Materialien für den bilingualen Unterricht

Biologie

Discover Biology

Cornelsen

Materialien für den bilingualen Unterricht: Biologie
Discover Biology

Erarbeitet von
Horst-Dieter Mathews und Simon Olmesdahl, Essen

Beratende Mitwirkung
Redaktion Naturwissenschaften, Cornelsen Verlag; Karl-Heinz Junge, Brakel

Redaktion: Marc Proulx
Redaktionsassistenz: Eva Baumgart, Merlene Griffin, Abbie Headon (extern), Mareike Seehaus, Dr. Ilka Soennecken
Umschlaggestaltung: Klein & Halm Grafikdesign, Berlin
Layout und technische Umsetzung: Rainer Bachmaier, Berlin; Heike Freund, Hameln;
werkstatt für gebrauchsgrafik, Berlin

www.cornelsen.de
www.cornelsen.de/bilingual

1. Auflage, 3. Druck 2014

Alle Drucke dieser Auflage sind inhaltlich unverändert
und können im Unterricht nebeneinander verwendet werden.

Druck: Himmer AG, Augsburg

ISBN 978-3-464-31844-7

Contents

Unit III

A message from the authors

With this book we think you will discover that learning about biology in English does not have to be hard. In fact, you may even find it more enjoyable than learning in your own language.
We have tried not to overload the texts with details and have kept the language as clear as possible for you. At the same time, we have taken great care to get the facts right and to give you all the information on plants, animals and the human body that you need at your age.
In *Discover Biology* you will find a number of helpful tools (see below) which will make it easier for you to understand the texts and topics. These tools include lots of full-colour pictures and diagrams that make the book more fun to use.
By doing the experiments and other practical tasks, you will learn to work like a real scientist who asks questions about nature and uses a variety of methods to find the answers.

Helpful tools in this book

Chapter objectives

Each chapter begins with a brief overview of the things you will learn about and the skills you will develop while working through that chapter. The list of chapter objectives gives you an idea of what to expect and what is of particular importance in a chapter.

Key terms

The structure of the bones
- blood vessel *Blutgefäß*
- bone marrow *Knochenmark*
- compact bone tissue *hartes Knochengewebe*
- head (of a bone) *Gelenkkopf*
- long bone *Röhrenknochen*
- marrow cavity *Markhöhle*
- nerve *Nerv*
- shaft (of a bone) *Knochenschaft*
- spongy bone tissue *schwammiges Knochengewebe*

Key terms[1]

At the beginning of each new chapter section you will find a box with 'Key terms'. These are important words used in biology which appear for the first time in the book. Always look at the key terms before reading the text, then watch out for them while you are reading. You can find the terms easily in the text because they are in **bold print**. At the end of a chapter all the key terms boxes are listed with page numbers so that you can go back and review[2] the key terms to make sure you have understood them (see also 'Chapter revision' below).

New vocabulary

Words and phrases which you might not know yet are numbered[3] in the text and explained in German on the same page.

Pronunciation help

If you are not sure how to pronounce a key term or a new word, you can find out by looking up the word in the appendix at the back of the book (see also 'Appendix' below).

Activate your English

Talking about causes and effects
Causes
Carbon monoxide can cause ... by ...
Tar affects the ... by causing the ... to ...
... is caused by ...
... results from ...
Effects
When nicotine/... enters / passes into / is carried to ..., it ...
... leads to ...
... can result in ...
As a result of this, ...
The result (of this) is that ...

Tasks and *Activate your English*

The tasks at the end of most sections help you to understand the information in the texts and pictures and make you think further about the material. They also give you practice in discussing topics and ideas with other people. To help you say what you want to say in English, you will find words and phrases under the heading 'Activate your English' either in a box next to a task or on the inside cover of the book.

Saying it in German

In each chapter you will find at least one task or activity with the heading: 'Say it in German', which asks you to speak or write in German about something you have just learned. These tasks and activities will not only help you find out if you have fully understood the material, but will also give you practice in using the equivalent terminology in German.

1 key term *Schlüsselwort*
2 review sth. *etwas überprüfen, nochmals durchsehen*
3 number sth. *etwas nummerieren*

Biology skills[1]

Within each chapter in a unit you will find coloured pages which introduce a biology skill that is relevant to the work in the chapter. These 'Improve your skills' pages offer you strategies and methods to help you learn better and more easily. Tasks and activities give you practice with the new skill. You can always find something useful here to improve the way you study biology.

Chapter revision[2]

There is a 'Revision' page at the end of each chapter. Here you can review and practise the things you have learned in the chapter. You can also find exercises to improve your English.

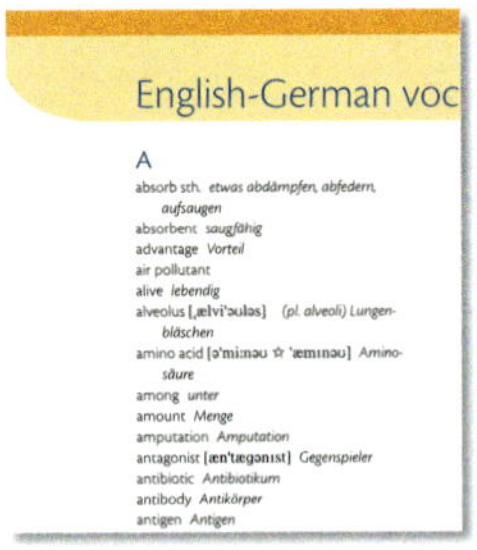

English-German voc

A

absorb sth. *etwas abdämpfen, abfedern, aufsaugen*
absorbent *saugfähig*
advantage *Vorteil*
air pollutant
alive *lebendig*
alveolus [ˌælviˈəuləs] *(pl. alveoli) Lungenbläschen*
amino acid [əˈmiːnəu ☆ ˈæmɪnəu] *Aminosäure*
among *unter*
amount *Menge*
amputation *Amputation*
antagonist [ænˈtægənɪst] *Gegenspieler*
antibiotic *Antibiotikum*
antibody *Antikörper*
antigen *Antigen*

Appendix[3]

The appendix is a section at the back of the book where you can find both English-German and German-English vocabulary lists. Here you can look up any English key term or annotation that appears in the book and find its German equivalent. Also, if you have trouble pronouncing a word in English, you can look up the phonetic spelling[4] of the word. The German-English vocabulary list can help you when you are asked to talk or write about a subject. If you know what you want to say in German, you can look up the English equivalent here.

We sincerely hope you enjoy working with *Discover Biology*!

Horst-Dieter Mathews and Simon Olmesdahl

1 skill *Fertigkeit*
2 revision *Überprüfung*
3 appendix *Anhang*
4 phonetic spelling *Lautschrift*

I. Biology is the study of living things

A tiger is a fascinating creature:
It can run extremely fast.
It is very strong.
It is beautiful, but dangerous.

We can all agree that the tiger in the picture is a living thing. But what is it about the tiger that makes it 'alive[1]'? You can ask the same question about yourself or about plants, insects or bacteria. It is not an easy question to answer.
People have asked questions like these for thousands of years, and they have spent a lot of time studying living organisms[2] to find the answers. This is where the word 'biology' comes from: *bios*, the Greek word for 'life', and *logos*, which means 'study'. And this is what this book is all about: finding out about living things and their characteristics[3].

Think about the factors that tell us that a tiger is alive. Make a list.

II. Life has certain properties[4]

Life on this planet has many different forms – think of a bird or a tree, for example. All living things have some qualities in common. These are called the properties of life. By these properties we can tell if something is alive or lifeless like a stone.

Here are five of the most important properties of living things:

Organisms move

This is very obvious[5] when you think about animals. A dog can run, a bird or a bat[6] can fly and fish and dolphins can swim. But what about a flower or a tree? Plants can move as well. You can see this clearly when you watch the flower of a crocus or a tulip[7], which opens up in the morning and closes in the evening.

Organisms grow

You know that your life began as a single cell[8] in your mother's womb[9] and that you have grown ever since. This growth[10] is controlled by the information stored in your genes[11]. Humans, animals and plants are small at the beginning of their lives and become bigger as they grow older.

1 alive *lebendig*
2 organism *Organismus*
3 characteristic *spezielles Merkmal*
4 property *Eigenschaft*
5 obvious *offensichtlich*
6 bat *Fledermaus*
7 tulip *Tulpe*
8 cell *Zelle*
9 womb *Gebärmutter*
10 growth *Wachstum*
11 gene *Gen*

Organisms reproduce

Life can only come from life. All plants, animals and humans produce new organisms, which are the same species as their parents. When males and females reproduce[1] sexually, they pass their characteristics on[2] to their offspring[3], which grow and develop into individual organisms that can one day reproduce. In this way life is passed on from one generation to the next.

Organisms respond[4] to their environment

There are many ways in which organisms find out what is going on around them. How animals and plants react[5] to influences from outside can also be very different. When a cat sees a mouse, it tries to catch it. A flower opens its petals[6] in the morning in response to[7] the light of the sun. A Venus Flytrap[8] closes its leaves very quickly when an insect touches the hair cells on the leaves.

Organisms need energy

For all life processes – movement[9], growth, reproduction[10] and response – living organisms need energy. Humans and animals get their energy from their food, where it is stored[11] in chemical[12] form. They can use this energy for tasks or activities, as you do, for example, when you play soccer. Green plants get their energy from the sun.

III. Organisms are made of cells

Your school building has many rooms which have particular[13] functions. The science rooms – the biology, physics and chemistry labs[14] – are set up so that you can do experiments there. Cells are arranged in your body like the rooms in a school building. Some cells work together to form your heart, while others work to form your brain.

Your life started as a single cell. Growth takes place by a process called cell division[15]: when a cell reaches a certain size, it divides[16] into two cells. These two then divide again to make four, and so on. All organisms grow in this way.

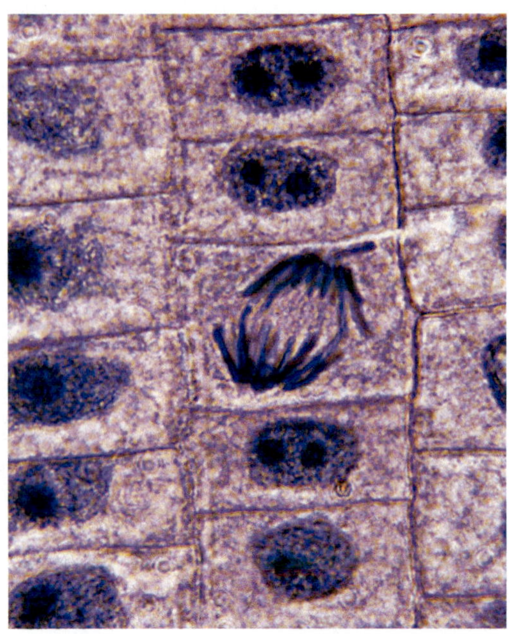

1 reproduce *(sich) fortpflanzen*
2 pass sth. on *etwas weitergeben*
3 offspring *Nachwuchs*
4 respond to sth. *auf etwas reagieren*
5 react to sth. *auf etwas reagieren*
6 petal *Blütenblatt*
7 in response to sth. *in Reaktion auf etwas*
8 Venus Flytrap *Venusfliegenfalle*
9 movement *Bewegung*
10 reproduction *Fortpflanzung*
11 store sth. *etwas lagern*
12 chemical *chemisch*
13 particular *besonders*
14 lab(oratory) *Labor*
15 cell division *Zellteilung*
16 divide *teilen*

Up to a certain point, all these cells look (more or less) the same. But then the cells become specialized[1]. When these cells divide, they form groups that have the same function. These cell groups are called tissues[2].
There are about 220 different types of tissues in your body. For example, some cells become specialized muscle tissue. They can contract[3] and relax[4], making the muscle shorter and longer. Others grow into nerve[5] cells. This tissue can pass along signals from one part of your body to another. Still others become the lens[6] of your eye. These cells are specialized to focus[7] the light that falls on your eye.

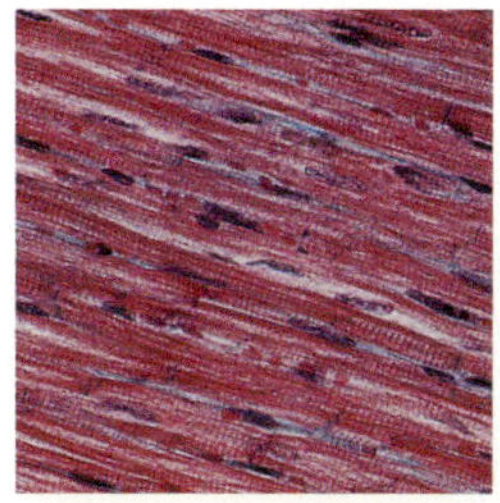
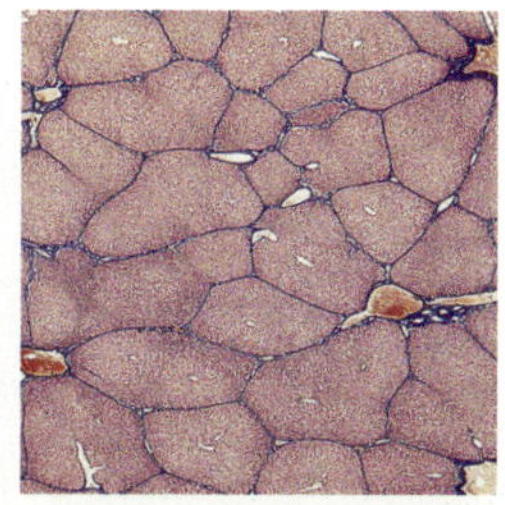
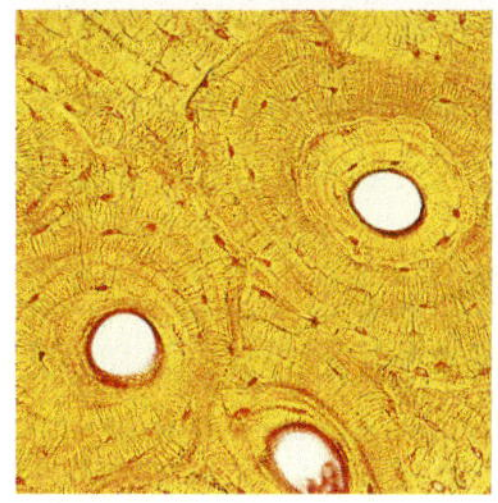
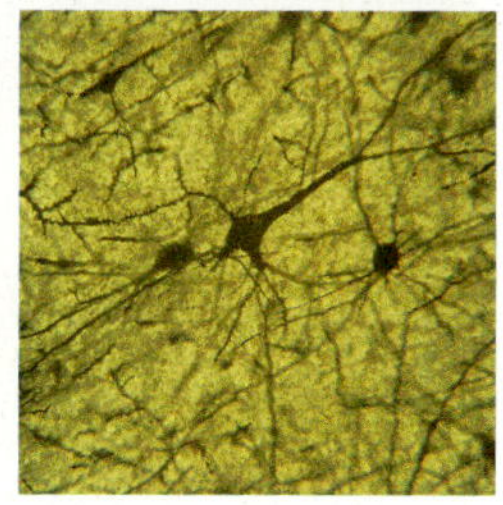

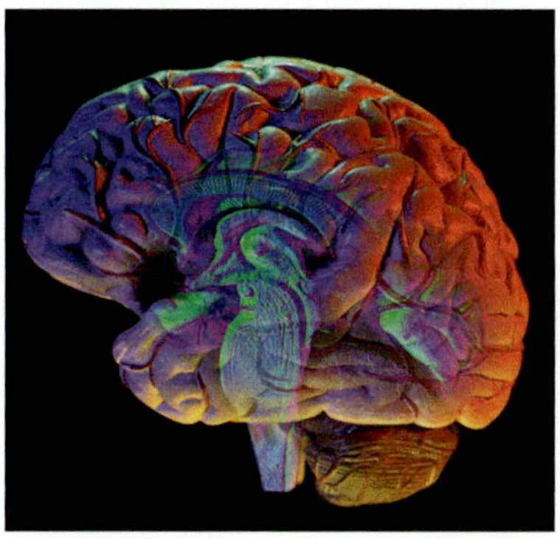
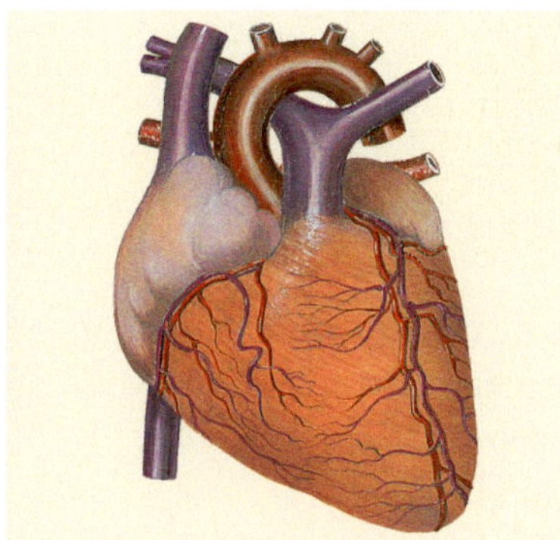

Different tissues combine to make organs[8].
Your heart is made mainly of muscle and nerve tissues.
Your nose consists of skin tissue, nerve tissue, bone and cartilage[9] tissues.
Your brain consists mainly of nerve tissue, covered by several layers[10] of skin.

Like the tiger at the beginning of this introduction, and like all other life forms, you are an organism. In a complex organism many organs work together. In fact, the organs and tissues of an adult person are made up of no less than 100 trillion[11] (= 100,000,000,000,000) cells.

IV. Useful study methods

Learning vocabulary for biology

In order to study biology, you need to know certain words which allow you to say exactly what you mean. Studying biology in English means that you have to learn these words in two languages. Learning both the English and the German words to describe something may take a bit more time, but it also trains your memory, so you get better the more often you do this.
In your English class you have been shown some good methods for studying vocabulary. You can use those same methods to learn biological terms. Here are just a few – try them out and find out which method works best for you.

Keep a vocabulary notebook:
The most common method is to keep a notebook with all the important words that are new to you, their translations, and a little note[12] if the pronunciation is difficult. Use a new page for each topic. Don't forget headings and dates!

1 specialized *spezialisiert*
2 tissue *Gewebe*
3 contract *sich zusammenziehen*
4 relax *sich entspannen*
5 nerve *Nerv*
6 lens *Linse*
7 focus sth. *etwas scharf einstellen*
8 organ *Organ*
9 cartilage *Knorpel*
10 layer *Schicht*
11 trillion *Billion*
12 note *Notiz*

Keep a vocabulary box:
Write down the English term on one side of a card and the German translation on the other side. Keep the terms in three groups: 'daily', 'weekly' and 'monthly'. Once you know a word from the 'daily' group well, put it in the 'weekly' section. Then once a week, go through all the 'weekly' words and put the ones you have learned well into the 'monthly' section, and revise[1] this section at the end of the month.

Software:
There are many software programs for training vocabulary. Some of them are freeware, which means you can download them for free from the Internet. It is important to choose a program which allows you to type in your own words, since most of them will not have much biology vocabulary.

Remember, any or all of these methods can help you with vocabulary, but *you* still have to take the time to learn it!

Keeping a biology notebook

You also need a place to write down other kinds of information – things which the teacher tells you to note down during lessons, sentences or drawings which you copy from the board, and pictures and texts which you complete for your homework.
Here are some basic[2] ideas for keeping a notebook:

Always have a heading that tells you what the text is about (don't just write 'Page 73, No. 4'!).

Use lots of paragraphs to make reading easier.

When you start a new topic, make sure you leave more space than between paragraphs.

1 May 2009

The heart has four chambers

A vein from the body enters the right atrium. From there the blood goes through a heart valve to the right ventricle.

Now the blood goes to the lungs, where it takes up oxygen and releases carbon dioxide. The pulmonary vein then goes to the left atrium, which opens into the left ventricle.

From the left ventricle an artery (the aorta) takes the blood to the rest of the body.

Blood vessels which take blood to the heart are called ▲ *veins, no matter if the blood is oxygenated or deoxygenated!*

Gas exchange takes place in the lungs and in the body cells.

Always have a date.

Underline the biological terms in a different colour from the text.

If something is really important, draw a little symbol next to it, like this one.

Finally, it is a good idea to reserve the first few pages of your notebook as a calendar for writing down the topic of each biology lesson.

Aug 22, 09:	Aug 24, 09:	Aug 26, 09:	...
– Parts of the respiratory system – Questions	– Revision – Breathing experiments (bell jar model)	– Reading: Chapters 1 and 2 – Quiz	...

1 revise sth. *etwas wiederholen*
2 basic *grundlegend*

Unit I Human Biology

Chapter 1

Movement

Chapter objectives

- To **name** the main parts of the human body and the bones of the skeleton
- To **describe** the structure and function of the main types of joints
- To **compare** and **contrast** the muscles of the skeleton, heart and inner organs
- To **explain** how muscles move a bone
- To **find out** what gives a bone its strength
- To **list** the most common injuries of the skeleton and muscles and **explain** how to prevent them

As you know, it is not easy for young people to sit in a classroom for hours and not be able to move around and get fresh air. And it is quite natural that you want to get up after a lesson to stretch your body and move your arms and legs. These signals from your body that it needs movement are a sign that you are healthy. Taking regular breaks during working hours is a necessary part of staying healthy for both children and adults.

1.1 Bones, joints and muscles are needed for movement

When you take part in a ball game and run after the ball, your legs have to absorb[1] thousands of pounds each time your feet hit the ground. This is possible because your bones are built in such a way that they are light and very strong at the same time. If you want to score a goal[2] in football, for example, you have to observe[3] what the other players do and follow the path of the ball with your eyes. You have to move your body quickly from one position to another, stop suddenly to avoid other players, turn quickly round, kick the ball[4] and start running again.

The reason your body can carry out these different movements is because you have **joints** between the bones and muscles which allow the bones to move in different directions.

Before you have a closer look at the bones and muscles inside your body, see if you can remember the names of those parts of your body which you can see from the outside. You will have learned many of them in your English lessons, but some, which you can see in Figure 2, may be new to you.

Figure 1 Playing a ball game

1 absorb sth. *etwas abdämpfen, abfedern*

2 score a goal *ein Tor schießen*

3 observe sth. *etwas beobachten*

4 kick the ball *den Ball schießen*

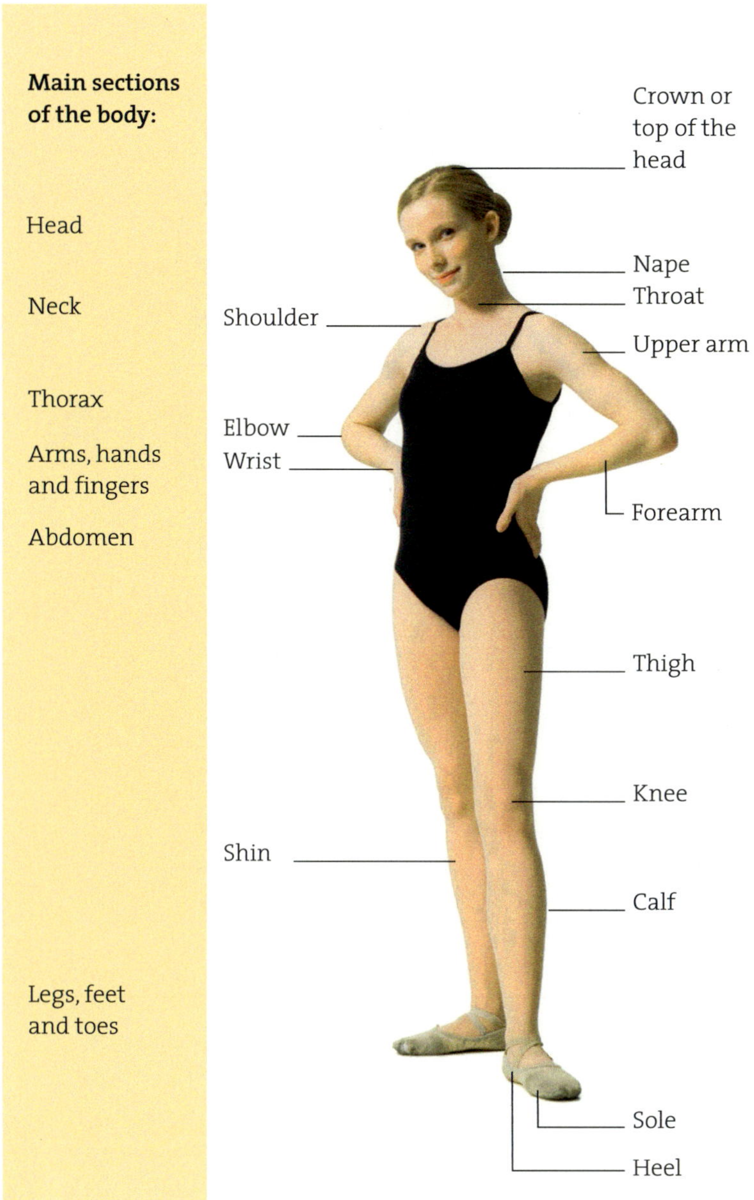

Figure 2 Parts of the body

Key terms

Parts of the skeleton

- backbone (vertebral column) *Wirbel-säule*
- breastbone *Brustbein*
- carpals *Handwurzelknochen*
- cartilage *Knorpel*
- collarbone (clavicle) *Schlüsselbein*
- fibula *Wadenbein*
- finger bones *Fingerknochen*
- humerus *Oberarmknochen*
- intervertebral disc *Bandscheibe*
- jawbone *Kieferknochen*
- joint *Gelenk*
- metacarpals *Mittelhandknochen*
- metatarsals *Mittelfußknochen*
- pelvis *Becken*
- radius *Speiche*
- ribcage *Brustkorb*
- ribs *Rippen*
- shin bone (tibia) *Schienbeinknochen*
- shoulder blade *Schulterblatt*
- skull *Schädel*
- spinal cord *Rückenmark*
- tarsals *Fußwurzelknochen*
- thigh bone (femur) *Oberschenkel-knochen*
- toe bones *Zehenknochen*
- ulna *Elle*
- upper/lower jaw *Ober-/Unterkiefer*
- vertebra (pl. vertebrae) *Wirbelkörper*

1.2 The skeleton carries and protects your body

The bones of your skeleton support[1] your body and give it shape. They also protect[2] the inner organs. The brain is kept safe inside the **skull**. The **ribs** and **breastbone** form a protective[3] cage around the heart and lungs, and the **pelvis** protects the organs of your abdomen. The delicate[4] **spinal cord**, which transports information from your brain to all the organs in the body and the limbs, lies inside the **backbone**.

The bones of your skeleton store minerals such as calcium[5] and phosphorus[6], which make your bones hard. Your nerves and your muscles also need calcium and phosphorus to function properly. If there are not enough of these minerals in your blood because you did not consume the necessary amount[7] with your food, calcium and phosphorus are gradually released[8] from your bones. You can imagine what effect this can have if it happens over a long period of time[9].

1 support sth. *etwas unterstützen*
2 protect sth. *etwas schützen*
3 protective *schützend*
4 delicate *empfindlich*
5 calcium *Kalzium*
6 phosphorus *Phosphat*
7 amount *Menge*
8 release sth. *etwas freisetzen*
9 period of time *Zeitraum*

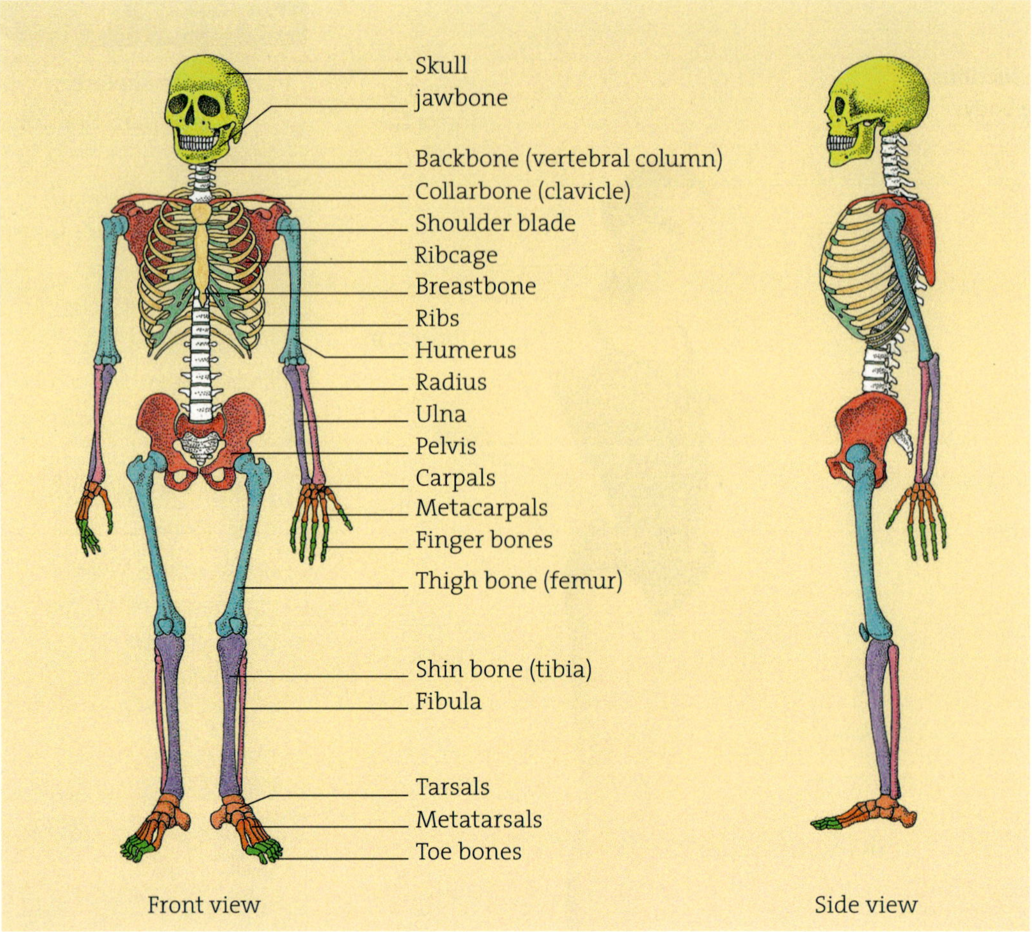

Figure 3 **Human skeleton (front and side view)**

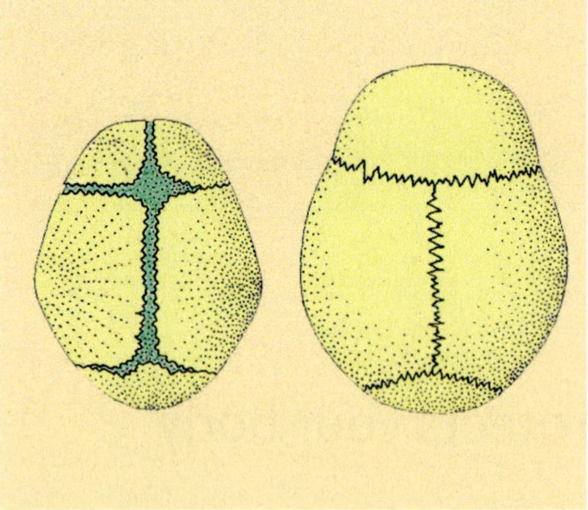

Figure 4 **Skull of a new-born baby (left) and of an adult (right)**

The human skull consists of a mosaic of bones, which in an adult are completely fused[1] together, but in a new-born baby are not fully fused on the top (see Figure 4). The spaces[2] between the bones gradually close during the baby's first year of life. This is why it is important to be very careful when you touch the top of a baby's head. In the human skull, the **lower jawbone** is the only bone that can be moved freely.

The **backbone** consists of 33 small bones called the **vertebrae**. The solid[3] part of each single vertebra has an oval shape. Located at the back of the vertebra is a bony ring with a hole through which the nerves of the spinal cord run from the skull to the pelvis. Three pieces of bone stick out of the ring and point to the back and to the side. This is where the muscles of your back are attached[4].

In Figure 6 you can see what a single vertebra looks like and how the vertebrae are placed one on top of the other. The vertebrae are separated from one another by discs of **cartilage** called **intervertebral discs**. Cartilage is a softer and more elastic tissue[5] than bone.

A baby's backbone is rather straight. But when the child gets older, the shape of its backbone changes. In a grown-up person it is shaped like a double-S (see Figure 5). This shape makes the backbone strong and flexible, and it is what allows you to move in an upright position.

1 fuse sth. *etwas verbinden*
2 space *Raum, Zwischenraum*
3 solid *fest*
4 attach sth. *etwas befestigen*
5 tissue *Gewebe*

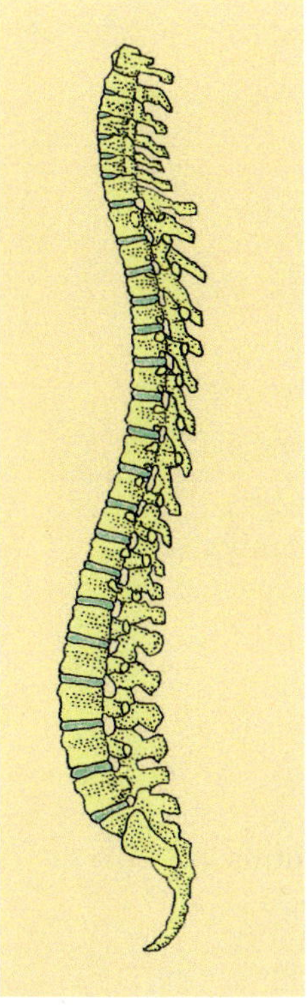

Figure 5 **The backbone seen from the left side**

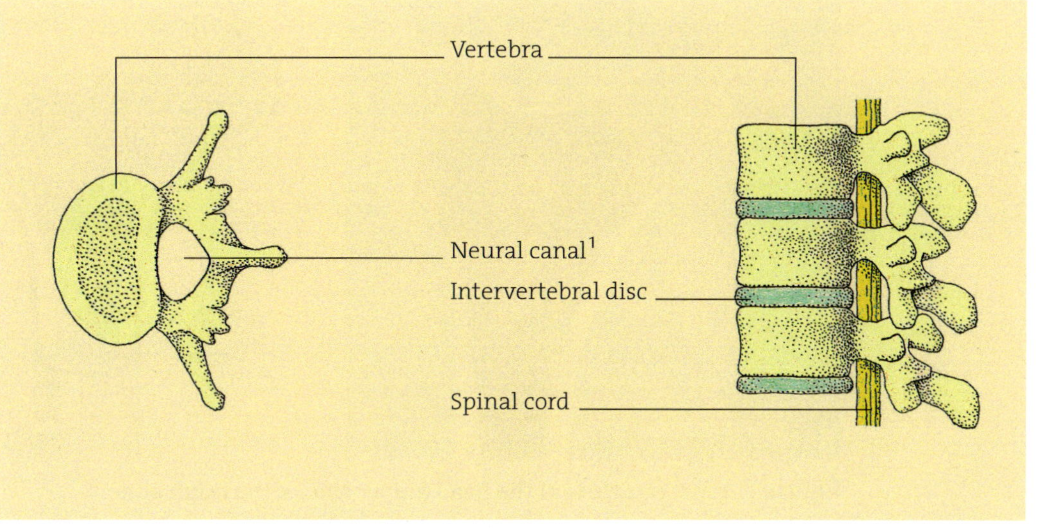

Figure 6 **A single vertebra (left) and a section of the backbone (right) showing the vertebrae, the intervertebral discs and the spinal cord**

The discs of cartilage between the vertebrae are like cushions, which absorb the shocks when you run or jump. They also prevent[2] the vertebrae from rubbing[3] against each other when you move.

Twelve pairs of ribs form a cage around the heart and the lungs. The **ribcage** can move up and down because pieces of cartilage, not bone, connect the ribs to the breastbone at the front and to the vertebrae at the back (see Figure 3).

1.3 All our bones are built in the same way

The bones of the human skeleton have different shapes. Some are long and thin, like the bones in the upper arms or upper legs, and others are broad[4] and flat, like the shoulder blades or the pelvis. But all bones have a similar inner structure[5].

They are covered with a thin but tough[6] membrane with lots of blood vessels and nerves running through it. Tiny blood vessels also run into the bone. As a living organ, a bone must receive[7] food and oxygen for growth as well as repair and it must be able to get and send out information.

Beneath this thin membrane is a layer of **compact bone tissue**, which is very hard and strong. It surrounds[8] a much lighter **spongy bone tissue**, which has lots of spaces in it. These spaces contain **blood vessels**, **nerves** and **bone marrow**. The bone marrow is very important because it produces red and white blood cells. The bone tissue contains[9] calcium salts, which make the bone hard, and elastic[10] protein[11] fibres[12], which prevent the bone from breaking.

The **long bones** in your upper legs carry the weight of your whole body, so they have to be very strong.

Key terms

The structure of the bones
- blood vessel *Blutgefäß*
- bone marrow *Knochenmark*
- compact bone tissue *hartes Knochengewebe*
- head (of a bone) *Gelenkkopf*
- long bone *Röhrenknochen*
- marrow cavity *Markhöhle*
- nerve *Nerv*
- shaft (of a bone) *Knochenschaft*
- spongy bone tissue *schwammiges Knochengewebe*

1 neural canal *Neuralkanal*
2 prevent sth. *etwas verhindern, vorbeugen*
3 rub *reiben*
4 broad *breit*
5 structure *Struktur*
6 tough *zäh*
7 receive sth. *etwas erhalten*
8 surround sth. *etwas umgeben*
9 contain sth. *etwas enthalten*
10 elastic *dehnbar, elastisch*
11 protein *Eiweiß*
12 fibre *Faser*

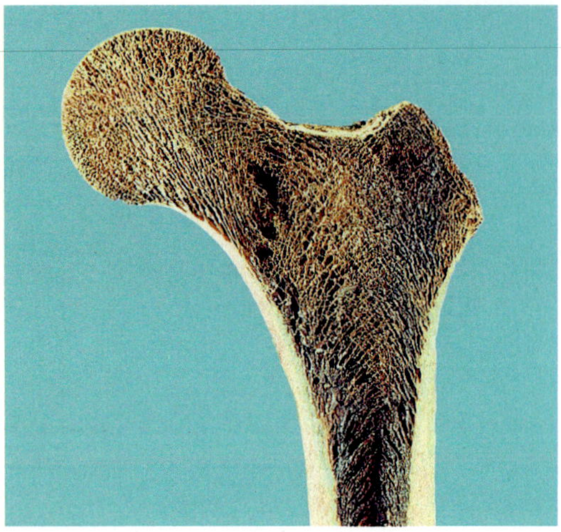

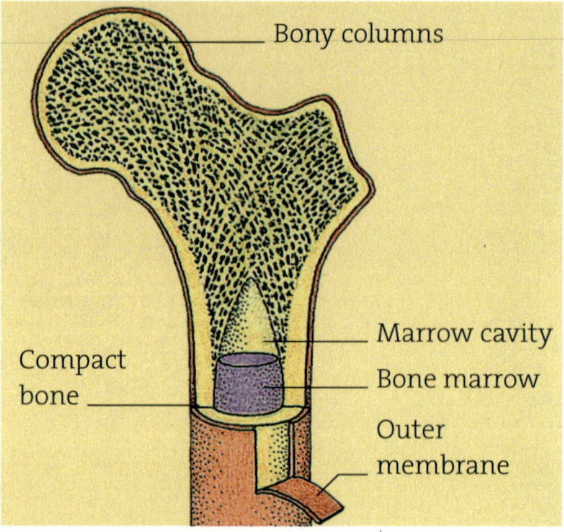

Figure 7 Inner structure of the head (upper end) of the thigh bone

But at the same time they must not be too heavy. Figure 7 gives you an idea what the upper end – the **head** – of the human thigh bone looks like inside.
Thin bony columns run from one side of the bone to the other. They are bent[1] like the arches[2] of a bridge and form a strong network, whose spaces are filled with bone marrow. This structure gives the thigh bone its great stability.

Did you know?

The body fully replaces the tissue at the end of the thigh bone in the knee every six months.

1 What are the main functions of the skeleton?
2 Explain the special shape of the backbone.
3 State the reasons why we have got intervertebral discs.
4 What do all bones have in common[3]?
5 Describe the inner structure of the thigh bone and explain why it is built in this way.

1.4 Joints fit bones together

Joints are found where two bones come together. Figure 8 shows the three types of freely moveable[4] joints, which allow different kinds of movements:

- At a **hinge joint** the bones can only bend in one direction. A good example is your knee joint (see Figure 10).
- A **ball and socket joint**, like your shoulder joint, allows the bones to move in many directions.
- A **saddle joint**, like at the base[5] of the thumb, enables[6] the bones to move in more than one direction, but it is not as freely moveable as your shoulder.

Key terms

The joints
- ball and socket joint *Kugelgelenk*
- bone membrane *Knochenhaut*
- cruciate ligament *Kreuzband*
- hinge joint *Scharniergelenk*
- joint capsule *Gelenkkapsel*
- knee cap *Kniescheibe*
- layer of cartilage *Knorpelschicht*
- ligament *Gelenkband*
- meniscus (pl. menisci) *Meniskus*
- saddle joint *Sattelgelenk*
- socket *(hier) Gelenkpfanne*
- synovial fluid *Gelenkschmiere*

1 bend sth. *etwas biegen*
2 arch *Bogen*
3 have sth. in common *etwas gemeinsam haben*
4 freely moveable *frei beweglich*
5 base *Basis, Grund*
6 enable sth. *etwas ermöglichen*

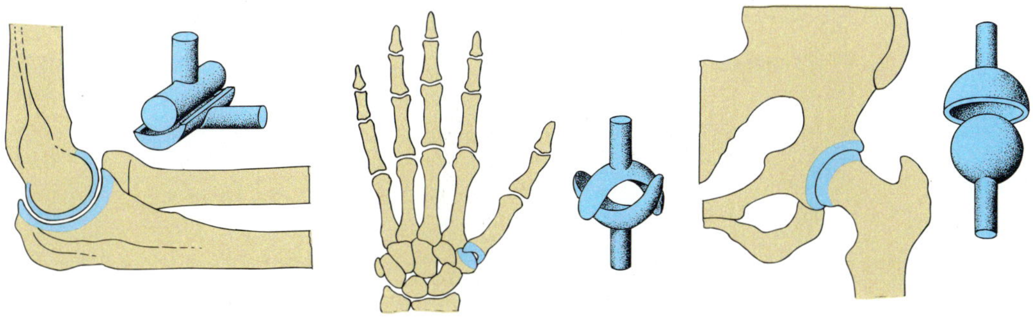

Figure 8 Three types of joints

Where the bones of a freely moveable joint meet, they are covered with a protective **layer of cartilage**. Between the bones is a thick liquid called **synovial fluid**, which makes the joint move smoothly. If the cartilage is worn down[1] or if there is not enough synovial fluid, then it becomes very painful[2] to move the joint.

Strong and elastic bands called **ligaments** hold the bones of the joint together. On the outside the joint is covered and protected by a **capsule** of tough skin.

The knee is a rather complex[3] joint which can cause a lot of trouble, especially in old age. It can also cause problems when a person is very overweight[4]. It is a hinge joint that is formed by three bones: the thigh bone, shin bone and fibula (see Figure 10 a). A small flat bone, the **knee cap**, protects the joint at the front.

The lower head of the thigh bone has two rounded parts which fit into **sockets** on the upper head of the shin bone. In between these are two elastic rings of cartilage, the **menisci**,

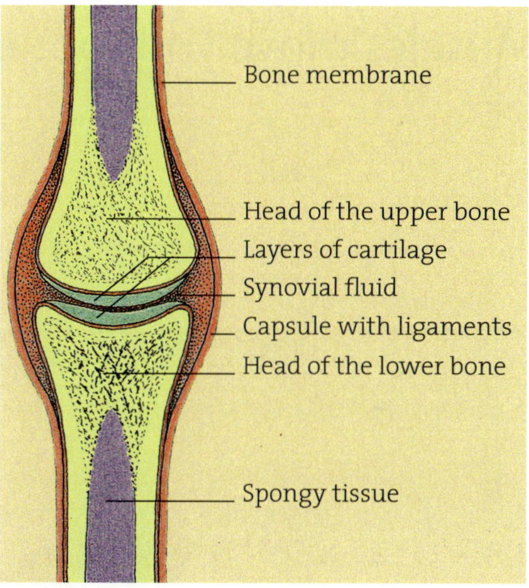

Figure 9 The parts of a joint

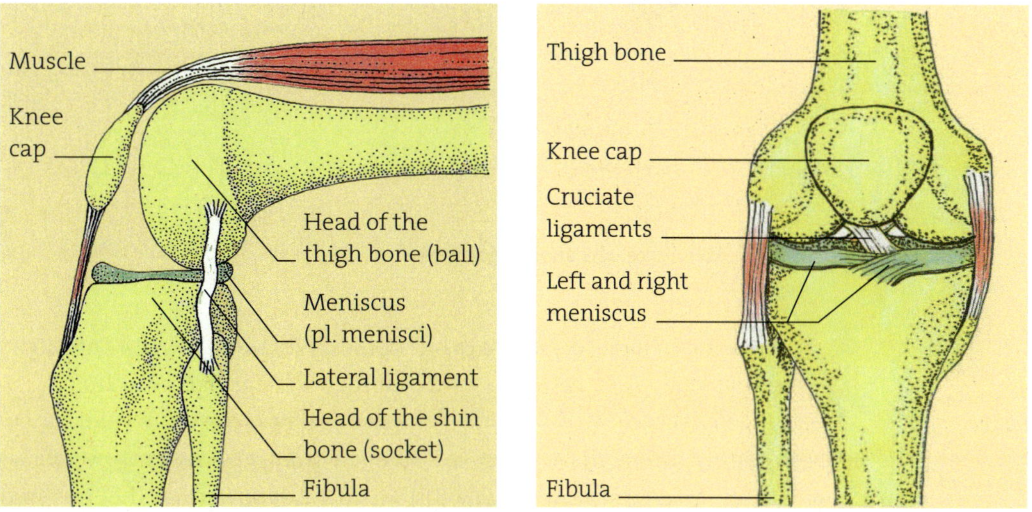

Figures 10 a/b Right and left knee joint (side and front view)

1 wear sth. down *etwas abnutzen*
2 painful *schmerzhaft*
3 complex *kompliziert*
4 overweight *übergewichtig*

which work like shock absorbers[1]. When you walk, bend down[2] or lift a heavy weight, they slightly[3] change their shape and position. In this way they add to the stability of the joint.
The **cruciate ligaments** are arranged[4] cross-wise and hold the two bones in the correct position. Two straight[5] ligaments on either side, the lateral ligaments, connect the head of the thigh bone with the shin bone and the much thinner fibula. They allow you to move the lower leg a little bit to the side as well.

1 Describe the main parts of a joint and their functions.
2 Give as many examples as possible for hinge joints in your body. What kind of movements are possible with such joints?
3 There is a special joint in your thumb which allows your hand to catch and hold a ball, but also to use a pen for writing or drawing. What do you know about this joint?
4 What kind of joint is your hip joint? Explain which bones form this joint and how the bones can move.
5 Describe the special structure of a knee joint. Explain how the different parts of the joint work together. You can use the illustrations in this chapter (see Figures 8–10).

1.5 Muscles move your bones and your blood

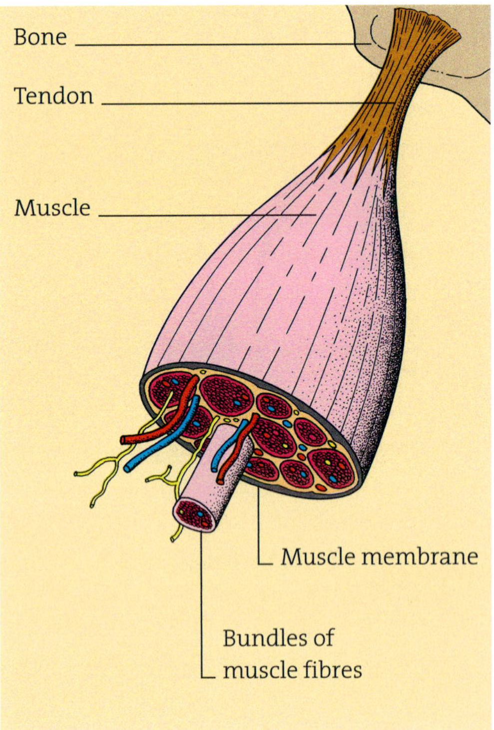

Figure 11 **Structure of a skeletal muscle**

Key terms

Muscles and their functions
- antagonist *Gegenspieler*
- bend sth. *etwas beugen*
- biceps *Armbeuger*
- cardiac muscle *Herzmuskel*
- connective tissue *Bindegewebe*
- contract *(sich) anspannen*
- extensor *Streckmuskel*
- flexor *Beugemuskel*
- involuntary muscle *unwillkürlicher Muskel*
- muscle fibre *Muskelfaser*
- relax *sich entspannen*
- skeletal muscle *Skelettmuskel*
- straighten sth. *etwas strecken*
- tendon *Sehne*
- triceps *Armstrecker*
- voluntary muscle *willkürlicher Muskel*

Muscles can be found everywhere in our bodies. We have about 600 of them altogether. There are three different types: **voluntary**, **involuntary** and **cardiac muscles**.
The muscles which move our bones – the **skeletal muscles** – are voluntary, which means that we can control their actions by our will. About a quarter of our body weight is made up of skeletal muscles, which can work quickly and with a lot of power. But they get tired after some time and need to rest.

1 shock absorber *Stoßdämpfer*
2 bend down *sich niederbeugen*
3 slightly *geringfügig, leicht*
4 arrange sth. *etwas anordnen*
5 straight *gerade*

1 gut *Darm*
2 bladder *(Harn)blase*
3 pacemaker *Schrittmacher*
4 fasten sth. *etwas befestigen*
5 spindle-shaped *spindelförmig*
6 branched *verzweigt*
7 network *Netzwerk*
8 agility *Gewandheit, Flinkheit*
9 flexibility *Elastizität, Biegsamkeit*
10 upright posture *aufrechte Haltung*

Our inner organs, like the stomach, gut[1] and bladder[2], normally work on their own. They have involuntary muscles, which work slowly and do not get tired.
The heart, which pumps blood through the body, is made up of cardiac muscle, which works automatically because it has its own natural pacemaker[3]. But when you are afraid or excited, your brain sends information to your heart and makes it beat more quickly.
The three types of muscle also differ in their inner structure. Skeletal muscles consist of bundles of thin **muscle fibres** covered over by a layer of tough **connective tissue**. Both ends of the muscle are fastened[4] by **tendons** to one or more bones of the skeleton. Tendons are strong, elastic bands of connective tissue. The muscle tissue in the walls of your inner organs, however, consists not of fibres but of spindle-shaped[5] cells. The heart muscle has muscle fibres similar to those of skeletal muscles, but they are branched[6] and form a complex network[7].

1.6 Skeletal muscles work in pairs

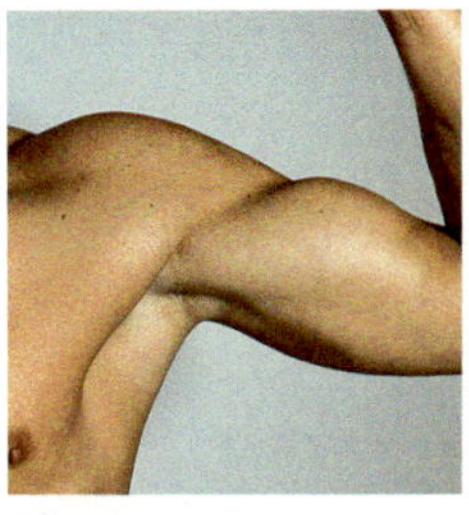

Figure 12 The biceps muscle in the upper arm contracts.

Muscles can pull at a bone but cannot push it. Therefore you always need one muscle to pull the bone in one direction and another to pull it in the opposite direction, so that it returns to the original position.
If you **bend** your arm, the **biceps** muscle in the front of your upper arm **contracts** and becomes shorter and fatter (see Figure 12). At the same time, the **triceps** muscle in the back of your upper arm **relaxes** and becomes longer and thinner. If you **straighten** your arm, the triceps muscle contracts and the biceps relaxes.
The biceps is a **flexor** muscle because it bends (flexes) the lower arm. The triceps is called an **extensor** muscle, because it straightens the forearm. Muscles like your biceps and triceps, which pull at the same bone and move it in opposite directions, are called **antagonists**.

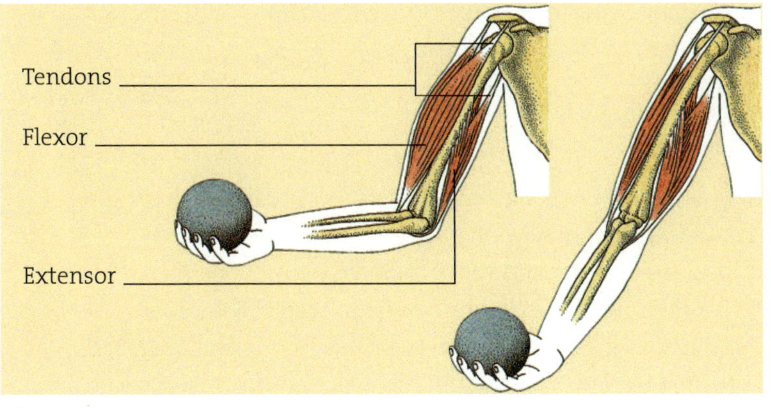

Figure 13 Flexor and extensor muscles in the upper arm work as antagonists when a weight is lifted or lowered.

The name *bi*ceps comes from the fact that *two* tendons attach the flexor muscle in the upper arm to the shoulder blade; the *tri*ceps is so named because the extensor muscle is attached with *three* tendons.
Muscles get stronger when they are used regularly. Therefore you can train your muscles by doing different sorts of sports activities. Exercise increases not only the strength and agility[8] of muscles, but also the flexibility[9] of joints. Strong muscles are necessary for a good upright posture[10].

1 Describe the inner structure of a skeletal muscle. Explain the functions of the different parts.
2 Point out where you can find voluntary and involuntary muscles in your body. How do they differ in form and function?
3 Use your knowledge about the function of biceps and triceps in your upper arm to explain how the lower leg is able to move.
4 Explain the difference between tendons and ligaments.
5 **Say it in German:** Work with a partner. At dinner your father or mother asks you what topic you discussed in your biology lesson that day. Tell him or her how skeletal muscles work and how antagonists function when you move your arms and legs.

Reading, understanding and memorizing a text

It is difficult to read a biological text like a story because it is full of facts and contains specialized terminology[1] which is not used in ordinary[2] everyday communication.
Here is how you can get as much information from a text as possible without wasting time:

Reading Skills

Prepare yourself:
Begin by reading the heading and subheadings[3], as well as the key words in the box. This will help focus your brain on what is important. You do not need to understand every word right away – the diagrams and the texts will help you later.

When reading:
- **Read carefully**, one paragraph at a time. After each paragraph, stop and ask yourself, 'What is the main message of this paragraph?' Look up the meaning of the new words and their pronunciation in the boxes and in the glossary.
- **Highlight[4] the important facts**, if the book belongs to you, by circling or underlining words or making marks (e.g. exclamation or question marks) in the margin[5]. Before you highlight anything, read the whole paragraph first – that will make it easier to decide what is most important. Make sure not to highlight too much – if everything is underlined, nothing is emphasized[6] – but rather, highlight only the most important phrases.
- **Make notes** on the main points: a good rule is to make one note per paragraph.

Task:
Use one of the sections in this chapter to practise with a friend: each of you writes down one question for each paragraph. Then exchange your questions and try to answer them. You will soon find out which paragraphs you have to go back to and read again.

Mind mapping
Another useful way of understanding a text is by making a mind map. You visualize[7] the text by arranging the key terms around a central word or phrase on a sheet of paper:

Step 1: Start with an empty, unlined piece of paper and some coloured pens.
Step 2: Write the topic of the text (or the title of the chapter) in the middle of the paper.
Step 3: Using different colours, draw various branches[8] or sub-branches[9] out from the central word or phrase and along them write key words or phrases – not complete sentences. Remember you can add more information at any time.
Step 4: Add symbols, pictures and icons to emphasize important points.

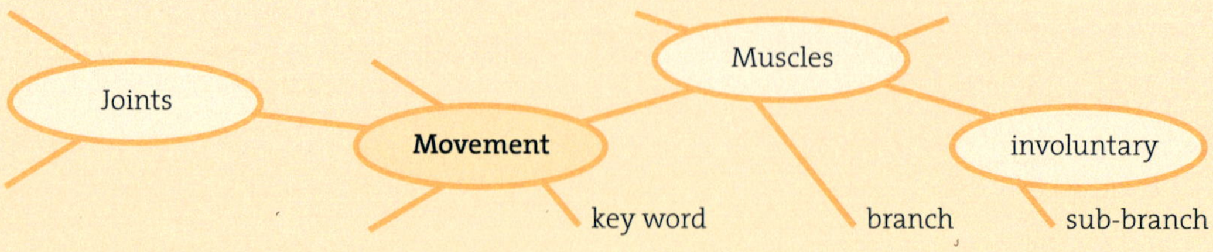

Task:
After you have read Sections 1.1–1.6, complete the branch 'joints' by adding sub-branches with the appropriate words or short phrases.

1 terminology *Fachsprache*
2 ordinary *normal*
3 subheading *Zwischenüberschrift*
4 highlight sth. *etwas hervorheben*
5 margin *Rand*
6 emphasize sth. *etwas betonen*
7 visualize sth. *etwas veranschaulichen, bildlich darstellen*
8 branch *Ast*
9 sub-branch *weiterer Abzweig*

Activity 1

Observe your body

Do the following exercises alone or with a classmate:

1 Write down the names of the joints which are involved in these actions and describe what they do:
a Move your foot up and down.
b Open and close your hand.
c Give the 'thumbs up' signal.
d Stand on tiptoes[1].
e Lift a book from a desk.

2 Bend your forefinger[2] until it touches the palm of your hand, then straighten it again. Repeat this movement as many times as you can. What does your finger feel like? Why do you have to stop after some time?

3 You can feel the tendons of some muscles using your forefinger. For example, touch the tendons on your hand that move your fingers. You can see them quite well when you straighten the fingers. Also touch the large Achilles tendon at the back of your ankle. What does this feeling tell you about the qualities of tendons?

Activity 2

Investigate the strength of a bone

Note:
When working with chicken bones make sure to clean your hands thoroughly[3] and everything you have touched afterwards in order to avoid an infection[4] from salmonella bacteria[5].

1. Remove the meat from two thigh bones of a fresh (uncooked) chicken. First try to bend both of the bones. You will find that they are hard and rigid.
2. Now place the bones in two 1000 ml beakers[6]. Add enough dilute hydrochloric[7] acid to one of the beakers to cover the bone and the same amount of water to the other beaker. Leave the bones in the solutions[8] for 3 to 4 days.
   Alternatively, you can put the first bone in a beaker with vinegar[9], in which case you have to leave it for about a week, changing the vinegar several times.
3. When taking the bone out of the hydrochloric acid solution, use a pair of tongs[10]. Rinse the bone carefully in clear water.
4. Now try to bend both bones again.

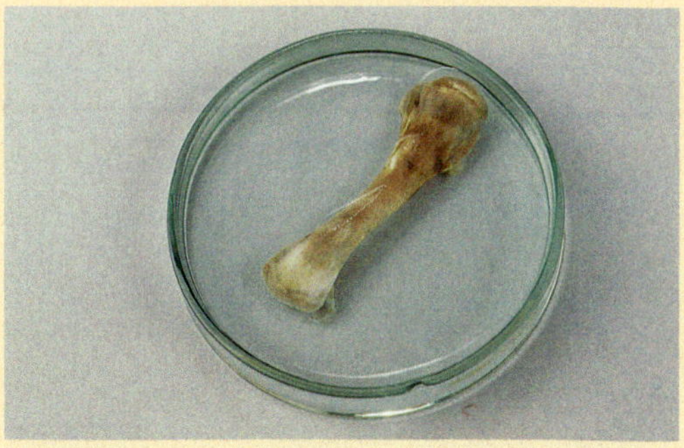

Figure 14

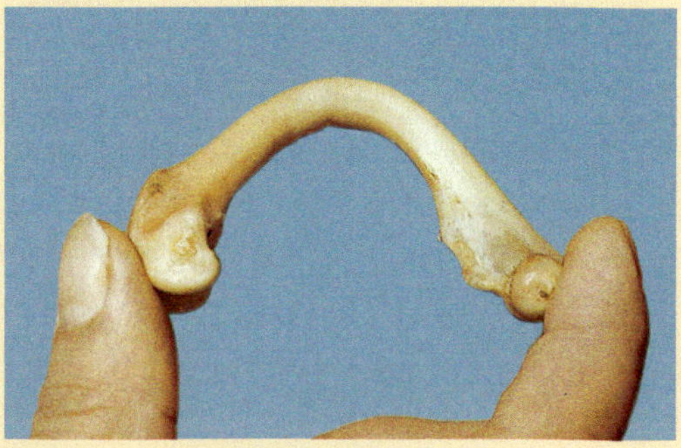

Figure 15

1 on tiptoes *auf Zehenspitzen*
2 forefinger *Zeigefinger*
3 thoroughly *gründlich*
4 infection *Ansteckung, Infektion*
5 bacterium (pl. bacteria) *Bakterie, Bakterium*
6 beaker *Becherglas*
7 dilute hydrochloric acid *verdünnte Salzsäure*
8 solution *Lösung*
9 vinegar *Essig*
10 tongs *Greifzange*

1 Describe what happens when you try to bend the bones just as you did at the beginning.
2 Speculate what you think could be the reason for the difference.
3 Explain what gives a bone its flexibility and strength.

Activity 3

Build a simple model of the backbone

Material:

- 12 discs[1] made of wood and 11 discs made of foam material[2], each about 0.5 cm thick and 3 cm in diameter[3], with a hole in the center
- piece of strong wire[4] about 15 cm long
- pair of pliers[5]
- wooden board about 10 cm x 10 cm, 1 cm thick

Procedure:

Drill a small hole in the center of the board, pull one end of the wire through it and use the pliers to bend it to one side. Then place the board on the table with the free end of the wire pointing upwards.

Now string the wooden discs and the foamy discs alternately[6] onto the wire (Figure 16). Begin and end with a wooden disc.

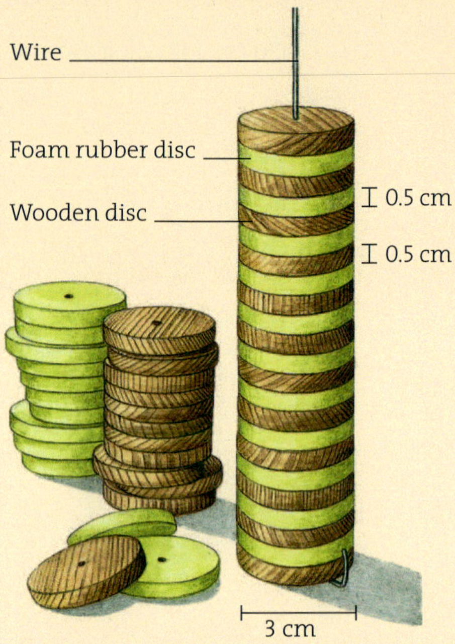

Figure 16 **Model of the backbone**

1 Point out which parts of the backbone are represented in your model.

2 Bend the wire to one side, then to the other.

a Describe what happens to the discs.

b What conclusion can you draw about the function of the foamy discs?

Activity 4

Build a cardboard[7] model to demonstrate how the biceps and triceps work

Draw the shapes of the upper arm and forearm on a piece of cardboard and cut them out.

1. Make holes in the two pieces of the model where the elbow joint should be, and attach the two pieces with a press stud[8] or a paper fastener[9].
   The cut-outs for the upper arm and forearm should now be movable.
   Then make two holes in each arm section where the tendons of the biceps and triceps are attached (at points T1, B1, T2 and B2). Pull a rubber band through holes B1 and B2 and another one through T1 and T2. Fasten them with knots at the back of the model.
2. Now move the forearm up and down by pulling at the rubber band between B1 and B2 and then between T1 and T2.

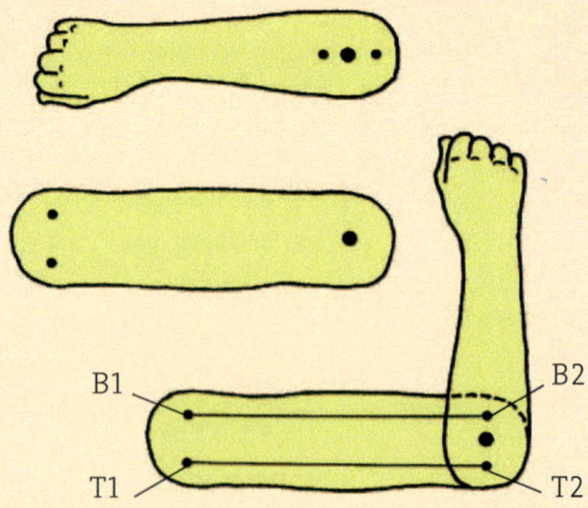

Figure 17 **Cardboard cut-out pieces for the upper arm and forearm**

1 disc *Scheibe*
2 foam material *Schaumstoff*
3 diameter *Durchmesser*
4 wire *Metalldraht*
5 pliers *Zange, Kombizange*
6 alternately *abwechselnd*
7 cardboard *Pappe*
8 press stud *Druckknopf*
9 paper fastener *Rundkopfklammer*

1 Define the special characteristics of the joint you have constructed.

2 What can you demonstrate with this model?

3 Discuss the usefulness of this model for the demonstration of the joint and muscle action in your forearm.

1.7 Defects in posture can be prevented

Statistics show that more than 50 per cent of all schoolchildren in Germany have postural defects[1] of one sort or another.

In Figure 18 you can see how wrong posture can develop. If a heavy school bag is carried on one side of the body, the backbone is permanently[2] bent sideways. This posture changes the position of the intervertebral discs. Some of them move sideways and press against the nerves inside the spinal cord. This can lead to back pain over a longer period of time. Placing the weight of the body on one leg while standing is a bad habit which can lead to similar problems. When the backbone is permanently curved sideways, the disease is called scoliosis[3]. The cause for this condition, which mostly develops during childhood[4], is still unknown. Another common defect of the backbone is 'round shoulders' or a round back[5], which can arise from sitting in an incorrect position. And if the lower backbone is curved forward too much, a hollow back[6] can result[7].

Figure 18 **Carrying a schoolbag the wrong way (left) and the correct way (right)**

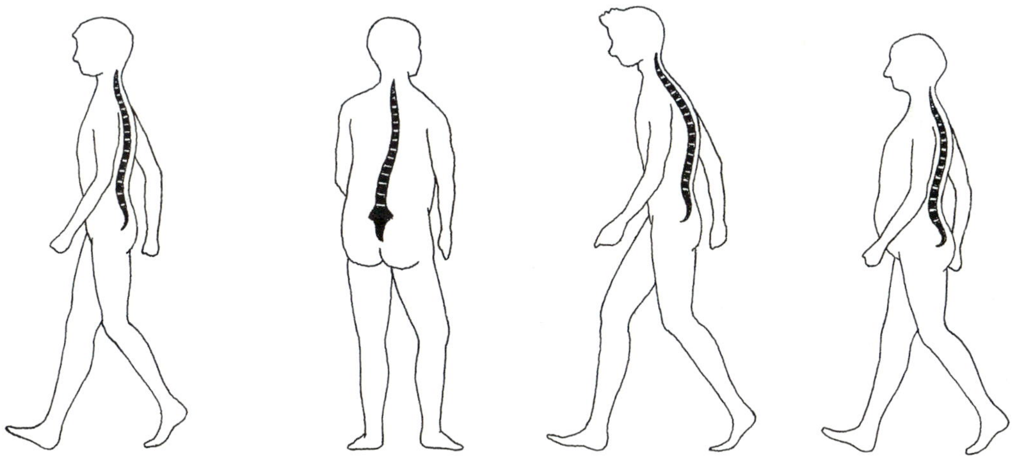

Figure 19 **Normal posture (left) and examples of defective posture**

The strain[8] on the intervertebral discs, the nerves, ligaments and muscles caused by wrong posture in childhood can lead to lasting defects and back pain in later life.

It is best to change your position often when you are working or resting. This way your muscles contract and relax, which, by turns, increases and reduces the stress on your joints and ligaments.

You should try to move as many parts of the body as possible, so activities like gymnastics, swimming, cycling and walking are especially good for you.

1 postural defect *Haltungsschaden*
2 permanently *andauernd*
3 scoliosis *seitliche Verkrümmung der Wirbelsäule, Skoliose*
4 childhood *Kindheit*
5 round back *Rundrücken*
6 hollow back *Hohlrücken*
7 result *auftreten*
8 strain *Belastung*

Your feet carry the weight of your whole body. If the foot muscles or ligaments are weak, then they cannot keep up the natural shape of the foot. Normally the sole of the foot is curved upwards and forms an arch[1]. But for some children the arches of the feet are not stable enough, so they sink down slightly when weight is put on the feet, resulting in fallen arches[2]. In the case of a child with flat feet[3], the soles completely touch the floor when the child is standing (see Figure 20).

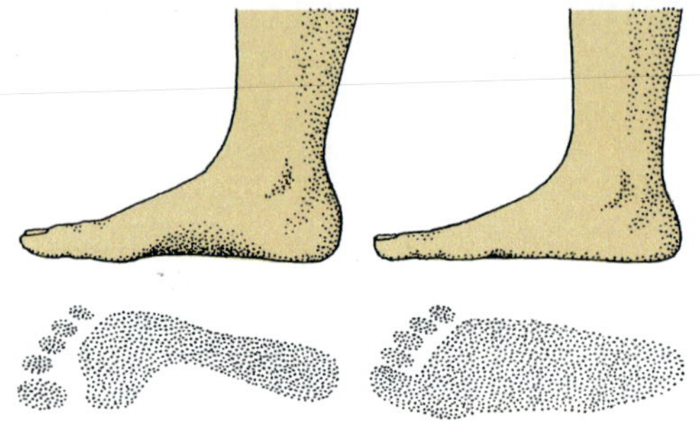

Figure 20 A normal foot and a flat foot

If a child complains of foot or ankle pain, he or she should see a doctor. The doctor may recommend a shoe insert[4].

Walking barefoot[5] on a natural surface and foot exercises can strengthen the foot muscles. Tight-fitting, fashionable[6] shoes as well as shoes with high heels can lead to foot problems and defects in posture. Choosing shoes which fit correctly can prevent these problems.

Rules for proper and healthy posture:

- Carry your school bag across both shoulders 'rucksack-style.' The alternative is to carry the bag in your arms in front of you.
- Sitting up straight in school and elsewhere can save you from back pain later in life.
- As people spend a lot of time in front of the computer nowadays, here are some tips for correct sitting posture that prevents strain and fatigue:

  - Sit directly in front of your keyboard and screen;
  - Set the top of the screen just below eye level;
  - Sit back in your chair and do not lean forward;
  - Sit upright, but try to keep your shoulders relaxed;
  - Use a chair and desk that allow you to keep the elbows and hands at the same height as the keyboard. Shift the keyboard back a bit from the edge of the desk to make space to rest your hands;
  - Position the mouse close to the edge of the keyboard;
  - Keep your feet flat on the floor.

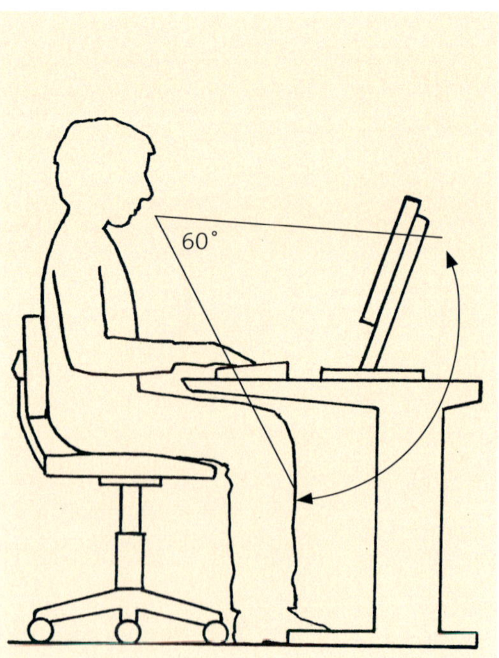

Figure 21 Correct posture when working at a computer

- Try not to sit in one place for a long time. Stand up, move around, stretch and bend your body, or do some gymnastic exercises, which relax your muscles and your backbone. Also try to move about during breaks in school. You will be able to concentrate and learn much better after getting some exercise.

1 arch *Wölbung*
2 fallen arches *Senkfuß*
3 flat feet *Plattfüße*
4 shoe insert *Schuheinlage*
5 barefoot *barfuß*
6 fashionable *modisch*

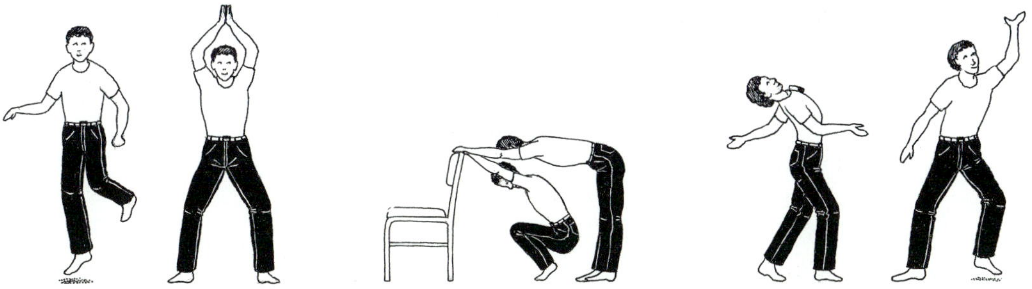

Figure 22 Some exercise ideas

1 Describe the normal shape of the backbone compared to the shape in a person with a round or hollow back. For language help, see **Activate your English** on the inside cover.
2 What kinds of physical activities[1] would you suggest for someone who works in a job where they have to sit a lot? Explain your answer.
3 Develop a poster with physical exercises for the prevention[2] of foot problems and defects in posture. You can get more information online and from doctors, hospitals, and health insurance[3] companies.

1.8 Treatment[4] of common injuries to bones and muscles

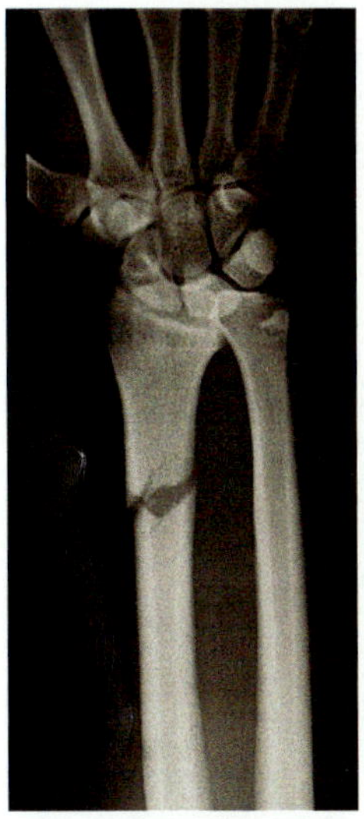

Figure 23 An x-ray of a broken radius

1 physical activity *körperliche Tätigkeit*
2 prevention *Vorbeugung*
3 health insurance *Krankenversicherung*
4 treatment *Behandlung*
5 damage sth. *etwas beschädigen*
6 pound sth. *auf etwas hämmern, schlagen*
7 endure sth. *etwas ertragen, aushalten*
8 injury *Verletzung*

Bones and muscles can withstand a lot without being damaged[5]. When you are angry, you may pound[6] a table very hard without **injuring** yourself. Remember that we have already talked about the forces that act on the muscles and bones during a football game. Or think of the weight that an athlete's bones have to endure[7] during a high jump.
The most common **injuries** that affect the skeletal system are **sprains**, **dislocations** and **fractures**. Sprains happen more frequently than any other joint injury[8]. A **muscle strain** is an injury that often affects the skeletal muscles of athletes, but can happen to anybody during the course of normal daily activities, especially when heavy objects are lifted quickly.
The following table gives you some information on these injuries. It is important that you know the most common ones and what can be done to help a person who is injured. If it is not clear what is wrong or if the injury hurts a lot, the person should see a doctor, who will decide on the type of **medication** or further medical treatment.

Key terms

Injuries and their treatment
- antibiotic *Antibiotikum*
- bandage *Verband*
- cast *Gipsverband*
- complete fracture *vollständiger Bruch*
- dislocation *Verrenkung*
- fracture *Knochenbruch*
- germ *Keim*
- heal *heilen*
- incomplete fracture *unvollständiger Bruch*
- injure yourself *sich verletzen*
- injury *Verletzung*
- medical care *medizinische Hilfe*
- medication *Medikament*
- muscle strain *Muskelzerrung*
- open/closed fracture *offener/geschlossener Bruch*
- plaster cast *Gipsverband*
- overstretch sth. *etwas überdehnen*
- regenerate *sich regenerieren*
- splint *Schiene*
- sprain *Verstauchung*
- sprain sth. *etwas verstauchen*

Table of common disorders of the skeleton and its muscles

Type of injury	What is wrong	What can be done
Muscle strain	If a cold muscle is **overstretched**, some of the muscle fibres can become damaged.	The injured part of the body must be rested. Then the damaged muscle fibres will quickly **regenerate**.
Sprain	A tendon or the ligaments of a joint have been overstretched or torn[1].	If the sprain is mild, it will **heal** by itself if it gets enough rest. But if it hurts a lot, you need **medical care**.
Dislocation	The ends of two bones have been pulled out of their normal positions in a joint.	The bones have to be put back into their proper positions. Do not try to do this yourself! They have to be held in place by a **bandage** or **plaster cast** until the joint is healed.
Fracture	With a **complete fracture** a bone has broken into two or more pieces. With an **incomplete fracture** there is only a crack in the bone. With an **open fracture** the bone pierces[2] the skin, whereas with a closed fracture it does not. When the skin is broken in an **open fracture**, **germs** can get into the body.	The broken ends of the bone are put together again and are held in position by **splints** and a **cast**, or in some cases by a bandage. In the case of an open fracture the doctor will prescribe[3] an **antibiotic** to kill the germs that may have entered the body.

1 tear sth. *etwas zerren*
2 pierce sth. *etwas durchdringen*
3 prescribe sth. *etwas verschreiben*
4 dislocate sth. *etwas verrenken*
5 accidentally *zufällig*

1 A football player runs into another player and has to be treated for a dislocated[4] shoulder. What has actually happened to the injured player?
2 Explain what a doctor does with someone who has **sprained** their elbow. What does the doctor tell the patient about moving the injured arm?
3 Point out what happens when you sprain your ankle.
4 Explain why a strained muscle is so painful.
5 Describe how an open fracture is treated.
6 What may be wrong when you feel strong pain in the calf of your left leg after being accidentally[5] kicked there during a game of football? What could you do to find out?
7 **Say it in German:** Your classmate has been ill for the past few days and wants to know what topic you worked on in the last biology lesson.
  a Tell him or her that you talked about defects in posture and explain what can be done to prevent them.
  b Explain all that you can remember about 'injuries to bones and muscles, and how to treat them'.

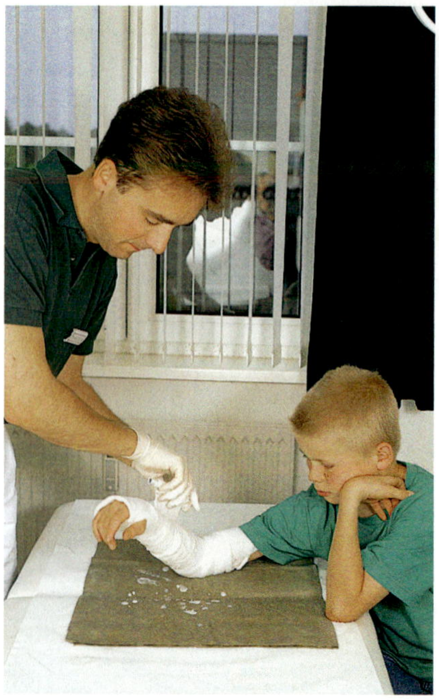

Figure 24 A broken arm is put in a plaster cast

Summary

The skeleton gives the body stability and protects the inner organs. The special shape of the backbone (double-S) is responsible[1] for the body's upright posture. Joints and muscles make the skeleton mobile. Skeletal muscles work in pairs and are voluntary, whereas the cardiac muscle and the muscles of the inner organs are involuntary. The muscles are attached to the bones by tendons. Regular use strengthens the muscles and keeps the joints flexible.

Key terms

See the boxes:

Parts of the skeleton .... p. 15
The structure of the bones .... p. 17
The joints .... p. 18
Muscles and their functions .... p. 20
Injuries and their treatment .... p. 27

A Fill in the gaps

Complete the sentences by using the correct words and phrases on the right. Write the sentences into your exercise book. Each word or phrase should be used only once.

consists of	are connected
are attached	runs through[2]
protect	covers
are made of	contain
is composed[3] of	is divided into

1. *One of the functions of your skeleton is to ... the inner organs.*
2. *The backbone ... a large number of vertebrae.*
3. *The spinal cord ... the vertebrae.*
4. *The muscles ... to the bones by tendons.*
5. *A tough membrane ... the outside of a bone.*
6. *A skeletal muscle ... bundles of muscle fibres.*
7. *Your food must ... enough calcium to make your bones hard.*
8. *Bones which ... by joints can be moved.*
9. *The head of the thighbone ... two rounded parts.*
10. *The menisci ... elastic rings of cartilage.*

B Bio-quiz

1. The only bone in your skull that can be moved is your ....
2. Involuntary muscles move your ....
3. Muscles which pull a bone in opposite directions are called ....
4. The bones of a joint are held in place by ....
5. ... prevent the vertebrae from rubbing against each other.
6. Blood cells are made in the ....
7. The muscle which straightens your forearm is called ....
8. The joint at the base of your thumb is a ....
9. The extensor muscle in your arm is also called the triceps because it ....
10. The bones in your body are moved by voluntary muscles, which means ....

1 responsible *verantwortlich*
2 run through sth. *durch etwas hindurchführen*
3 be composed of sth. *aus etwas bestehen*

C Critical thinking

1 Why do you think it is an advantage that some muscles in your body work without the steady control of the brain?

2 Do you think it is useful that the brain makes your heart beat faster when someone or something frightens you? Explain your answer.

3 Explain why there are spaces between the bones on the skull of a new-born baby.

Chapter 2 **Respiration**

Chapter objectives

- To **name** the parts of your respiratory system and **describe** their function
- To **explain** the role of the diaphragm and the intercostal muscles in breathing
- To **explain** how the gases in the air are exchanged in your lungs
- To **find out** the effect of exercise on your breathing rate
- To **analyse** the dangers of smoking to a person's health

Since the moment you came into this world, you have been breathing in and out, day and night, even when you sleep. This process is called **respiration**: your body takes in **oxygen** from the atmosphere and releases **carbon dioxide**. But why does your chest and your belly move when you breathe? Why do we have to breathe at all? And where does the oxygen go that you take in? This chapter will help answer these questions.

2.1 The organs used for breathing make up your respiratory system

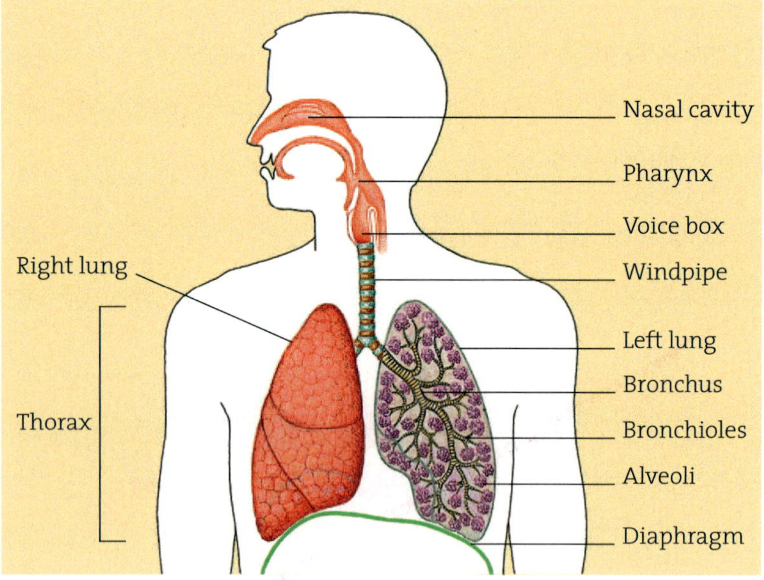

Figure 1 The respiratory system

Key terms

Anatomy of the lungs
- alveolus (pl. alveoli) *Lungenbläschen*
- bronchus (pl. bronchi) *Bronchie*
- bronchiole *Bronchiole*
- capillary *Kapillare*
- cilium (pl. cilia) *Wimperhärchen*
- diaphragm *Zwerchfell*
- left and right lungs *Lungenflügel*
- mouth cavity *Mundhöhle*
- nasal cavity *Nasenhöhle*
- pharynx *Rachen*
- pleural membranes *Lungen- und Rippenfell*
- voice box *Kehlkopf*
- windpipe (trachea) *Luftröhre*

Breathing (1)
- carbon dioxide *Kohlenstoffdioxid*
- oxygen *Sauerstoff*
- respiration *Atmung*
- respiratory system *Atmungssystem*

Let us first look at how air gets into your lungs. When you breathe in, the air enters your body through your nose or your mouth. In your **nasal cavity** the air is warmed, moistened[1] and cleaned. The inside of your nose is covered in mucus[2]. Here dust particles and germs[3] are caught and millions of tiny hairs (**cilia**) brush them towards your throat. By sneezing and coughing, your body forces them back out.

The air then passes into your **pharynx** and through the **voice box**, which helps us produce sounds for speaking and singing. It flows down your **windpipe (trachea)**, which is 11 to 14 cm long, and divides[4] into two tubes called **bronchi**. Rings of cartilage protect the windpipe and hold it open all the time. The bronchi carry the air into both lungs. The lungs are spongy[5], shaped like sacks and lie inside the ribcage, which protects them (see Figure 2).

1 moisten sth. *etwas anfeuchten*
2 mucus *Schleim*
3 germ *Bazille, Bakterie*
4 divide into sth. *sich in etwas aufteilen*
5 spongy *schwammig, weich*

The bronchi then divide into smaller and smaller tubes, or **bronchioles**, like the branches[1] and twigs[2] of a tree. At the ends of the bronchioles are clusters of tiny balloons called **alveoli**. Each individual alveolus is smaller than a grain of salt. Around each one is a close net of blood **capillaries** (see Figures 3 and 4). The purpose and function of the alveoli and capillaries will be explained in the next section.
Between the lungs and ribs are two layers[3] of skin called **pleural membranes**. They allow the lungs to move smoothly along the ribs. A thin sheet of muscle, the **diaphragm**, separates your thorax (the part of the upper body which includes the ribs, the heart and lungs) from the organs of your abdomen.

1 Explain why it is better to breathe through your nose than through your mouth.
2 What do you think would happen if there were no rings of cartilage round the windpipe?
3 Describe the function of the pleural membranes.

2.2 The lungs cannot work without muscles

Your lungs are soft and cannot move by themselves. So how does breathing work?
When you watch somebody run or play sports, you can see that his or her chest moves up and down. This is because there are muscles between the ribs called **intercostal muscles**, which **contract** to lift the ribcage. This makes more space inside the ribcage so that the lungs can expand and fill with air. When the intercostal muscles **relax**, the ribcage sinks, the space inside becomes smaller, and the air is pushed out of the lungs.

Key terms

Breathing (2)
- contract *sich zusammenziehen*
- intercostal muscles *Zwischenrippenmuskeln*
- relax *sich entspannen*

Another muscle plays an important role during breathing: the diaphragm. It is normally dome-shaped[4], but when you breathe in, it contracts and flattens, making the room inside your chest bigger and sucking air into your lungs. When the diaphragm relaxes, it becomes dome-shaped again and the space for the lungs becomes smaller. This forces the air out of the lungs.

1 branch *Ast, Zweig*
2 twig *kleiner Zweig*
3 layer *Schicht*
4 dome-shaped *gewölbt*

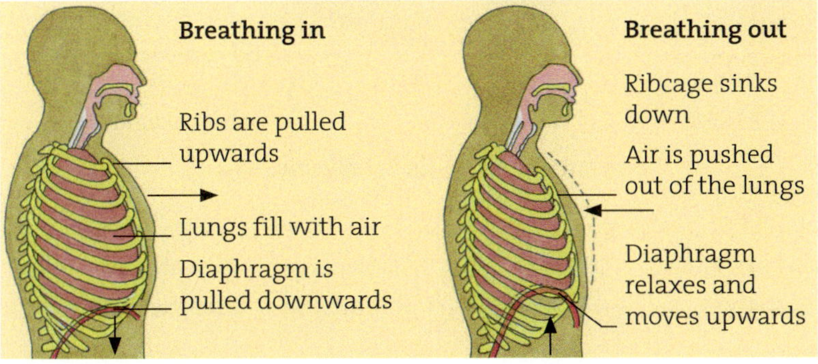

Figure 2 The movement of the ribcage and diaphragm

When you look at your classmate sitting next to you, you can hardly see any breathing movement in his or her chest. So how do we breathe when we are at rest?
In this case, the contracting and relaxing of the diaphragm muscle brings enough air into your lungs to give the organs the oxygen they need.
You will discover more about the working of your diaphragm when you do the bell jar experiment (see 'Improve your skills', p. 36).

1 Stand up straight and place your left hand on your lowest ribs and your right hand flat on your abdomen. Then breathe in and out normally. Describe and explain how your ribcage moves.
2 How do you breathe when you are sleeping?
3 **Say it in German:** Explain to a friend how his or her lungs work at rest and during sport.

2.3 Gas exchange takes place in the alveoli and in the cells

Alveoli and capillaries both have very thin walls so that gases can pass through them easily. The air you breathe in has a lot of oxygen in it, which passes through the capillary walls and enters the bloodstream[1]. At the same time, the blood coming from the organs carries carbon dioxide. So while oxygen is being taken into the body, carbon dioxide is being moved in the opposite direction: it passes from the bloodstream into the alveoli and is then breathed out. This process is called **external respiration**.

The oxygen coming in is taken up by the red blood cells in the capillaries, which carry it to the different cells of your body. Those cells then use the oxygen to get **energy** from food substances which you have eaten and which the blood has transported there. The oxygen is needed to burn these substances in a process which produces carbon dioxide. Since this process takes place inside your cells, it is called **internal respiration**, also known as cellular respiration.

Key terms

Exchange of gases
- energy *Energie*
- external respiration *äußere Atmung*
- internal respiration *innere Atmung*

Did you know?

There are about 300 million alveoli in your lungs. If they were placed side by side, their combined surface would form an area of about 70 square metres – nearly the size of half a tennis court.

1 bloodstream *Blutbahn, Blutkreislauf*
2 close-up *Nahaufnahme*
3 nitrogen *Stickstoff*

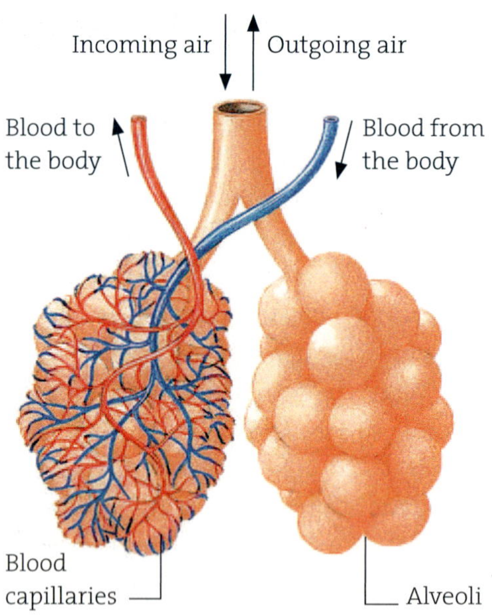

Figure 3 Alveoli with capillaries

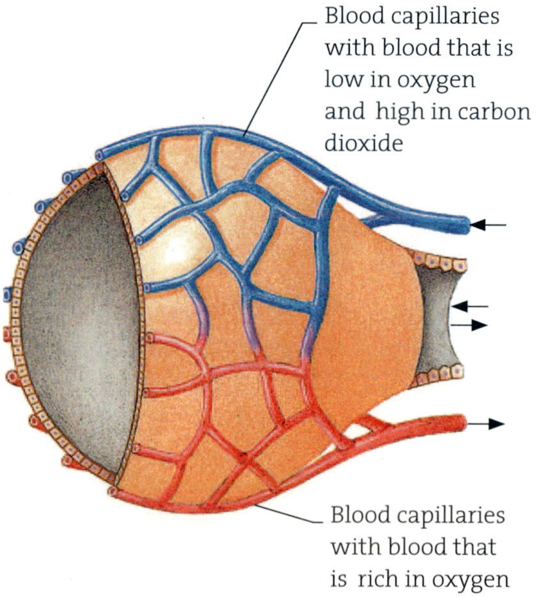

Figure 4 Close-up[2] of alveolus

1 Describe where gases are exchanged in the body and in which direction this exchange takes place.
2 Explain the advantage of having such a large number of small alveoli.
3 Look at the table on the right and answer these questions:
   a How does the amount of oxygen and carbon dioxide change in your lungs?
   b Why do we breathe out more carbon dioxide than we breathe in while the amount of nitrogen stays the same?

Gas	% in air you breathe in	% in air you breathe out
Oxygen	21	16–17
Carbon dioxide	0.03	3.5
Nitrogen[3] and other gases	78	78

2.4 Smoking is bad for your health

Health organizations have found that children have their first contact with cigarettes between the ages of 8 and 14. Children and teenagers start smoking for different reasons. Some start because their friends smoke, while others want to impress people or be accepted in a group. The earlier young people start smoking, the greater are the dangers to their health.
In Germany more than 112,000 people die each year from diseases connected with smoking. This is about the number of deaths that would occur if a large plane crashed every day and killed all its passengers. According to the World Health Organization, about half of all regular cigarette smokers are killed by their habit[1].

Figure 5 He wants to belong to the group

What makes tobacco smoke so dangerous?

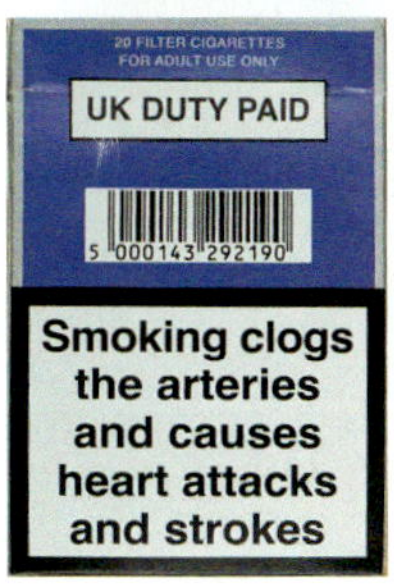

In the European Union every packet of cigarettes and tobacco must display a health warning in bold letters and information on how much tar, nicotine and carbon monoxide are contained in one cigarette. This is important information because these three substances are the reason why smoking is so dangerous.
Nicotine is a poisonous[2] **drug**, which passes quickly from the lungs into the blood and is carried to all the organs in the body. It has two particularly negative effects on the body: First, nicotine is **addictive**. When the body gets used to[3] nicotine, it does not want to be without it. This is why many people find it hard to give up smoking. Second, the drug causes your blood vessels[4] to contract, which makes it difficult for your blood to circulate properly. This can lead to **heart attacks** and **strokes**. The risk of having a heart attack is two or three times higher in smokers than in non-smokers. Also, most leg **amputations** in Germany are carried out on smokers. The tissues in the arms and legs can die when they do not get enough oxygen-rich blood.
Carbon monoxide, which is one of over 4000 chemical compounds[5] in tobacco smoke, is also absorbed into the blood. This gas not only reduces the amount of oxygen that the blood can transport, but also affects body cells so that they cannot take in as much oxygen as they normally do. This is why smokers are often short of breath[6] when they are active during work or sports. Carbon monoxide is also produced by car engines, so it not only affects smokers.

Key terms

Smoking and its effects
- amputation *Amputation*
- be addictive *abhängig oder süchtig machen*
- bronchitis *Bronchitis*
- cancer *Krebs, Krebserkrankung*
- carbon monoxide *Kohlenstoffmonoxid*
- drug *Droge, Arzneimittel*
- emphysema *Emphysem*
- headache *Kopfschmerz*
- heart attack *Herzanfall*
- inflammation *Entzündung*
- lung cancer *Lungenkrebs*
- nicotine *Nikotin*
- passive smoking *Passivrauchen*
- stroke *Schlaganfall*
- tar *Teer*
- tumour *Geschwulst, Tumor*

1 habit *(hier) Sucht*
2 poisonous *giftig*
3 get used to sth. *sich an etwas gewöhnen*
4 blood vessel *Blutgefäß*
5 chemical compound *chemische Verbindung*
6 short of breath *kurzatmig*

Tar is the third dangerous substance and the reason why the lungs of heavy smokers are coloured black on the inside. When tobacco smoke cools in the mouth, pharynx and lungs, the tar in the smoke settles[1] on the inner surface of these organs. Someone who smokes 20 cigarettes a day breathes a full cup of tar (210 g) into his or her lungs in one year.

Normally your lungs are kept clean by mucus[2], which covers the inner lining[3] of the lungs. The mucus catches and holds dust particles and germs. The cilia (see p. 30) then transport the mucus out of the lungs. Tar makes the cilia stick together so that they stop working properly. The mucus then stays in the lungs and becomes an ideal breeding ground[4] for bacteria and viruses. It can also lead to a disease called **emphysema** when the affected person has to cough repeatedly to get rid of[5] the mucus. This damages the delicate alveoli. Holes then develop in the lung tissue (see Figure 6) and it becomes more difficult for the person to breathe.

Figure 6 Lung tissue showing signs of emphysema

Tar and smoke also irritate the lining of the respiratory organs. When the bronchi become swollen and painful this **inflammation** of the bronchi is called **bronchitis**. Finally, tar contains over 400 different chemicals, and more than 30 of them can cause **cancer**.

Lung cancer is the most serious disease which can result from long-time smoking. Cancer begins when cells of the inner lining of the bronchi start to divide in an uncontrolled way. Later **tumours** develop in the lungs. Once lung cancer starts, it may spread to other parts of the body via the blood and lymph system. Lung cancer is difficult to detect at an early stage, when it can still be treated. If smokers go to see their doctor when they feel a pain in their chest, it may be too late and the chances of surviving are relatively small. In Germany, Britain and the USA, lung cancer is one of the major causes of death. In Germany, 54,000 people die of lung cancer every year. Although some non-smokers may get lung cancer, 80 per cent of all people who die of this disease are smokers.

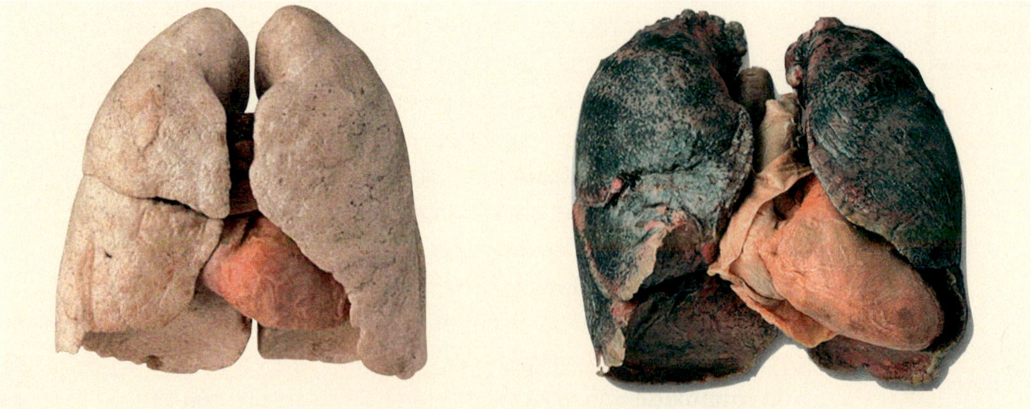

Figure 7 A healthy lung (left) and the lung of a heavy smoker (right)

Smoking is especially dangerous during pregnancy[6]. If a woman smokes during her pregnancy, the chances are high that her new-born baby will be smaller than normal because the develop-

1 settle *sich niederlassen*
2 mucus *Schleim*
3 inner lining *Schleimhaut*
4 breeding ground *Brutstätte*
5 get rid of sth. *etwas loswerden*
6 pregnancy *Schwangerschaft*

ing foetus[1] will get less oxygen and food. Brain cells are especially sensitive to carbon monoxide, so the risk of damaging the foetus's brain is greater.
Even people who do not smoke at all can be affected by tobacco smoke if they are around smokers too often. **Passive smoking** not only irritates your eyes and causes sore throat[2] and **headaches**, it can lead to the same diseases that smokers get. Today more and more countries are introducing laws which forbid smoking in public buildings, hospitals, offices and restaurants, in order to protect the health of non-smokers.

1 foetus *Fötus*
2 sore throat *Halsschmerzen, Halsentzündung*

State	Bars	Restaurants	Private workplaces other than bars or restaurants	Government buildings
Alabama			▲	▲
Alaska		▲	▲	▲
Arizona	●	●	●	●
California	●	●	●	●
Florida		●	●	●
Nevada		●	●	●
New York	●	●	●	●
Texas				
Wyoming				▲
England	●	●	●	●

● Smoking is prohibited ▲ Smoking is restricted

Table 2 **Anti-smoking laws in some US states and in England** *(From: http://ash.org/smokingbans.html, 5 February 2010)*

1 Name the three most dangerous substances in tobacco smoke. In small groups, put the information on these substances from pp. 33–34 into cause-and-effect tables as shown below (one table for each substance). Then take turns describing the effects of the substances to each other. See **Activate your English** below for language help.

Cause	*Effect(s)*
Nicotine enters the blood …	…

2 Explain why there is even more reason for pregnant women not to smoke.

3 In Germany the average age when people start smoking is 13 ½. Why do you think teenagers take up smoking? Make a list of reasons why it is better not to start.

4 **a** Look at Table 2 and compare and contrast the laws in two US states or in one of the states and England. The following words and phrases will help you: … *is more/less strict than* … – … *is not as strict as* … – … *is (not) prohibited/restricted in* … – … *are (not) allowed to smoke in* … – *Although,* … – *However,* …

b Research: Find out what the current anti-smoking laws are where you live and compare them with the laws in England.

Activate your English

Talking about causes and effects

Causes

Carbon monoxide can cause … by …
Tar affects the … by causing the … to …
… is caused by …
… results from …

Effects

When nicotine / … enters / passes into / is carried to …, it …
… leads to …
… can result in …
As a result of this, …
The result (of this) is that …

Working with models

Models are helpful in biology for understanding complicated processes. They are effective because they only show the most important features of what is being described or simulated. By leaving less important aspects out, models make things easier to understand. This also means, however, that they do not fully represent reality. We need to keep this in mind when we compare models with 'the real thing'. The following examples demonstrate this.

1 Using a bell-jar[1] model

Materials:
- plastic bell jar (see Figure 8 below)
- glass tube[2] branched in two at one end
- rubber sheet[3]
- rubber ring for fastening the rubber sheet
- rubber bung[4] with one hole
- two balloons

1 bell jar *glockenförmiges Gefäß*
2 glass tube *Glasrohr*
3 rubber sheet *Gummihaut*
4 rubber bung *Gummistopfen*

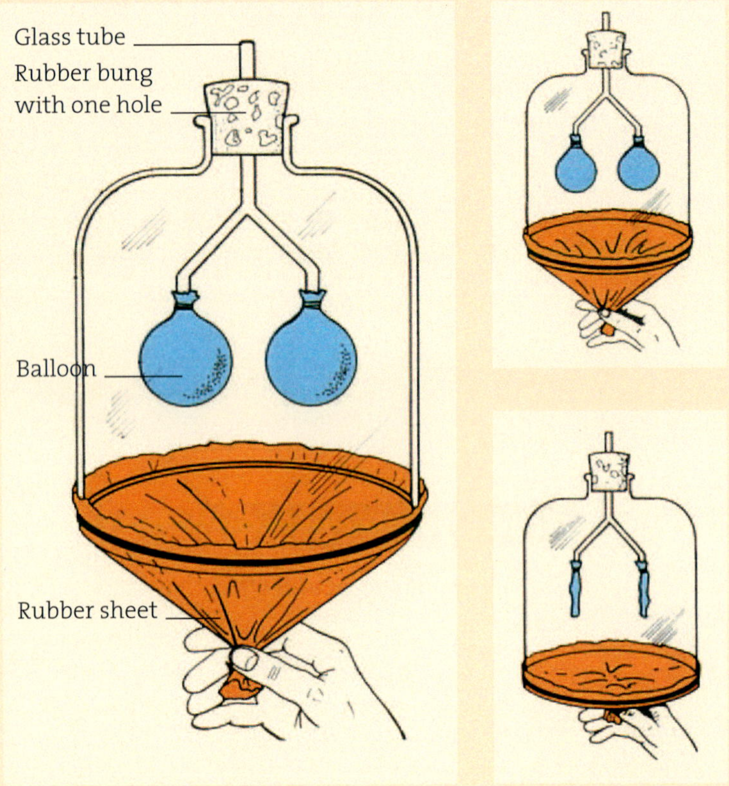

Figure 8 The bell-jar model

Procedure:
1. Pull the rubber sheet downwards (see Figure 8). What happens inside the bell jar?
2. Now push the rubber sheet upwards. Describe what happens.
3. Explain why the balloons inside the bell jar behave this way.
4. Now think about your lungs. Compare the model with your respiratory system by completing the following table.

Part of the model	*Respiratory organ*
Bell jar	...
...	...

Table 3 A comparison

1 Based on what you have observed from the bell-jar model, explain how we breathe in and out. Make sure you mention the respiratory organs in the second column of the table above.
2 Explain why the model cannot show exactly how breathing works in our body.
3 How is our breathing during physical exercise different from the breathing demonstrated by the bell-jar model?

2 Making a smoking apparatus

The aim of this activity is to better understand what smoke does to your lungs.

Materials:

- Erlenmeyer flask[1]
- Glass tubes (two)
- Rubber bung with 2 holes
- Cigarette without a filter
- Rubber tubing[2]
- Suction bulb[3] or suction pump[4]
- Cotton wool[5]

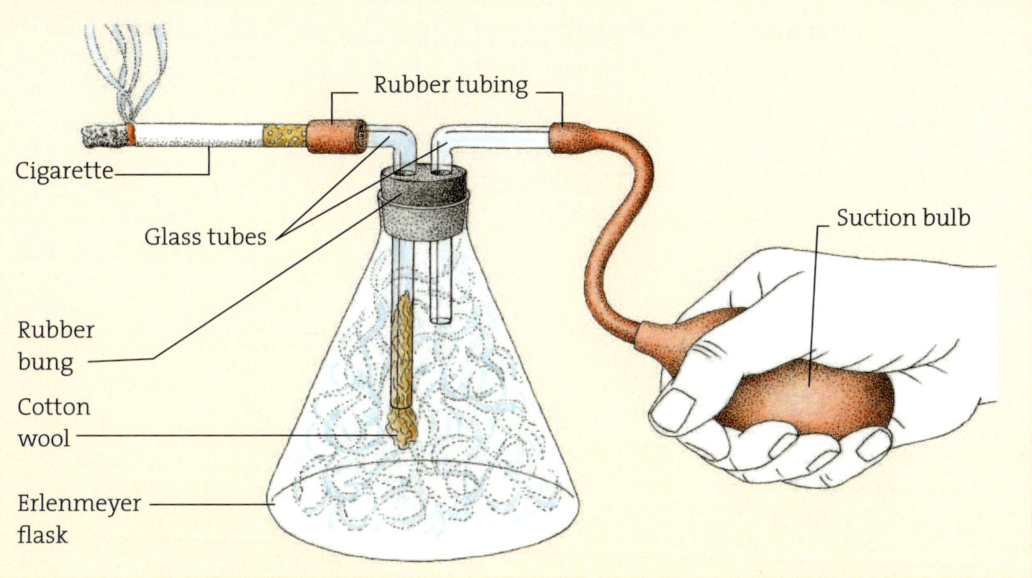

Figure 9 **A smoking apparatus**

Figure 10 **A cigarette filter does not absorb all dangerous substances**

Procedure:

1. Set up the apparatus as seen in the picture above. Use a short rubber tubing to connect the cigarette to the glass tube.
2. Light the cigarette. Then draw air[6] through it by squeezing the suction ball. Repeat the action for about one minute.
3. Describe what happens to the model as you work the suction ball.
4. Write down your observations.

1 Erlenmeyer flask *Erlenmeierkolben*
2 rubber tubing *Gummischlauch*
3 suction bulb *Saugball*
4 suction pump *Saugpumpe*
5 cotton wool *Watte*
6 draw air through sth. *Luft durch etwas hindurchziehen*

1 As in all models, certain parts of the apparatus stand for certain things in reality. Make a table and match each part with the corresponding organ in your body (see Table 3, p. 36). Note: the rubber tubing and bung are needed for the model and do not represent parts of the body.

2 Use your observations and what you have learned from the chapter so far to explain the effect that smoking has on your organs.

Activity 1

Investigating[1] how exercise affects the breathing rate[2]

Before you begin, read all the instructions. Then do the whole activity without stopping.

1. a Sit quietly and relaxed for about *two minutes*.
   b Now count for *one minute* how many times you breathe in.
   c Record[3] the number in your exercise book.
2. a Do some physical exercise[4], for example bending the knees or stepping up onto a chair and down again, for *two minutes* without stopping.
   b Now sit down quickly and count again for *one minute* how many times you breathe in.
   c Record the number in a table as shown in Table 4 below.
   d Relax for *one minute*, then record your breaths per minute again. Repeat this four times.

This diagram gives you an overview. Starting at the top, follow the wheel in a clockwise direction for one round.

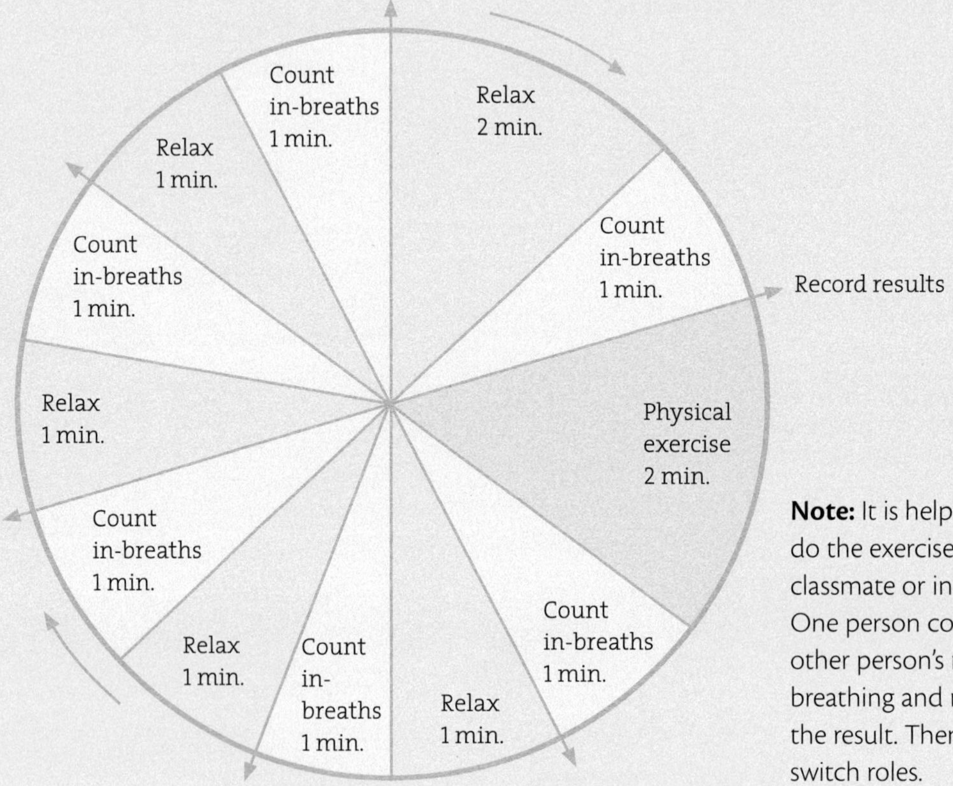

Note: It is helpful to do the exercise with a classmate or in a group. One person counts the other person's rate of breathing and records the result. Then you switch roles.

Figure 11 Experimenting with your breathing rate

Time passed after activity	*Number of breaths per minute*
1 min.	…
3 min.	…
5 min.	…

Table 4 Activity and breaths per minute

1 investigate sth. *etwas untersuchen*
2 breathing rate *Zahl der Atemzüge pro Zeiteinheit*
3 record sth. *etwas aufschreiben, notieren*
4 physical exercise *körperliche Aktivität(en)*

1 Compare your results with those of your classmates.
2 Give reasons for the quick rise in the number of breaths per minute during the exercise.
3 Why do you think it takes so long after exercising for your breathing rate to return to normal?

2.5 Breathing, swallowing and speaking

When you breathe in through your nose or mouth, the air passes first through the pharynx and then the windpipe or trachea. At the opening of the trachea is a small piece of tissue called the **epiglottis**, which directs air to the voice box. When you breathe in or out, the epiglottis is in an upright position so that air can pass through the trachea. When you swallow food, it automatically folds down and closes the entrance to the trachea so that the food goes into the gullet[1].
This is why it is a good idea not to talk or laugh while you are eating or drinking, as food or liquid can go 'down the wrong pipe'. When this happens, a reflex action in the voice box makes you cough until the food or liquid is forced out of the windpipe.

Key terms

Breathing and speaking
- epiglottis *Kehldeckel*
- vocal cords *Stimmbänder*

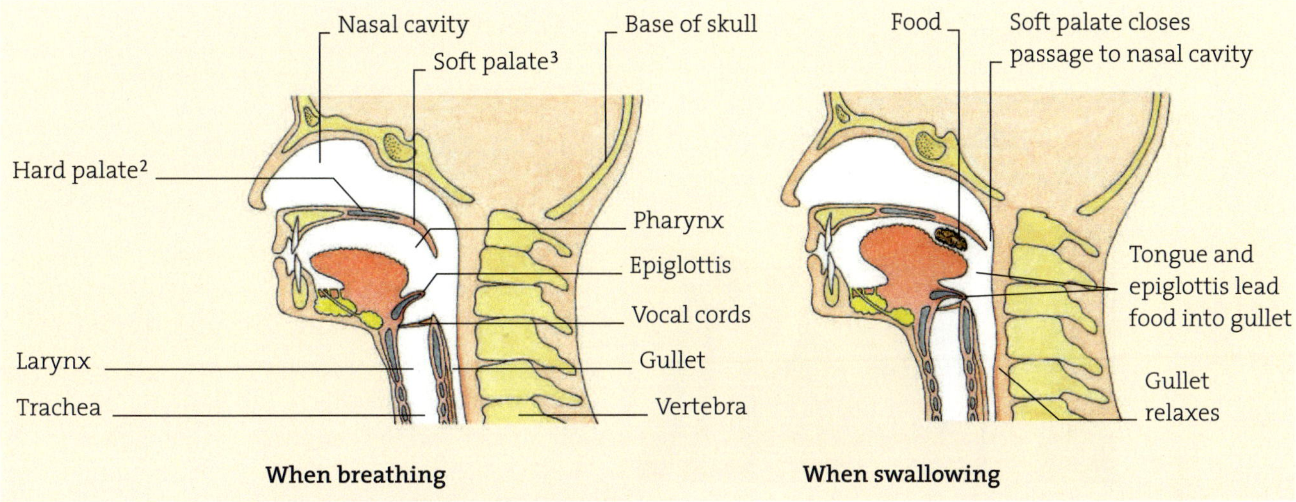

Figure 12 Position of the epiglottis

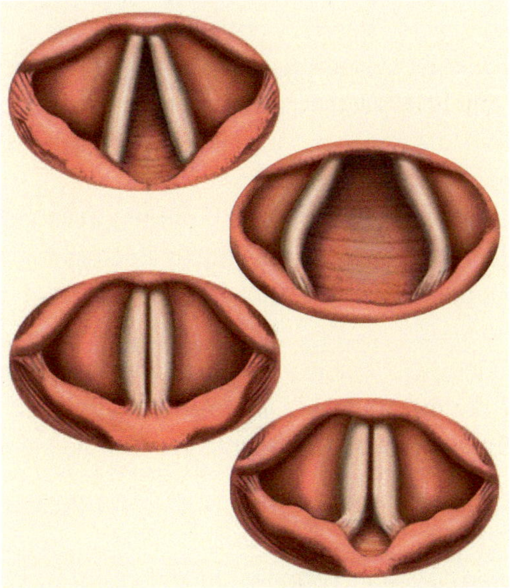

Figure 13 How vocal cords produce different sounds

Further facts

Your voice
At the front and back of the voice box are two flat folds of skin called the **vocal cords**. When air passes over them, they vibrate[4] and produce sounds. The pitch[5] and the quality of the sounds are varied when certain muscles change the tension[6] in the cords and the distance between them. As the vocal cords are slightly longer in males than in females, men normally have deeper voices than women. The human voice is produced by a complex coordination[7] of the lips, the tongue, the jaws, the breathing movements and the tension and position of the vocal cords.

1 gullet *Speiseröhre*
2 hard palate *harter Gaumen*
3 soft palate *weicher Gaumen, Gaumensegel*
4 vibrate *vibrieren*
5 pitch *Tonhöhe*
6 tension *Spannung*
7 coordination *Abstimmung, Koordination*

2.6 Diseases of the respiratory organs

Many diseases of the respiratory system are caused by air pollutants[1] from the burning of fossil fuels[2] in industrial plants[3], households, cars and other motor vehicles. These substances damage the breathing passages and make it easier for **germs** to **infect** the mucous membranes[4] in the nose, mouth, windpipe and lungs.

Common cold: This **infection** of the nasal passages[5] is caused by a **virus**. Typical symptoms[6] are a running nose, light fever, headache, and loss of appetite and sense of smell. While no medication[7] can stop a cold virus, some can relieve[8] the unpleasant symptoms. When you have a cold, you should keep your body warm and drink lots of liquids. After two or three days your immune system[9] will normally win the fight against the germs and you will start feeling better.

Key terms

Germs and infections
- bacterium (pl. bacteria) *Bakterie, Bakterium*
- bronchitis *Bronchitis*
- common cold *Schnupfen, Erkältung*
- cough *Husten*
- germ *Krankheitserreger*
- infect sb./sth *jdn./etwas infizieren, anstecken*
- infection *Infekt, ansteckende Krankheit*
- influenza, flu *Grippe*
- pneumonia *Lungenentzündung*
- sore throat *Halsschmerzen, Halsentzündung*
- tuberculosis *Tuberkulose*
- virus *Virus*

Figure 14 **Inhalations help dissolve mucus**

Cough: This condition[10] often arises in connection with a cold. When the lining of your throat and, in some cases, the upper part of your windpipe become infected, you get a **sore throat**. By coughing, the body throws out mucus and the harmful viruses and **bacteria** that gather there. If the cough does not go away after a few days, you should see your doctor.

Influenza: Better known as (the) **flu**, this viral infection has symptoms similar to a cold. Sore throat and aching muscles are also typical. The treatment[11] is basically the same: rest and drink lots of liquids, preferably water. A flu can be dangerous and even deadly for old or weak people. It can also sometimes lead to more serious diseases such as pneumonia.

Pneumonia and **tuberculosis:** Both are dangerous infectious diseases of the lungs which are caused by bacteria. They can be treated[12] with antibiotics.

Bronchitis: In a person with bronchitis the lining of the bronchi is inflamed[13] and secretes a thick and sticky mucus. The body tries to clear out the mucus by coughing. Inhalations[14], e.g. with dried chamomile flowers, wet packs[15] and the drinking of special teas helps dissolve[16] the mucus and relieve the inflammation.

To protect yourself against infections: Do not work or play in rooms that are overheated, because the air dries out the mucous membranes of your nose and throat. Avoid breathing in tobacco smoke and the exhaust fumes of cars. Spend a lot of your free time outside, get exercise and eat lots of fresh fruit and vegetables every day.

1 air pollutant *Luftschadstoff*
2 fossil fuel *fossiler Brennstoff*
3 industrial plant *Industriebetrieb*
4 mucous membrane *Schleimhaut*
5 nasal passage *Nasengänge*
6 symptom *Symptom, Krankheitserscheinung*
7 medication *Behandlung/Einnahme von Medikamenten*
8 relieve *erleichtern*
9 immune system *Immunsystem*
10 condition *Befinden, Zustand*
11 treatment *Behandlung*
12 treat sb./sth. *jdn./etwas behandeln*
13 inflamed *entzündet*
14 inhalation *Inhalieren*
15 wet pack *feuchter Wickel*
16 dissolve *auflösen*

1 Why do you think it is good to open classroom windows during breaks?
2 Compare the common cold with influenza.
3 When did you last have a cold? Describe the symptoms and how you got rid of it.

Activity 2 Measuring[1] the vital capacity[2] of your lungs

Materials:
- large bottle (5–6 litres) with measurements (see Figure 15 below)
- rubber bung[3]
- piece of rubber or plastic tubing
- glass or plastic container filled with water
- glass tube for a mouthpiece
- glass tube for the other end of the rubber tubing
- stand with a clamp[4] to hold the bottle[5]

Procedure:
1. Fill the large bottle completely with water and close it with the rubber bung. Place the bottleneck under water in the container, clamp the bottle to the stand and loosen the rubber stopper. Then stick one end of the rubber tube into the bottle and push it up to the top.
2. Now take a deep breath[6], place your lips round the free end of the glass tube (the mouthpiece) and breathe out as much air as you can.

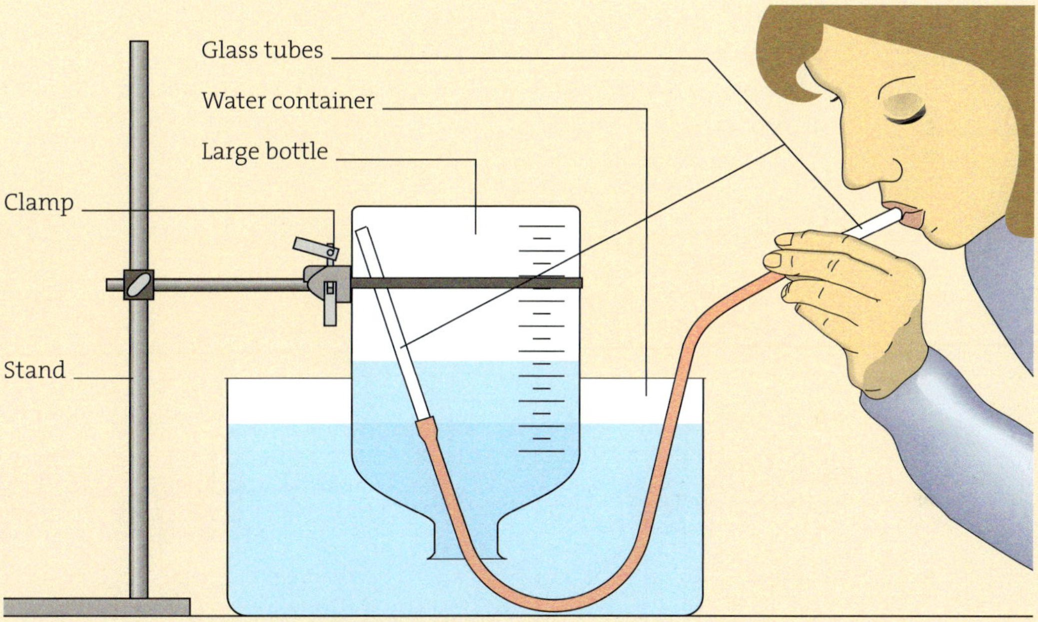

Figure 15 **Collecting the exhaled air**

3. Note what happens to the water in the bottle.
4. When you cannot exhale[7] any more, seal the open end of the glass tube with your finger while pulling the tube out of the bottle. Now close the bottle with the bung.
5. Record how much air the bottle contains. Keep the neck of the bottle under water the whole time.

1 measure sth. *etwas messen*
2 vital capacity *maximaler Lungeninhalt*
3 stopper *Stöpsel*
4 clamp *Klammer*
5 clamp sth. to sth. *etwas zusammenklammern*
6 breath *Atemzug*
7 exhale *ausatmen*

1 Describe what happened when you exhaled into the bottle.

2 Your teacher may ask different people in your class to carry out the experiment. Among them there should be a good athlete and, if possible, a pupil who plays a wind instrument (e.g. a clarinet, a saxophone or a trumpet), boys as well as girls. The mouthpiece should be exchanged or cleaned before each new person does the experiment.
Collect the results of everyone's vital capacities in a simple table and try to explain the differences.

3 Discuss what you could do to increase the capacity of your lungs.

Activity 3

Finding out how exhaled air differs from inhaled[1] air

For this activity you need to know that lime water[2] is an indicator for carbon dioxide (CO_2). The clear solution[3] becomes milky white when carbon dioxide dissolves in it.

Materials:
- 2 Erlenmeyer flasks[5] (250 ml)
- 2 glass tubes, 15 cm long
- 2 glass tubes, 10 cm long
- short glass tube for a mouthpiece
- 2 rubber bungs, each with two holes
- T-tube
- 2 pieces of rubber tubing to connect the flasks with the T-tube
- longer piece of rubber tubing to connect the mouthpiece with the T-tube
- freshly prepared lime water (about 300 ml)

1 inhale (sth.) *(etwas) einatmen*
2 lime water *Kalkmilch, Kalkwasser*
3 solution *Lösung*
4 flask *Kolben*
5 assess sth. *etwas bewerten, feststellen*

Procedure:
1. Connect the two Erlenmeyer flasks with glass tubes and rubber tubing exactly as shown in Figure 16.
2. Fill each flask with 50 ml of clear lime water.
3. Now place your lips round the mouthpiece and slowly breathe in and out 15 to 20 times. Do not breathe too hard!

Figure 16 Investigating the CO_2 content of inhaled and exhaled air

1 In which flask did you see bubbles when you breathed out and in which did you see them when you breathed in?
2 Describe the special way in which the two flasks are connected.
3 In which flask did the lime water change?
4 Assess[5] what the results of the experiment tell you about inhaled and exhaled air.
5 Explain the reasons for the difference between the air you breathe in and the air you breathe out.

Summary

Your respiratory system is made up of all the organs you use for breathing. When you draw air into your lungs, gases are exchanged in the alveoli: oxygen from the air passes into the blood while carbon dioxide from the organs passes from the blood into the lungs and is breathed out. The oxygen you take in travels to the organs, where it is used in a burning reaction to gain energy from food substances. In this process, carbon dioxide is formed and passes into the bloodstream.

Two different groups of muscles are used for breathing: the diaphragm and the intercostal muscles. When you are relaxed, you mostly use the diaphragm to expand your lungs. But when you need more oxygen, for example for physical exercise, the intercostal muscles help expand the chest so that the lungs can take in more air.

A Study tasks

1. Make a mind map based on the information in the summary above (see 'Improve your skills', p. 22).
2. Write down all the words from the Key terms boxes on cards. With a partner, take turns drawing cards and explaining the terms. If you are not sure how to express something in English, check the text for words that can help you and copy them into your vocabulary notebook.
3. Draw a diagram showing the path of oxygen and carbon dioxide through your body. Copy it onto a transparency and present it to the class.

B Bio quiz

1. The following text has a lot of mistakes. Find and correct them.

> Air enters your body through your nose, where oxygen is taken up into the blood. It then flows through the windpipe, which is held open by rings of bone. In your lungs, the air goes through an artery until it arrives in the alveoli. Their walls are very thick, so oxygen passes out of the blood into the alveoli. The carbon dioxide is then transported to the organs in the red blood cells. Here it helps to get energy from sunlight, a process which is called external respiration.

2. Complete the sentences using Key terms from this chapter (see below).
   a) When we breathe in, the ... inside our noses helps protect us against harmful substances.
   b) When you are sleeping, the movement of your ... gives you enough air to breathe.
   c) The cells in our bodies use ... to burn the food we eat, which produces ....
   d) Once you start smoking regularly it is hard to stop because the ... in tobacco is very ....
   e) Laws against smoking in public places make it easier for non-smokers to avoid ....
   f) The most serious disease caused by smoking is ..., which kills large numbers of people every year.

C Critical thinking

1. When you smoke, the cilia in your lungs stop moving for a while or at least slow down. What is the consequence of this?
2. When a rib is broken, there is a danger that the broken bone will tear the lung behind it. Explain what you think the result of this might be.
3. Research: The table on page 35 shows you non-smoking laws in some US states and in England. What is the current situation in Germany and how effective are the laws?

Key terms

See the boxes:

Chapter 3

Blood and Blood Circulation

Chapter objectives

- To **name** the different components of blood and compare their functions
- To **explain** the process of blood clotting
- To **describe** the anatomy of your heart
- To **compare** the three main types of blood vessels and their functions
- To **trace** the path of blood as it circulates in your body
- To **explain** how heart disease develops and how it can be prevented

As you read this, your heart is pumping blood around your body. This continuous flow of blood is essential, because blood carries oxygen and nutrients[1] to your cells. But this is not the only function blood has. It also keeps all the parts of your body at nearly the same temperature. If you cut your finger, blood closes the wound[2] after a short time. And if the wound gets infected[3], your blood helps fight the germs[4].

In this chapter you will learn how blood does all these things, and how the heart manages to pump all of your blood around your body in less than a minute.

3.1 Blood forms a liquid organ[5]

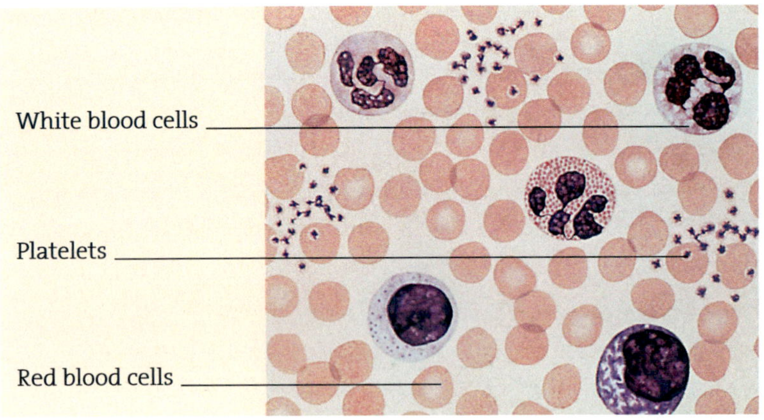

Figure 1 **Human blood under the microscope with the three types of blood cells**

Your body contains about four to six litres of blood, depending on your height and body weight. Blood is not simply a red-coloured liquid but also consists of different cells which float in a yellowish substance called plasma. In fact, about 55 per cent of your blood is **plasma**, which is mostly water in which digested[6] food (glucose[7], amino acids[8] and fats[9]), waste substances[10], plasma proteins, hormones[11], vitamins[12] and **antibodies** are dissolved.

Red blood cells make up about 44 per cent of the blood. They can be easily recognized because they are wheel-shaped and red. The colour comes from a substance called **haemoglobin**, which contains iron[14]. Haemoglobin becomes bright red when the iron combines with oxygen in the

Key terms

Your blood (1)

- antibody *Antikörper*
- blood circulation *Blutkreislauf*
- bone marrow *Knochenmark*
- haemoglobin *Hämoglobin, roter Blutfarbstoff*
- plasma *Plasma*
- red blood cell (erythrocyte) *rotes Blutkörperchen*

Did you know?

- Your bone marrow produces more than 2 million new red blood cells every second.
- If you could place all your red blood cells one after the other, they would form a string which reaches five times round the globe.

1 nutrient *Nährstoff*
2 wound *Wunde*
3 infected *infiziert*
4 germ *Keim, Krankheitserreger*
5 liquid organ *flüssiges Organ*
6 digested *verdaut*
7 glucose *Traubenzucker*
8 amino acid *Aminosäure*
9 fat *Fett*
10 waste substance *Abfallstoff*
11 hormone *Hormon*
12 vitamin *Vitamin*
14 iron *Eisen*

lungs. Since red blood cells have no nucleus[1], they cannot move by themselves, which means they have to be carried through the body via the blood stream[2].

Red blood cells are produced in the red **bone marrow**[3] and after about 120 days are broken down in the liver[4], the spleen[5] and the bone marrow. Their haemoglobin is stored[6] and used for the building of new red blood cells.

The blood of a healthy person contains on average[7] between five and six million red blood cells in one cubic millimetre[8]. People who live in high mountain regions, where the air contains less oxygen than in lowland areas, have a much higher concentration of red blood cells.

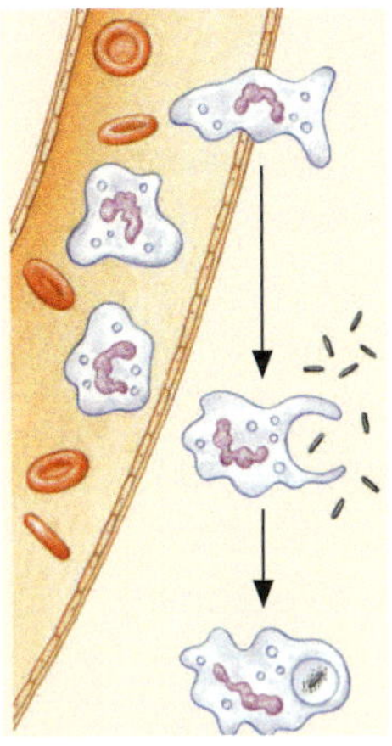

Figure 2 A phagocyte leaves a capillary and eats up a bacterium

White blood cells make up less than one per cent of your blood. Although they are fewer in number than red blood cells, they are larger, have a nucleus and can change their shape while they actively move around. There are about 4,000 to 10,000 white blood cells in one cubic millimetre of blood.

White blood cells can even squeeze[9] through very small spaces in the **blood capillaries**. These are the smallest type of **blood vessel** in your body – most of them are thinner than your hair. This makes it possible for white blood cells to reach nearly every part of the body, where they destroy[10] germs and break up dead body cells.

When germs – for example bacteria[11], viruses[12] or foreign cells – enter the body through a mucous membrane or a wound in the skin, they are first attacked by a type of white blood cell called a **phagocyte**, which is produced in the bone marrow. Phagocytes come out of the capillaries in the area of the wound, surround the germs and consume them together with any damaged cells.

Key terms

Your blood (2)
- blood capillary *Blutkapillare*
- blood vessel *Blutgefäß*
- lymphocyte *Lymphzelle, Lymphozyt*
- phagocyte *Fresszelle, Phagozyt*
- platelet (thrombocyte) *Blutplättchen*
- tonsils *Mandeln*
- toxin *Giftstoff*
- white blood cell (leucocyte) *weißes Blutkörperchen*

Lymphocytes are another type of white blood cell and are also produced in the bone marrow. They are found in the lymph nodes[13], for example in the **tonsils**. When your tonsils hurt and swell up, it is a sign that your body is fighting against an infection. Lymphocytes fight against germs and their poisonous products (**toxins**) by producing specific antibodies. Antibodies are proteins which can either destroy the germs and other dangerous substances directly or make them stick together, so that they can be eaten by phagocytes.

Platelets, which make up less than one per cent of your blood, are so small that one cubic millimetre of blood contains between 250,000 and 400,000 of them. Platelets are not complete cells, but small pieces of specialized cells from the bone marrow, so they have no nucleus and have an irregular[14] shape[15]. They help close wounds. When you cut your finger, blood flows out of the damaged[16] blood vessels. The body first reacts by contracting[17] the walls of the blood vessels. This makes the opening smaller. As a second step, the platelets, which stick to the rough edges of the wound, form a kind of patch[18] over it. Then the platelets and damaged cells release

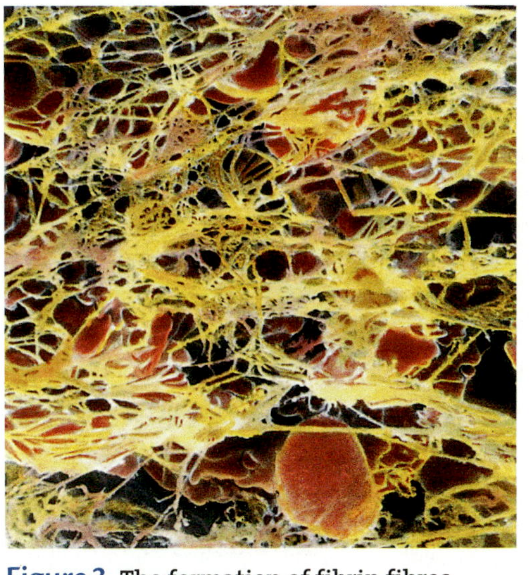

Figure 3 The formation of fibrin fibres during blood clotting

1 nucleus *Zellkern*
2 blood stream *Blutstrom*
3 bone marrow *Knochenmark*
4 liver *Leber*
5 spleen *Milz*
6 store sth. *etwas speichern*
7 on average *im Durchschnitt*
8 cubic millimetre (mm^3) *Kubikmillimeter*
9 squeeze through sth. *sich durch etwas hindurchzwängen*
10 destroy sth. *etwas zerstören*
11 bacterium (pl. bacteria) *Bakterie*
12 virus *Virus*
13 lymph node *Lymphknoten*
14 irregular *unregelmäßig*
15 shape *Form*
16 damaged *beschädigt*
17 contract *sich kontrahieren*
18 patch *Pflaster*

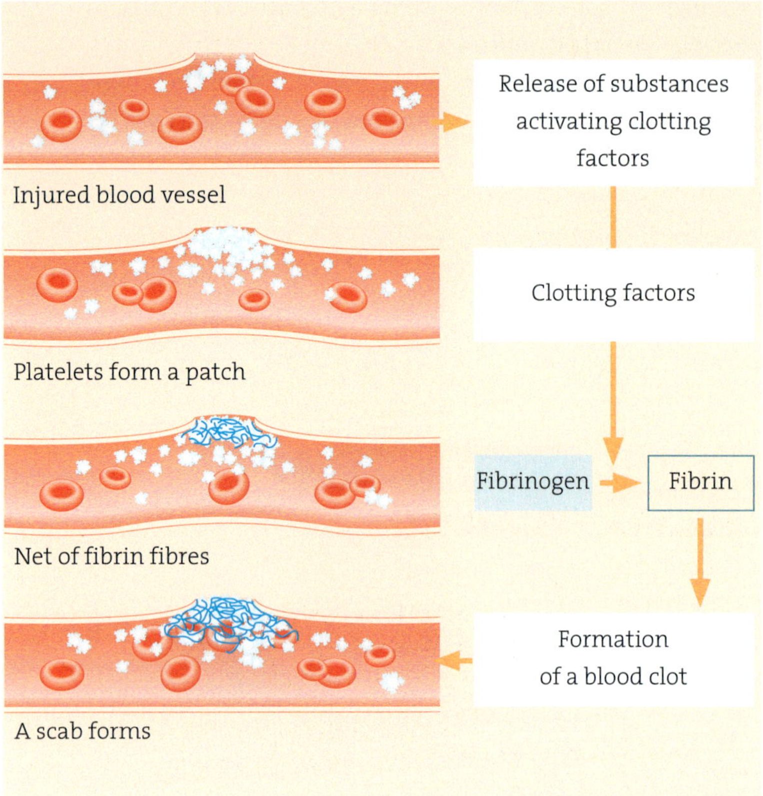

Figure 4 **The closing of a wound**

Key terms

Your blood (3)
- blood clot *Blutgerinnsel*
- fibrin *Fibrin*
- fibrinogen *Fibrinogen*
- oxygenated blood *sauerstoffreiches Blut*
- scab *Kruste*

substances which start a complicated process in the blood, in which a special fluid protein (**fibrinogen**) changes its form to become solid[1] fibre[2] (**fibrin**). This fibre creates a net across the wound, enclosing[3] damaged tissue, white and red blood cells and also platelets. The result is a **blood clot**, which stops the bleeding[4]. The blood clot hardens to become a **scab**. Beneath the scab a new layer[5] of skin slowly grows and after several days the scab falls off.

1 **Say it in German:** Imagine that your younger brother broke his leg while playing football. He now has his leg in plaster and is constantly complaining that his leg hurts and that it is taking too long to heal. Tell him that the healing process is a long and complicated one, and explain to him what happens during the process.

3.2 Blood is an efficient transport system

Your blood plays an important role in transporting gases, nutrients and other valuable substances around your body. The two main gases it carries are oxygen and carbon dioxide (see Chapter 2, p. 32).

When the blood flows through the lungs, it slows down and the red blood cells come into contact with the thin walls of the capillaries. Across these walls carbon dioxide passes from the blood into the lungs, while oxygen from the lungs enters the capillaries. Here the oxygen is taken up by the red blood cells, where it combines with haemoglobin, changing the colour of the blood to bright red.

The **oxygenated blood** is then transported to all parts of your body, where oxygen is taken up by the organs and carbon dioxide is passed into the blood. But oxygen and carbon dioxide are not the only things your blood carries:
- Digested food is transported from the gut to the liver and to tissues in different parts of the body.
- Waste products are carried to the kidneys[6], where they are filtered out of the blood.
- Hormones – chemical messengers[7] which regulate important body functions – are released into the blood by special hormone glands[8] and reach the places where they are needed through the blood plasma.

1 solid *fest*
2 fibre *Faser*
3 enclose sth. *etwas umschließen*
4 bleeding *Blutung*
5 layer *Schicht*
6 kidney *Niere*
7 chemical messenger *chemischer Botenstoff*
8 gland *Drüse*

- Heat[1] is generated[2] by large organs in your body like the muscles and liver when they work, and your blood spreads this heat all around your body, even into the tips of your fingers and toes. In this way, the temperature is nearly the same in all parts of your body.

1. Blood has been called a 'liquid organ'. Explain why.
2. Describe the function of the red blood cells.
3. Point out why plasma is an important component of your blood.
4. What is the advantage[3] of the specific shape of red blood cells and why have we got such a large number of them?
5. Explain what happens to the blood when it slowly passes through the capillaries in the muscles and the lungs.

3.3 Your heart pumps your blood around the body

Blood could not move around your body without the help of your heart. This vital organ is about the size of your fist and lies right behind your breast bone, between your lungs. The heart is made of muscle tissue, so it does what all muscles do: it contracts and relaxes. This contraction squeezes blood out of the heart and into the body. With each **heartbeat** about 90 ml of blood are pumped into your blood vessels, which then transport it throughout your body.

When you are at rest, your heart beats about 70 times a minute. During this time more than six litres of blood pass through your heart (remember, the average adult body contains no more than five or six litres). When you exercise, your heart beats much faster.

Key terms

The heart
- aorta *große Körperarterie, Schlagader*
- atrium (pl. atria) *Vorkammer, Vorhof*
- coronary artery *Herzkranzgefäß*
- heartbeat *Herzschlag*
- longitudinal section *Längsschnitt*
- lung (pulmonary) artery/vein *Lungenarterie/-vene*
- main vein (vena cava) *Hauptvene*
- valve *Ventil*
- ventricle *Hauptkammer*

1 heat *Hitze*
2 generate sth. *etwas erzeugen*
3 advantage *Vorteil*

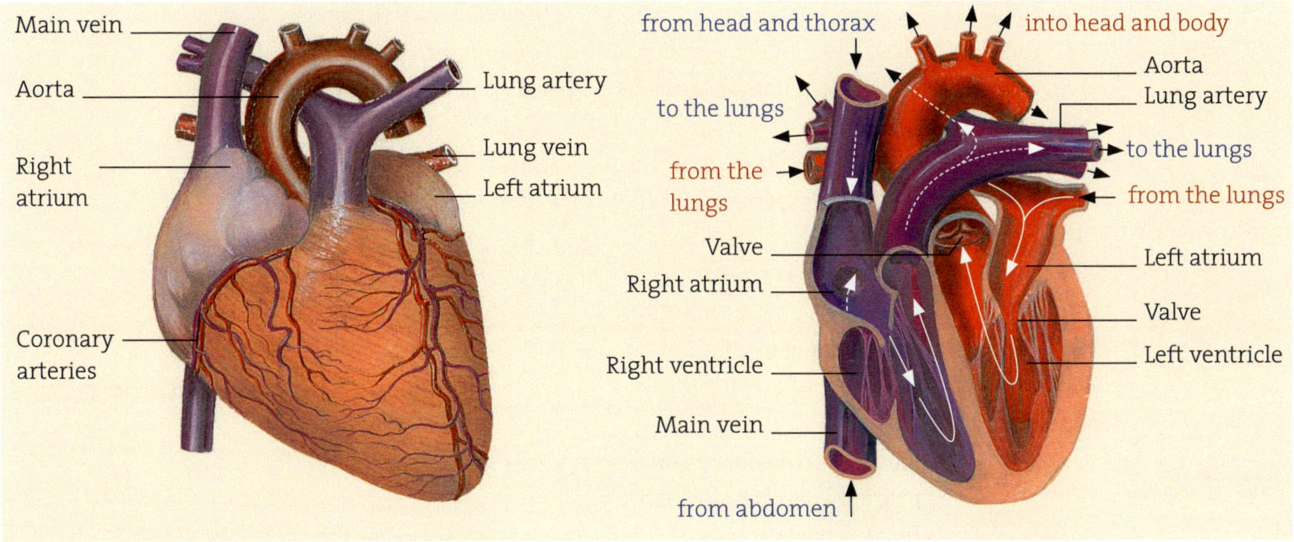

Figure 5 Outer and inner structure of the heart

The heart is divided into a right and a left side. Each side consists of a small chamber[1] called an **atrium**, and a large chamber called a **ventricle**. The right side of the heart pumps blood into the blood vessel leading to the lungs, while the left side supplies[2] blood to the vessel leading to the other parts of your body.

The heart works in three phases. In phase 1 the atria and ventricles are relaxed while blood returning from the body and the lungs pours into the right and left halves of the heart. Phase 2 begins when both atria contract and empty their contents into the ventricles. In phase 3 the two ventricles then contract, pressing the blood they contain into the **aorta** and the lung artery.

Blood passes through the heart in one direction only. **Valves** keep the blood from flowing backwards. They are like doors which open in only one direction. They are found between the atria and ventricles, but also between the ventricles and the arteries.

When the ventricles contract, the valves leading back to the atria close while the valves leading to the blood vessels open. In this way blood is forced out of the heart and into the blood vessels.

Did you know?

The heart of an average adult weighs between 300 and 350 g. In long-distance runners and in racing cyclists[3] the heart can have a weight of up to 500 g.

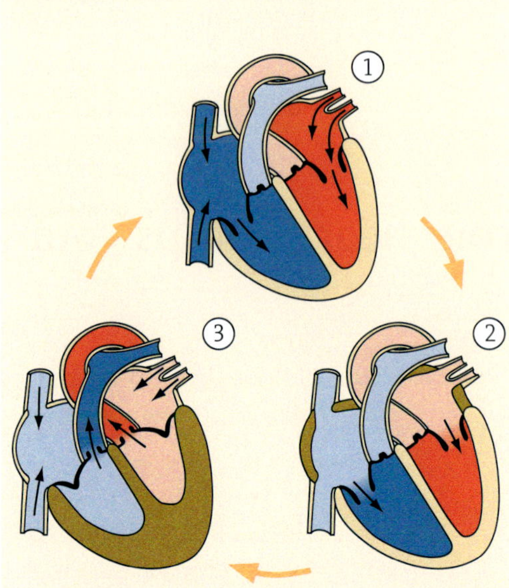

Figure 6 The cardiac cycle[4]

Every muscle needs energy, which it gets from nutrients in the blood. The heart muscle gets its nutrients and oxygen not from the blood which it pumps through its chambers, but from vessels which are on its surface and inside the muscle tissue. These are called **coronary arteries**. As you will see later, these arteries often cause heart problems.

1 chamber *Kammer*
2 supply sth. *etwas versorgen, liefern*
3 racing cyclist *Rennradfahrer*
4 cardiac cycle *Herzzyklus*
5 disadvantage *Nachteil*

1. Describe to a classmate how the heart is constructed. You can use a model of the human heart or Figure 5.
2. Use Figure 6 to explain the functioning of the valves as the heart goes through the cardiac cycle.
3. Look at Figure 5 and compare the size of the cardiac muscle (around the ventricles) on the right and left sides of the heart. Describe the differences. What might be the reason for this?
4. Why do you think heart patients are given medication which makes it harder for the blood to clot?
5. Compare a heart model with Figures 5 and 6 in your book. What are the advantages and disadvantages[5] of each in helping us understand how the heart works?

Activate your English

Discussing advantages and disadvantages

While/Whereas the model is ..., the illustration is ...

The advantage/disadvantage of the ... is that ...

... give/provide a clearer idea/impression of how ...

two-/three-dimensional

easier/harder to understand

Though the drawings are only ... they have the advantage that ...

3.4 Blood is transported in arteries, veins and capillaries

Your heart is a very strong muscle and when it contracts, it builds up a lot of pressure[1]. Therefore the blood vessels that carry the blood away from the heart have to be quite strong too. That is why their walls are made of thick muscle tissue. These vessels are called **arteries**.

Arteries take the blood from the heart to all parts of the body. They branch off[2] into smaller and smaller vessels until finally they are thinner than a single hair. The walls of these tiny vessels, which are now called **capillaries**, have to be very thin because gases and other substances have to pass through their walls into the cells of the organs. Equally, substances from the cells can pass into the capillaries, which are then transported by the blood back to the heart.

Along the way from the organs to the heart, the capillaries join together to build **veins**. Veins are the blood vessels that carry the blood to the heart. The muscle layer in their walls is much thinner than in the arteries and the blood flows with very low pressure. The veins have valves that allow the blood to move only in the direction of the heart. Since the sucking[3] effect of the heart is not strong enough to move the blood quickly through the veins, these vessels mostly lie close to arteries or skeletal muscles. When the muscles contract, the blood in the veins is pushed forward. As veins often run parallel to arteries, the pulse-wave[4] which moves through the artery also moves the blood forward in the vein.

Key terms

Blood transport (1)
- artery *Arterie*
- blood pressure *Blutdruck*
- capillary *Kapillare*
- vein *Vene*

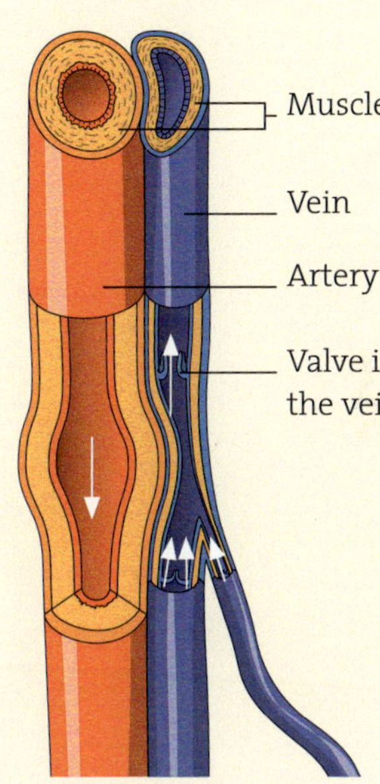

Figure 7 **Blood flow in an artery and a vein**

1 pressure *Druck*
2 branch off *abzweigen*
3 suck sth. *etwas saugen*
4 pulse-wave *Pulswelle*
5 cross-section *Querschnitt*

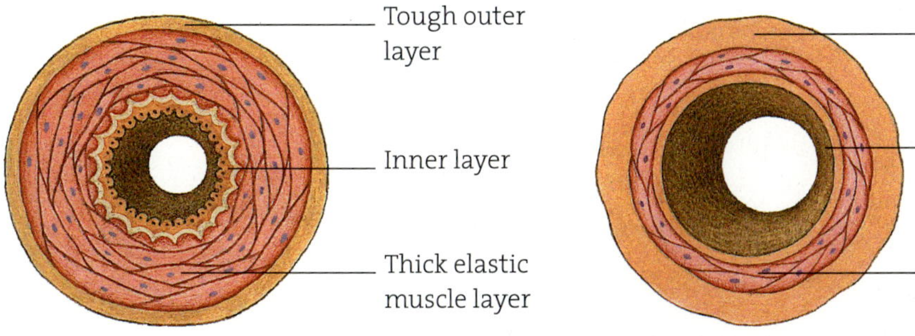

Figure 8 **Cross-sections[5] of an artery and a vein**

1 Using Figure 7, describe how the structure of the arteries and veins is responsible for the way in which the blood is transported from the heart to the organs and back to the heart.
2 What is the function of the capillaries in the system of blood circulation?
3 Explain how lack of exercise or movement affects the blood flow in your body.

3.5 Blood flows through your body in two separate circulations

The blood that arrives in your heart after bringing food and oxygen to your organs carries almost no oxygen. It is **deoxygenated**. Instead, it is loaded with another gas, carbon dioxide, which it has absorbed from the organs. This blood flows into the right side of the heart. When the heart contracts, the blood enters the **lung circulation** as the right ventricle pumps it into the **lung artery**.

Key terms

Blood transport (2)
- body circulation *Körperkreislauf*
- deoxygenated blood *sauerstoffarmes Blut*
- lung artery/vein *Lungenarterie/-vene*
- lung circulation *Lungenkreislauf*
- oxygenated blood *sauerstoffreiches Blut*

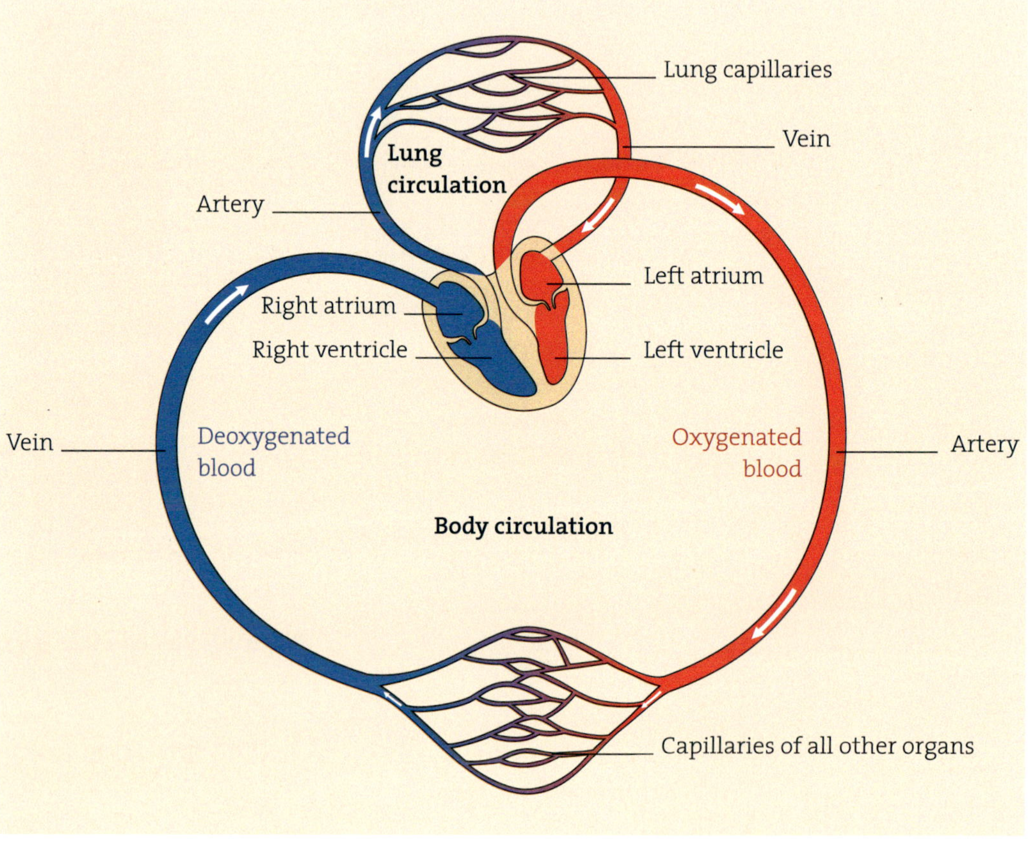

Figure 9 Lung and body circulation

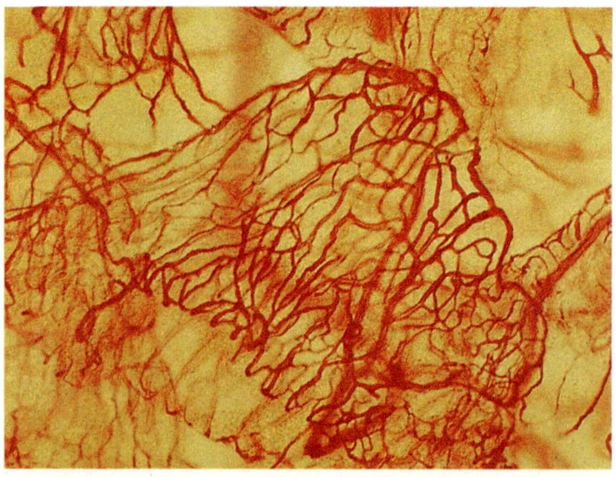

Figure 10 Blood capillaries under the microscope

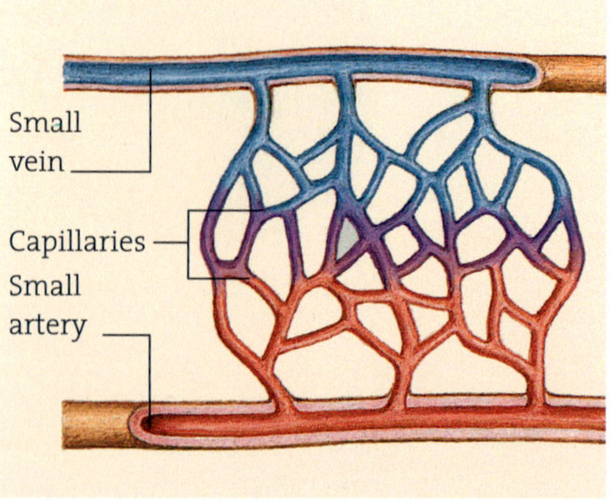

Figure 11 Illustration of blood capillaries

In the lungs the carbon dioxide is released and fresh oxygen is taken up (see also Chapter 2, p. 32). The blood is now **oxygenated** and has a bright red colour.
The **lung vein** transports the oxygenated blood back to the left side of the heart. The blood enters the **body circulation** as the left ventricle forces it into a large artery called the aorta, which takes the blood to all parts of the body.
Where the small arteries turn into capillaries, the organs take up oxygen from the blood and release carbon dioxide. Veins then transport the carbon dioxide back to the heart and the body circulation is completed.

1 Name a vein that carries oxygenated blood and one that carries deoxygenated blood. Figure 9 will help you. Explain how the two types of blood are different and why both blood vessels are veins.

2 Explain how the lung and body circulation are connected through the functioning of your heart.

3.6 Blocked arteries can cause heart disease[1] and strokes

As you have learned, your heartbeat forces blood into the arteries with a lot of pressure. This means the arteries need to be elastic so that they can expand[2] under the pressure. This elasticity is very important for keeping a constant flow of blood throughout the body. However, certain factors reduce the elasticity. Nicotine, high **blood pressure**, stress, a high-**cholesterol**[3] diet and **diabetes**, among other things, can make the walls of the arteries become harder and narrower. When cholesterol hardens, it makes the walls rough and increases the danger that platelets will cause the blood to clot within the blood vessels. Doctors call this **thrombosis**. A blood clot inside a coronary artery is dangerous, as it can block the artery and stop part of the heart from doing its job. The result is a **heart attack**. Patients often first experience a stinging pain in the left side of the chest and in their left arm. In such cases, a doctor must be called immediately. The person should sit quietly with his or her chest raised.

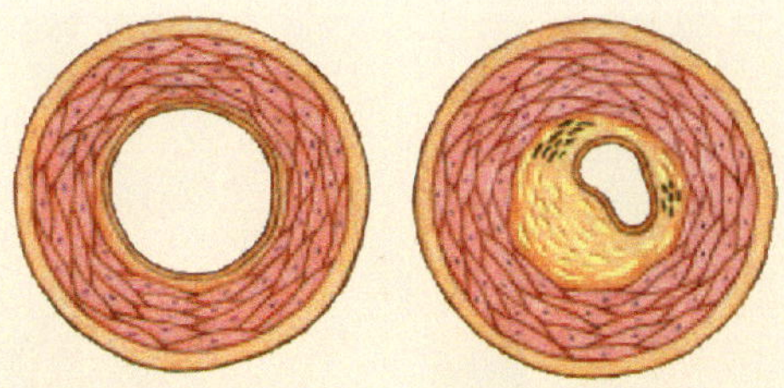

Figure 12 Coronary arteries

Key terms

Blocked arteries
- cholesterol *Cholesterin*
- diabetes *Zuckerkrankheit*
- heart attack *Herzanfall*
- medical treatment *medizinische Behandlung*
- paralysis *Lähmung*
- stroke *Gehirnschlag*
- thrombosis *Thrombose*
- blood pressure *Blutdruck*

Did you know?

How you can call for help in an emergency[5]:
Dial[6] 112 and say
1. What has happened
2. Where it has happened
3. How many people are injured
4. What kind of injuries you recognize
5. What your name is.

Speak slowly and clearly.
Ask for instructions.

1 heart disease *Herzerkrankung*
2 expand *sich ausdehnen*
3 high-cholesterol *mit viel Cholesterin*
4 deposit *Ablagerung*
5 emergency *Notfall*
6 dial sth. (a number) *(am Telefon) etwas (eine Nummer) wählen*

If an artery in the brain becomes blocked, the person suffers a **stroke**. Possible symptoms of a stroke are:
- blurred vision[1]
- a feeling of numbness[2] in a part of the body
- a loss of speech
- **paralysis** on one side of the body.

These symptoms can go away again if the patient gets **medical treatment** quickly. If help does not come in time, a stroke can cause permanent damage or even death.

1 Why is it important that arteries are elastic?
2 What kind of lifestyle reduces a person's chances of getting heart disease? Explain why.
3 Arteries can lose their elasticity and get blocked. Find out about the disease arteriosclerosis in your library or on the Internet and present your findings in class.

First Aid for bleeding

When trying to control severe[3] bleeding in a person who has had an accident, it is important that you avoid direct contact with the person's blood so that you do not get any diseases the injured[4] person might have. If you do not have latex gloves, use more than one piece of absorbent[5] cloth while following these steps:

1. Press the edges of the wound together or, using the palm[6] of your hand, press down lightly on the wound with a thick, clean absorbent cloth.
2. If possible, raise the site of the wound above the level of the heart.
3. If the wound is small, apply[7] a plaster[8]. Take care not to touch the gauze[9] in the centre of the plaster.
4. Apply a bandage using overlapping turns[10] with a roll of gauze. Most bleeding can be stopped this way, especially when for example the arm or leg is raised above the level of the head.
5. If an artery has been hurt and there is strong bleeding, apply a pressure bandage[11], place a thick pad over the bandaged wound and fasten the bandage firmly with a roll of gauze as in Figure 13.
   In such a case, call the emergency number for an ambulance immediately (see 'Did you know?', p. 51).

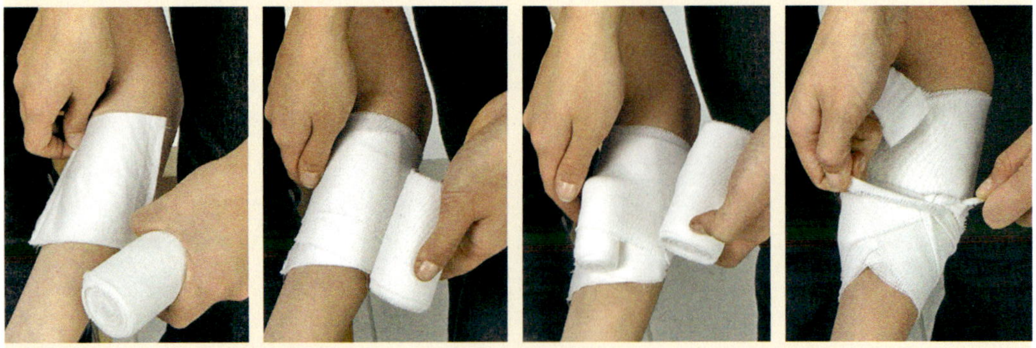

Figure 13 Applying a pressure bandage

For nosebleed[12]:
Sit still with your head slightly forward. Put a tissue or other cloth in front of your nose to absorb[13] the blood. Do not block the blood from coming out. A cold towel placed around your neck can help stop the bleeding.

1 blurred vision *verschwommene Sicht*
2 numbness *Gefühllosigkeit*
3 severe *schwer, heftig*
4 injured *verletzt*
5 absorbent *saugfähig*
6 palm *Handfläche*
7 apply sth. *etwas anlegen*
8 plaster *Pflaster*
9 gauze *Gaze*
10 overlapping turns *(hier) Überkreuzungen*
11 pressure bandage *Druckverband*
12 nosebleed *Nasenbluten*
13 absorb sth. *etwas aufsaugen*

Working with tables[1] and line graphs[2]

Putting information into the form of a table or diagram is an effective way of understanding a science text better. It makes it easier to see and interpret the important facts.

1 Table

A table is often used to summarize the information in a text and arrange it in a clearer way. It always has vertical columns[3] and horizontal rows[4], as in Table 2 below.
The information in Section 3.1 (p. 44) on the different types of blood cells can be used as the basis for such a table. The text gives details about the quantity, shape and function of the different blood cells, so these three qualities can be represented in further columns with corresponding headings, as shown below.

1 Copy the table below into your exercise book and fill in the information from the text.
2 Now look at the completed table and make statements about the qualities of the different blood cells.
3 With a partner, take turns quizzing each other on the different blood cells.

Name of blood cells	*Percentage*	*Quantity in 1 mm³ of blood*	*Shape*	*Function*
...	...	...	...	...

Table 1 Components of the blood

2 Line graph

The information in a table can often be represented in a line graph. This is useful for showing trends or changes over time, or changes in size, amount or intensity in relation[5] to a second factor. An example would be the change in heart rate[6] when the body weight increases[7] or decreases[8].
This table refers to mammals of different species:

Type of mammal	Body weight (kg)	Heart rate (beats per min)
Mouse	0.025	670
Rabbit	2	205
Human	70	72
Horse	450	38

Table 2 Relationship between body weight and heart rate in mammals

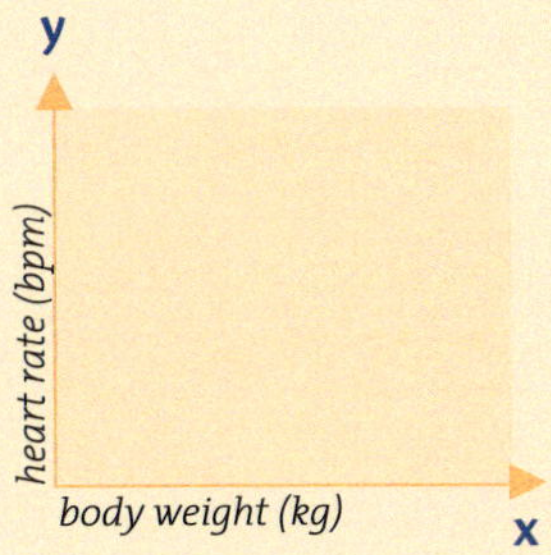

Figure 14 Changes in heartbeat in different mammals

Follow these steps to present the information in Table 2 in a line graph:
1. In your exercise book, draw an x-axis[9] (a horizontal line) and a y-axis (a vertical line) as in Figure 14 using a ruler and a sharp pencil.
2. Decide which data you want to plot[10] on the x-axis and which on the y-axis. In this case it is best to plot the different body weights on the x–axis and the heart rates on the y-axis. Label the axes.
3. Then connect the dots in your diagram to get a line graph. The graph helps describe and understand the connection between the two factors.
4. Finally, formulate conclusions about the information in your line graph. For language help, see **Activate your English**, 'Talking about a line graph', on the inside cover.

1 table *Tabelle*
2 line graph *Kurvendiagramm*
3 column *Spalte*
4 row *Zeile, Reihe*
5 in relation to sth. *im Verhältnis zu etwas*
6 heart rate *Herzfrequenz*
7 increase *zunehmen, steigen*
8 decrease *abnehmen, sinken*
9 x-/y-axis (pl. axes) *x-/y-Achse*
10 plot sth. *etwas (Zahlenwerte in ein Diagramm) eintragen*

Activity

Measuring changes in your pulse rate

Materials (for each group):
- stopwatch or wristwatch with a second hand
- some graph paper or squared paper

Procedure:

1. Form groups of four. Practise checking your pulse using your index and middle fingers[1] (see Figure 15 below).

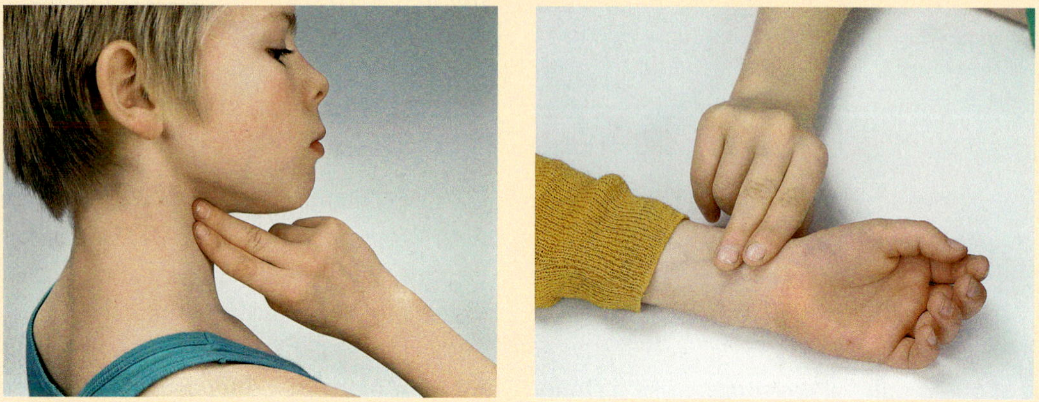

Figure 15 **Measuring the pulse rate on the neck and on the wrist**

2. Construct a data table like this in your notebook:

Name	*1. At rest*	*2. After walking for 1 min*	*3. After running for 2 min*	*4. After resting for 1 min*	*5. After resting for 2 min*
1st pupil	...	...	...	...	...
2nd pupil	...	...	...	...	...
3rd pupil	...	...	...	...	...

Table 3 **Table showing pulse rates (number of pulse beats per minute)**

3. Choose one person in your group to keep the time in this and the following experiments. The other three people follow these steps:
   a) Locate your pulse while sitting relaxed and count how many times your pulse beats in 15 seconds. Multiply this number by four to calculate your pulse rate for one minute. Write your pulse rate at rest[2] into the table.
   b) Walk in place[3] for one minute. Then stop and immediately take your pulse (as in step a). Write the number into column 2.
   c) Do the same for the tasks in columns 3, 4 and 5.
4. Use the data to construct a line graph for each member of your group. Use a different colour for each graph to make them easy to recognize. When plotting the time, you can ignore the 15 seconds you need each time to take your pulse.

1 index finger *Zeigefinger*
2 pulse rate at rest *Ruhepuls*
3 in place *auf der Stelle*

1 Describe how your pulse changes during the experiments.
2 Explain the effect that walking and running has on the pulse rate.
3 Comment on the pulse rates of the different members of your group.
4 Using the Internet and other sources, find out what effect regular exercise has on the heart.

3.7 Blood groups and blood transfusions

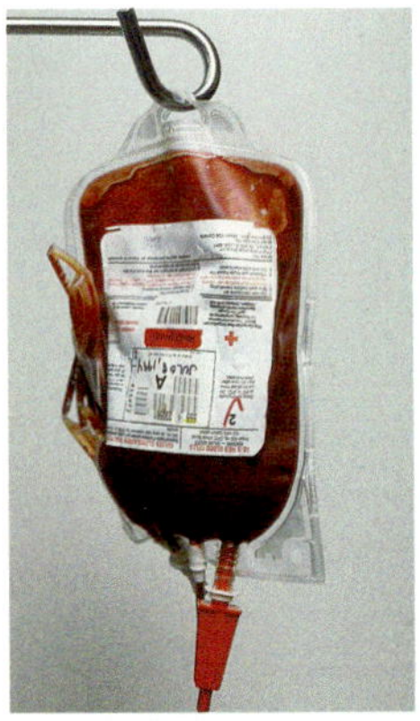

Figure 16 Blood transfusion bag

During an operation or after an accident, a patient sometimes needs a **blood transfusion**. In such a case it is important that the blood of the **donor**, i.e. the person whose blood is used, is the same type as the blood of the **recipient**, i.e. the person who gets the blood.

There are four major **blood groups or types:** A, B, AB and O. The differences between them have to do with two different protein molecules (**antigens**) on the surface of the red blood cells. They are called A and B.

Key terms

Blood groups and transfusions
- antigen *Antigen*
- blood group/type *Blutgruppe*
- blood transfusion *Bluttransfusion*
- donor *Spender*
- recipient *Empfänger*
- rhesus factor *Rhesusfaktor*

If you have red blood cells with molecule A, then your blood is group A. If your blood cells have molecule B, then B is your blood group. If the red blood cells carry A and B molecules, the blood group is AB, and when there are neither A nor B molecules on the surface of a person's red cells, the blood is type O.

But this is not all. In the plasma there are antibodies. The plasma of type A blood contains anti-B antibodies and type B blood carries anti-A antibodies. You may already have guessed correctly that type O blood holds anti-A and anti-B antibodies and blood of type AB has no antibodies at all.

A blood transfer is always safe when the blood types of donor and recipient are identical. As soon as a person receives the blood of another blood group than his or her own, the antibodies in the plasma attack the received red blood cells and start to clump them together[1]. It is as if the red blood cells were treated like germs. The clumps[2] of red blood cells can block capillaries and cut off the blood flow. This can be fatal[3] for the patient.

Blood group	Antigen on red blood cells	Antibody in plasma
A	A	anti-B
B	B	anti-A
AB	A and B	none
O	none	anti-A and anti-B

Table 4 Antigens and antibodies in people of different blood types

There are other blood groups beside the ABO system. Many people have the so-called **Rhesus factor**, a specific antigen, on their red blood cells. These people are Rh-positive (Rh+). People without this factor are Rh-negative. There are no natural antibodies against the Rhesus-factor in the blood of an Rh- person. So the first transfusion of Rh+ blood will not cause any harm, but the person's blood will start to develop and accumulate anti-Rh antibodies. In any subsequent[4] transfusion of Rh+ blood these antibodies will clot the received red blood cells together and cause them to burst[5]. Instead of helping the patient, the substances from the destroyed red blood cells can cause allergic shock reactions and even death. This is why it is important that the blood group of a patient is known before a transfusion is carried out.

ABO system		Rhesus system	
A	43 %	Rh+	85 %
B	14 %	Rh–	15 %
AB	6 %		
O	37 %		

Table 5 Distribution of blood groups in central Europe

1 clump sth. together *etwas verklumpen*
2 clump *Ballen, Klumpen*
3 fatal *tödlich*
4 subsequent *nachfolgend, später*
5 burst *platzen*

1 Explain why it is important to find out the blood group of a patient before he or she gets a blood transfusion.

2 Can a person with blood group 0/Rh– safely receive blood from a donor with blood group 0/Rh+? Explain.

3 a Using the information in the text and in Table 4, construct a table showing the compatible[1] blood types for each of the four major blood groups.

b Explain the information in the table. The words and phrases in the **Activate your English** box will help you.

Activate your English

Giving and receiving blood

Group A (blood) / (Blood) type A is (not) compatible with …

If your blood is group A / you are (blood) group A / you have type A (blood), then you can receive …

A donor with group/type A (blood) can/ cannot give/donate blood to sb. with …

A recipient with group/type B (blood) can/ cannot receive blood from sb. with …

3.8 The lymphatic system

As blood moves through the capillaries, some of the plasma liquid passes through the thin walls into spaces between the cells. This liquid is called **lymph**. As it moves round the cells, it keeps them moist[2], provides them with oxygen, nutrients[3] and minerals and carries away carbon dioxide and other waste substances produced by the cells. As the main **lymph vessels** join veins near the heart (see Figure 17), waste products[4] get into the blood circulation and are excreted[5] through the kidneys.
Like the blood system, the **lymphatic system** is a network of vessels which spreads throughout your body. But, unlike the blood's circulatory system, it has no pump[6] which pushes it along. The lymph vessels lie between muscles, and when the muscles contract, they squeeze the lymph forward. Like veins, the larger lymph vessels have one-way valves, which make the liquid flow in one direction only.
But the lymph system has another important function. As you can see in Figure 17, the lymphatic system has hundreds of small bulges[7] called **lymph nodes**. In some parts of your body the lymph nodes form larger **lymph glands**. You can probably remember a situation when you had an infection and your tonsils – the lymph glands in your neck – were swollen and hurt. Lymph glands are located wherever special white blood cells are made and stored. They filter out and destroy bacteria and other dangerous substances in the lymph. The lymphatic system is part of the body's defence system against diseases.

Key terms

The lymphatic system
- lymph *Lymphe*
- lymphatic system *Lymphsystem*
- lymph gland *Lymphdrüse*
- lymph node *Lymphknoten*
- lymph vessel *Lymphgefäß*

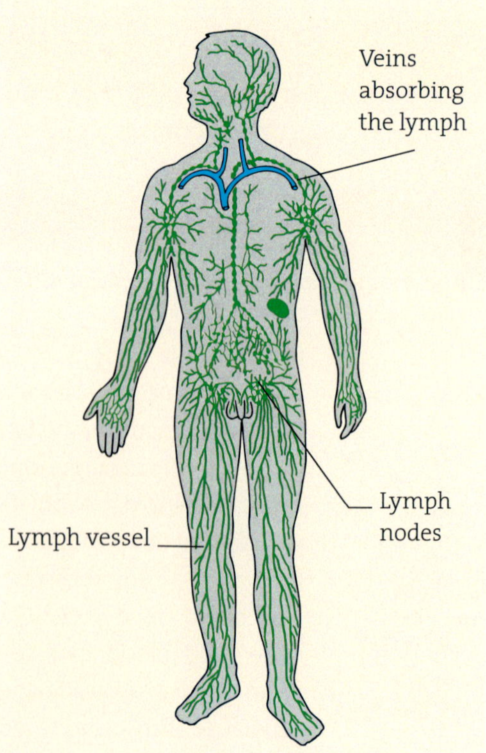

Figure 17 Lymphatic system

1 compatible *verträglich, passend*
2 moist *feucht*
3 nutrient *Nährstoff*
4 waste product *Abfallprodukt*
5 excrete sth. *etwas ausscheiden*
6 pump *Pumpe*
7 bulge *Anschwellung*

1 Describe the main functions of the lymphatic system.
2 What is lymph?
3 Why do you think your tonsils become swollen when you have a sore throat? Discuss.

Summary

The blood consists of plasma and three types of cells. The cells have different functions: red blood cells carry oxygen, white blood cells fight germs, and platelets help to stop the bleeding after a cut. Plasma takes carbon dioxide back to the lungs and also transports nutrients, waste products and hormones. The blood also keeps the body at a constant temperature by spreading heat from the organs and muscles.
The heart is a muscle which consists of four chambers: two atria and two ventricles. The right side of the heart pumps deoxygenated blood from the tissues to the lungs. The left side pumps oxygenated blood from the lungs into the body's circulatory system.
Among the blood vessels, arteries have strong elastic walls that carry the blood away from the heart, veins carry blood back to the right side of the heart, capillaries are tiny blood vessels with walls so thin that substances can pass through them. If arteries in the heart or brain get blocked, the result can be a heart attack or stroke.

Key terms

See the boxes:

Your blood 1, 2, 3 .... pp. 44, 45, 46
The heart .... p. 47
Blood transport 1, 2 .... pp. 49, 50
Blocked arteries .... p. 51
Blood groups and transfusions ...... p. 55
The lymphatic system .... p. 56

A Fill in the gaps

Complete the sentences below using each word or phrase in the box on the right only once. Write the sentences in your exercise book.

plasma	capillaries
oxygen	haemoglobin
platelets	carbon dioxide

1. *Blood arriving in the lungs has a low concentration of ….*
2. *Most of the carbon dioxide in the blood is carried by the ….*
3. *In contrast[1], oxygen is carried by a substance called ….*
4. *Blood which leaves the muscles has a high concentration of ….*
5. *The thin walls of the … allow gases to pass through.*
6. *In arteries that have lost their elasticity, … can cause the blood to clot.*

B Study tasks

1. Explain how the processes of respiration, blood circulation and movement are connected.
2. Why are athletes[2] often sent to mountainous areas for training before they take part in international competitions[3]?
3. Prepare and present a short talk on how your blood flows through your body. Make sure you use all the words and phrases listed on the right.

artery with oxygenated blood
vein with oxygenated blood
right/left ventricle
capillaries
lungs
artery with deoxygenated blood
vein with deoxygenated blood
right/left atrium
the other organs
body circulation

1 in contrast *im Gegensatz dazu*
2 athlete *Sportler, Leichtathlet*
3 competition *Wettkampf*

C Critical thinking

1 Why should everyone, but especially women, make sure they get enough iron with their food?
2 Blood samples are normally taken from the veins, not the arteries. Explain why.
3 In well-trained athletes, the number of heartbeats per minute goes up much more slowly under heavy stress compared to untrained people of the same age. What may be the reason?

Chapter 4

Food and Nutrition

Chapter objectives

- To **list** the three main functions of food
- To **name** the important groups of substances in our food and **explain** their functions
- To **use** the five groups of food to determine how to eat in a healthy way
- To **understand** how much energy the body needs
- To **describe** different eating disorders and **explain** how to avoid them

One of the first things you probably do when you come home from school is to search the fridge for something to eat. Eating comes naturally to us, but knowing what is good for our bodies does not. In order to find out what healthy[1] eating is, you will learn in this chapter about the different substances in our food, the right proportions of these substances in a healthy **diet**, and how much you should eat to get the **energy** you need.

4.1 Food serves three major functions

Figure 1 What you put in your mouth is important

Key terms

Aspects of food

- carbohydrate *Kohlenhydrat*
- diet *(hier) Ernährung*
- digest sth. *etwas verdauen*
- digestion *Verdauung*
- energy *Energie*
- fat *Fett*
- protein *Protein*

As you sit here reading, there are hundreds of things going on in your body: your stomach is **digesting** food, your skin is making new cells[2], your bone marrow[3] is producing blood cells, your brain is sending signals to all parts of your body, and much more. For all these things we need energy, which we get from our food.

It is important to realize that *energy cannot be made*. It can only be converted[4] from one state into another, and for this, both respiration and **digestion** are needed. Your digestive system breaks down your food, so that the chemical energy[5] that is in **carbohydrates**, **fats** and **proteins** can be converted and stored in ATP, a substance that your body uses in all processes that require energy. This conversion can only take place if your cells have enough oxygen and if, during the process, carbon dioxide is produced. This is why you have to breathe.

But energy alone is not enough to keep your body working properly: you also need building blocks[6] for growth and repair. This is the second function of food, which is fulfilled mostly by proteins.

Finally, food provides our bodies with substances which keep us healthy, for example vitamins for our immune system.

Did you know?

During digestion, your body creates 40 kg of ATP in the course of one day and uses most of it, too. This is a total of 48×10^{23} molecules of ATP – an unimaginably large number.

1 healthy *gesund*
2 cell *Zelle*
3 bone marrow *Knochenmark*
4 convert sth. *etwas umwandeln*
5 chemical energy *gebundene Energie*
6 building block *Baustein*

4.2 A healthy diet includes seven essential groups of substances

Our diet is the food we eat each day. The different substances that our body needs to maintain[1] its structure and functions belong to seven groups, as shown in Figure 2.

Figure 2 A typical breakfast

Comparative language note:

In English these seven groups of substances that are essential for life are called **nutrients**, whereas in the German language the word *Nährstoffe* is usually only used for carbohydrates, proteins and fats.

A closer look at the seven groups

Carbohydrates (sugars): These often come in the form of **starch**, which is a long chain of single **glucose** molecules. Your body can break this nutrient down very quickly and convert the energy for its needs. This is why footballers sometimes eat pasta before a match. You find carbohydrates in all products that contain flour[2], rice or potatoes.

Figure 4 Foods rich in carbohydrates

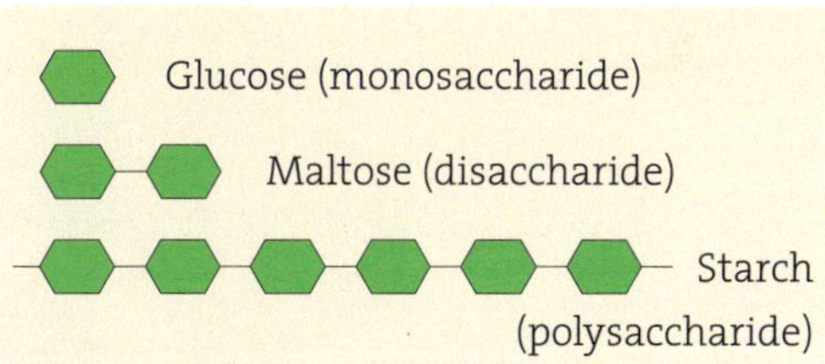

Figure 3 Different structures of carbohydrates

Key terms

Nutrition (1)
- amino acid *Aminosäure*
- dairy product *Milchprodukt*
- dietary fibre / roughage *Ballaststoffe*
- fatty acid *Fettsäure*
- glucose *Traubenzucker, Glukose*
- glycerol *Glycerin*
- iron *Eisen*
- magnesium *Magnesium*
- mineral *Mineral*

1 maintain *aufrechterhalten*
2 flour *Mehl*

Proteins: Almost everything in your body is made from proteins, so it is important that we eat enough foods rich in protein. These building blocks are very long chains of 20 different **amino acids**. We can also gain energy from proteins. Meat, fish and **dairy products** are the main sources of proteins, but they also occur[1] in peas and beans.

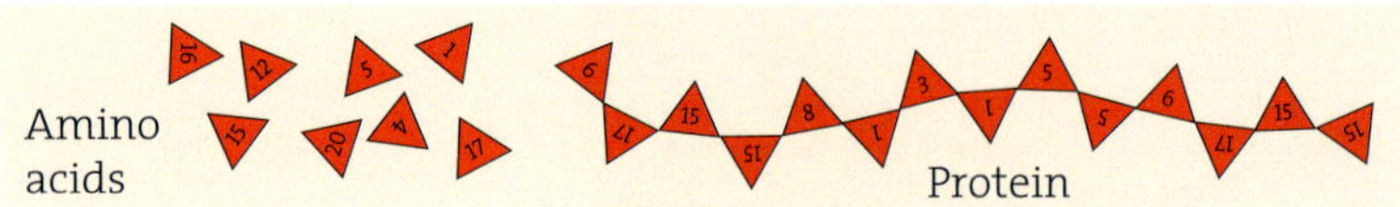

Figure 5 A chain of amino acids in a protein molecule

Figure 6 Foods rich in proteins

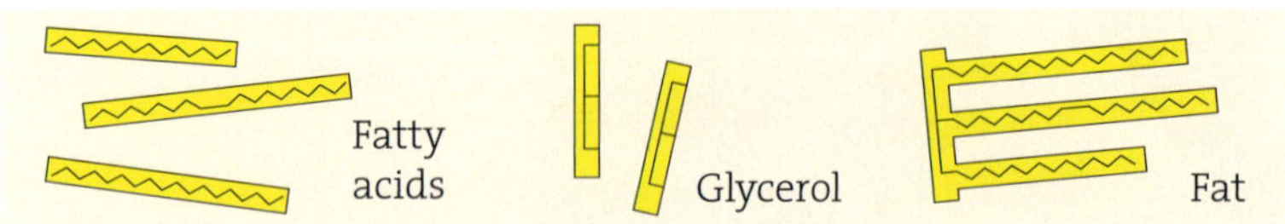

Figure 7 The structure of a fat molecule

Key terms

Nutrition (2)
- nutrient *Nährstoff*
- protein *Eiweiß, Protein*
- saturated fatty acid *gesättigte Fettsäure*
- starch *Stärke*
- unsaturated fatty acid *ungesättigte Fettsäure*
- vitamin *Vitamin*
- water *Wasser*

Fats: The fats in your food consist of three **fatty acids** which are attached to a molecule of **glycerol**. When the body digests fats, it cuts through the bonds[2] between the fatty acids and the glycerol, so that the four parts are separated. Processing fats takes a little longer than digesting carbohydrates, but fats contain a lot more energy per gram. There are two different kinds of fats, those with **saturated** and those with **unsaturated fatty acids**. Eating too much of the former can cause heart disease, so you should try to eat foods which are rich in unsaturated fatty acids. These can be broken down and digested more easily by your body. They occur in vegetable oils, for example olive oil and sunflower oil.

Figure 8 Foods rich in fats

Figure 9 Fruit and vegetables contain vitamins and minerals and also lots of fibre

Minerals: Although we only need a few grams per day, minerals play a vital role in all kinds of processes. For example, we need **iron** to build our red blood cells and **magnesium** for our muscles. Our body also needs iodine[3], potassium[4], sodium[5], calcium[6], phosphorus[7] and chloride[8].

Vitamins: There are 15 different kinds of vitamins, which have a variety of functions but are only required in small amounts. Your immune system cannot work without vitamin C, and vitamin D helps to build strong bones. B vitamins play an important role in carbohydrate and protein metabolism[9]. Your body cannot

1 occur *vorkommen*
2 bond *Bindung*
3 iodine *Jod*
4 potassium *Kalium*
5 sodium *Natrium*
6 calcium *Kalzium*
7 phosphorus *Phosphor*
8 chloride *Chlor*
9 metabolism *Stoffwechsel*

produce vitamins by itself, so it needs to get enough of them from other sources, mainly from the food you eat. Fruit and vegetables are the most important sources.

Dietary fibre: Fibre, also called **roughage**, cannot be digested by your body, which is why it is a great 'cleaning agent[1]' for your gut and helps prevent constipation[2]. You find dietary fibre in vegetables, fruits and wholegrain products[3].

Water: About three quarters of your body consists of water. It is the universal solvent[4] for all processes that take place inside you. Without water no chemical reactions could take place.

Did you know?

A person can live without food for several weeks in an emergency, but without water they can only survive a few days.

Figure 10 Your body is made up mostly of water

1 cleaning agent *Reinigungsmittel*
2 constipation *Verstopfung*
3 wholegrain product *Vollkornprodukt*
4 solvent *Lösungsmittel*
5 survive *überleben*

1 Give some examples of the functions of vitamins in the human body and list the main sources of vitamins in our diet.
2 Point out which of the seven nutrient groups provide energy. Write down the key information about each of these substances in your exercise book.
3 Explain what is important to keep in mind when buying foods that contain fat and why.
4 Why do you think human beings need to eat in order to survive[5], while plants live without eating? Explain your answer.

4.3 Eating right is important for a healthy lifestyle

Figure 11 Exercise is essential for a healthy lifestyle

Key terms

Food and lifestyle
- exercise *Bewegung*
- balanced diet *ausgewogene Ernährung*
- cereal *Getreideprodukt*

In order to enjoy a healthy lifestyle two things are essential: **exercise** and a **balanced diet**. Exercise does not have to mean joining a football club or running a marathon, but you do need regular physical activity to stay healthy. Taking the stairs instead of using lifts or escalators, walking instead of taking the bus, or cycling instead of asking someone to drive you somewhere is a good start.

A balanced diet – eating the right kinds of foods in appropriate amounts – can also be easier than you think. In fact, you may already be eating most of the right things. In the previous section you learned about the different groups of substances in our food. The next step is to find out *how much* of these substances your body needs in order to stay healthy.

The five main groups of valuable foods

The picture shown in Figure 12, which was published by the British government's Food Standards Agency, gives you a good overview of the most important food groups and the proportion of your diet that each of these food groups should represent.

Figure 12 The five main food groups

Most of the foods you eat should come from the four food groups on the left and right of the 'eatwell plate'. The last group at the bottom of the plate, foods containing fat and sugar, should be a very small part of your diet and cannot replace[1] foods from the other groups.
The following is a more detailed outline of the individual food groups. The recommendations[2] for a balanced diet are based on the leaflet issued by the British Nutrition Foundation called *Healthy Eating – A Whole Diet Approach*.

A Fruit and vegetables: eat about five servings[3] each day

This group contains all kinds of fruit and vegetables: fresh, frozen, dried and canned[4]. Pure fruit juice also counts, but only as one serving, no matter how much of it you drink. A serving is about 80 g, e.g. a medium-sized apple. Try cutting up some fruit or vegetables and eating them as snacks during the day, instead of crisps or sweets. This food group is rich in fibre and vitamins.

B Bread, cereals and potatoes: they should make up a third of your daily diet

Besides breakfast **cereals** and potatoes, this group consists of pasta, rice and maize products. You should try to include at least one type of food from this group in each meal, because these foods give you a lot of energy without using much fat. Also, if available, choose wholegrain cereals and noodles, which will give you an additional amount of fibre.

1 replace sth. *etwas ersetzen*
2 recommendation *Empfehlung*
3 serving *Portion*
4 canned *aus der Dose*

C Milk and dairy products[1]: try to eat 2–3 servings a day

All products that contain milk, such as cheese or yoghurt, and of course milk itself, belong in this group. Butter and cream are not in this category because they are high in fat (see group E). A serving in this group consists of, for example, 200 ml milk, 150 g yoghurt, or 30 g cheese. If available, use low-fat versions. These foods give you a lot of minerals and proteins.

D Meat, fish and alternatives: eat moderate[2] amounts

This group includes meat, poultry[3], fish, eggs and alternatives (see 'Further facts' below). Since some meat products are high in fat, make sure to remove[4] visible[5] fat and/or choose low-fat versions where possible. Also, it is a good idea to eat fish twice a week as it is low in fat but contains a lot of protein, minerals and essential omega-3 fatty acids. These foods provide you with protein, iron and magnesium as well as B and D vitamins.

E Foods containing fat and sugar: eat less of these

Butter, all kinds of oils, cream, fried foods and chocolate all belong in this group because they contain fat. You can find sugar in soft drinks, sweets, jam, cakes and ice cream. In fact, most of the foods that we eat for snacks and treats[6] are high in sugar or fat (or both). A small amount of fat is essential for a healthy diet, as some vitamins can only be absorbed into our cells when fat is present as a solvent[7]. It is important, however, not to eat too many foods with fat as they contain more energy than the body can use. Choose low-fat versions whenever possible, and choose products which are rich in unsaturated fats instead of those with high amounts of saturated fats.

Further facts

What about vegetarians and vegans?

Vegetarians do not eat meat and fish, while vegans go further, excluding all animal products including eggs, dairy products and honey from their diet. So they need to get the nutrients found in animal products from other sources. For example, nuts are high in fat, tofu is high in protein, and beans have a lot of iron. However, unlike meat and fish, these foods do not provide much zinc or vitamin B12, so vegetarians and vegans have to make especially sure that they get enough of these nutrients. Sometimes a nutrient is added to a food product which would normally be a poor source of it. Some fortified[8] breakfast cereals, for example, have a lot of vitamin B12.

1 dairy product *Milchprodukt, Molkereiprodukt*
2 moderate *mäßig*
3 poultry *Geflügel*
4 remove sth. *etwas entfernen*
5 visible *sichtbar*
6 treat *Leckerbissen*
7 solvent *Lösungsmittel*
8 fortified *angereichert*

1 Which of the food groups should make up the largest proportion of your diet? Explain why.
2 Make a 'food diary' by writing down everything you eat during a normal day. Then check how balanced your diet is by using the 'eatwell plate'. Make notes about what was healthy or unhealthy in your intake for the day.
3 Working with a partner, list the reasons why you think most people like to eat sweet and fatty foods. Then list the reasons why you should not eat too much of these types of foods. Compare your lists with those of your classmates.
4 Make a list of snacks that do not belong in group E. Share your ideas with the class and make a poster advertising 'healthy snacks'.
5 Why is fish particularly suitable for a healthy balanced diet?
6 Teenage girls who choose to be vegetarians need to pay extra attention to what they eat and to make sure that they have enough iron in their diet. Why?

4.4 How much energy do you need?

How much energy you need depends a lot on your job, your age and your lifestyle. An athlete needs more energy than a bank clerk[1], while a teenager needs more energy than a baby. Table 1 shows how much energy you need on average at different ages.

Energy is measured in joules or kilojoules (kJ), where 'kilo' means 1000. One joule is the energy it takes to lift one kilogram one metre high. For your diet this means in simple terms: if you take in more energy than you need through the food you eat, you will gain weight; if you take in less, you will lose weight.

Figure 13 a/b A baby and a teenage athlete have different energy needs

Age	Males (kJ)	Females (kJ)
0–3 months	2280	2160
4–6 months	2890	2690
7–9 months	3440	3200
10–12 months	3850	3610
1–3 years	5150	4860
4–6 years	7160	6460
7–10 years	8240	7280
11–14 years	9270	7720
15–18 years	11,510	8830
19–50 years	10,600	8100
51–59 years	10,600	8000
60–64 years	9930	7990
65–74 years	9710	7960
75+ years	8770	7610

Table 1 Estimated average energy requirements per day

1 bank clerk *Bank-angestellte/r*

Of course how active you are also makes a big difference. When you rest, the amount of energy you need to keep all your organs running is called the **basal metabolic rate**. This differs from person to person depending on factors such as age, gender[1], body fat, diet, etc. As soon as you start moving around, your energy requirements rise. Figure 14 shows you how many kilojoules you use when you do various activities for one hour.

Key terms

Energy needs
- basal metabolic rate *Grundumsatz*

Lying down 350 kJ

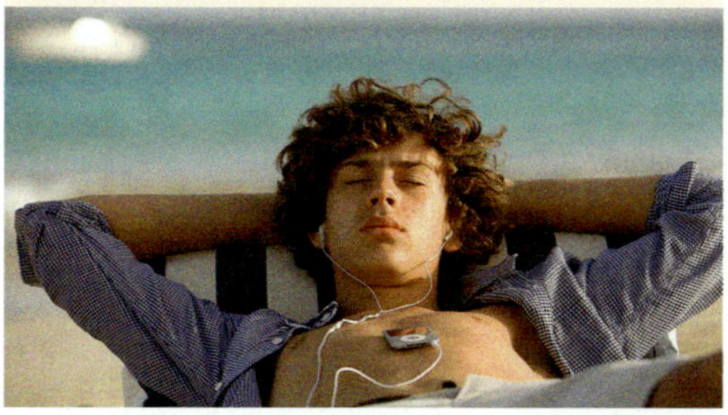

Cycling 1200 kJ

Cross-country skiing 3 200 kJ

Dancing 4 200 kJ

Figure 14 Energy requirements for different activities (in kilojoules per hour)

1 Name the two most common units for measuring energy.

2 Most people are hungry when they wake up in the morning although they have only been sleeping quietly. Explain why.

3 According to Table 1 on p. 64, how much energy does a boy aged 11–14 need? How much does a 23-year-old woman need? Give possible reasons for the difference. Discuss in class.

4 Look at Figure 14 and describe the energy requirements for the activities shown.

5 Make a list of your favourite activities and guess how high their energy requirements are. Then find out the real figures on the Internet. The following website may help you: *http://primusweb.com/fitnesspartner/calculat.htm* (for this website you will need to find out your weight in pounds, which can be done easily by using one of the many online conversion[2] sites). Share your results with a partner. Which of your findings surprised you most?

Did you know?

Besides kilojoule, you will also come across the term 'calorie'. It is an older unit for measuring energy, but a lot of people still use it. One calorie is the amount of heat you need to raise the temperature of one gram of water by one degree Celsius. Food energy is measured in large calories, also called kilocalories, which can be written capitalized as 'Calories' (Cal) or as 'kcal'.
1 Cal = 4.186 kJ

1 gender *Geschlecht*
2 conversion *Umrechnung*

4.5 Eating habits[1] can go wrong

We have seen how to choose a healthy diet and learned that we need regular exercise in order to maintain a healthy body weight. Although these guidelines may seem simple, it is not always easy for us to follow them. Many people have trouble managing their eating habits and some develop diseases, fatigue[2] and depression as a result.

While it is perfectly normal for people the same age to differ in size and weight, there is a way of measuring whether or not their body weight can be considered healthy. This is called the **body mass index (BMI)**. You can calculate your BMI using the following formula:

$$BMI = \frac{\text{body weight (kg)}}{\text{height (m)}^2}$$

Key terms

Body weight and eating habits
- anorexia nervosa *Magersucht*
- anorexic *magersüchtig*
- binge eating attack *Essattacke*
- body mass index (BMI) *Körpermassezahl*
- bulimia *Bulimie, Esssucht*
- eating disorder *Essstörung*
- obese *fettleibig*
- obesity *Fettleibigkeit*
- underweight *Untergewicht*

Less than 20	**Underweight**
20 to 25	Desirable[3] weight
25 to 30	Overweight
30 and over	**Obese**

Table 2 BMI categories for adults
From: World Health Organization, 2003

In general, BMIs *for adults* are classified in four categories, as shown in Table 2. Note that these values only apply to adults aged 20 and over, not to teenagers. A teenager's body changes rapidly, so this simple BMI classification cannot adequately[4] describe healthy ranges for young people. If an adult's BMI is outside the healthy range[5], they need to start paying attention to their eating habits, as being underweight or overweight can have serious consequences for your health. Normally your body knows when you are hungry and when you have to stop eating, so listening to your body's signals is important. Sometimes, however, the body develops bad eating habits, which can lead to **obesity** or **eating disorders**.

Eating and health problems

The three most common health problems associated with eating are obesity and the eating disorders **bulimia** and **anorexia nervosa**.

Obesity

In the Western world many people eat much more than their bodies require, which causes them to gain a lot of weight. A study in Germany in 2007 showed:
- 53 per cent of adult men and 35.6 per cent of adult women are overweight (i.e. they have a BMI between 25 and 30)
- 22.5 per cent of men and 23.3 per cent of women are **obese** (their BMI is 30 or more).

Being overweight or obese increases the risk

Figure 15 Obesity affects a lot of adults in the Western world

1 habit *Gewohnheit*
2 fatigue *Schwäche, Müdigkeit*
3 desirable *wünschenswert*
4 adequately *angemessen*
5 range *Spannbreite*

of developing heart disease and diabetes, causes difficulties during pregnancy and makes wounds heal more slowly. Obesity can also increase the danger of getting cancer.

Bulimia

Some teenagers who have a perfectly normal BMI sometimes have **binge eating attacks**. This means that they eat lots of food in a very short period of time, but feel bad afterwards and make themselves vomit[1]. Besides the emotional pressure which these people feel (having to hide their condition from their family and friends), the frequent vomiting can damage their vocal cords and teeth, and the lack of nutrients can lead to all kinds of deficiencies[2] in their bodies. Most people with bulimia need a doctor to help them break this habit.

Anorexia nervosa

Other teenagers do not eat enough food to supply their body with what it needs. They forget how to listen to their body's signals that tell them when they have to eat. Adults who eat very little food and have a BMI of under 17.5 are considered anorexic and need help from a doctor. Anorexia puts the body under enormous pressure. **Anorexic** teenagers may experience difficulty growing, while adults become very weak. In 2006 a Brazilian model died of anorexia – she had a BMI of 13.21. In the same year Spain banned models from fashion shows who had a BMI of under 18.

Figure 16 A case[3] of self-deception[4]

If you think that you or a friend might have one of these eating disorders, talk about it with someone you trust as soon as possible. Your biology teacher can also help you. While we all know it is normal to see a doctor about an obvious problem like a broken leg, some people feel embarrassed or ashamed about talking to anyone about eating disorders. Still, just like a broken leg, eating disorders are a serious matter and need to be treated. So don't worry about what others might think. It is best to get help quickly.

1 vomit *sich erbrechen*
2 deficiency *Mangel*
3 case *Fall*
4 self-deception *Selbsttäuschung*

1 Describe the consequences of obesity, bulimia and anorexia. Make a table comparing the three conditions.
2 Calculate your own BMI using an online calculator (e.g. *http://pediatrics.about.com/cs/usefultools/l/bl_metric_calc.htm*). It is important to choose a calculator for teenagers (not adults) and to use metric units of measurement when entering your height and weight. How reliable do you think the result is?
3 Why do you think teenagers in particular often feel they are too fat although they have a perfectly normal weight? Discuss this question with a partner and make a list of your ideas. Then, in small groups, discuss possible responses to this. How can teenagers become more satisfied with their bodies?
4 Using the Internet, the telephone book and any other resources, find out how people with eating disorders can be helped and where in your area people can find this help. Do you think there is enough help available for teenagers with eating disorders where you live? If not, where else could you get help?

4.6 Tips for healthy snacks

Healthy food does not have to be boring, and it can be very easy to prepare. Here are a few recipes[1] for snacks that are not only tasty but also good for you. Give them a try!

Smoothies[2]

This drink is the right choice for a hot day. It is quick, easy, and you can try out a lot of different ways of making it.

Ingredients[3] for two servings:

- 200 g natural low-fat yoghurt
- 2–3 medium-sized bananas
- 1 handful strawberries or other fresh fruit
- 400 ml fruit juice (e.g. orange, apple or cranberry)

Procedure:

Put all the ingredients into a blender[4] and mix them until smooth. You can also try flavoured yoghurt and any other mix of fruits. Bananas, though, make the drink especially smooth. For an extra boost of coolness, add some crushed ice.

Fruit muffins

Even cakes can be part of a healthy diet if you do not eat them all the time. This recipe makes six large or 12 medium-sized muffins.

Ingredients:

- 75 g white flour
- 75 g wholemeal[5] flour
- 2–3 teaspoons baking powder[6]
- 50 g caster sugar[7]
- 1 egg
- a pinch[8] of salt
- 75 ml milk
- 75 ml vegetable oil
- 50–75 g fresh or frozen fruit, e.g. raspberries, blueberries
- 50 g chopped[9] walnuts[10]

Procedure:

1. Heat the oven[11] to 190 °C. Using a special baking tray[12] for muffins, grease[13] the muffin tins[14] with a little oil.
2. In a large bowl[15] mix together the flour, baking powder, salt and sugar. In another bowl beat the egg, milk and oil lightly with a fork. Then pour all of the liquid ingredients into the dry ingredients and add the fruit and walnuts. Stir[16] with a tablespoon[17] until the ingredients combine, but no longer. Make sure to include the sides and bottom of the bowl.
3. Using a tablespoon, fill the muffin tins. Make sure to fill the tins only ¾ full. Bake for 15–20 minutes in the preheated oven. The muffins are ready when they are lightly browned.

1 recipe *Rezept*
2 smoothie *cremiges Getränk mit frischen Früchten*
3 ingredients *Zutaten*
4 blender *Mixer*
5 wholemeal *Vollkorn*
6 baking powder *Backpulver*
7 caster sugar *Streuzucker*
8 pinch *Prise*
9 chopped *gehackt*
10 walnut *Walnuss*
11 oven *Backofen*
12 baking tray *Backblech, Kuchenblech*
13 grease sth. *etwas einfetten*
14 muffin tin *Muffinform*
15 bowl *Schüssel*
16 stir sth. *etwas rühren*
17 tablespoon *Esslöffel*

Making a bar chart

Table 3 below shows the energy values for different foods and the amounts of nutrients these foods contain. A table like this provides concrete information but does not give you a quick overview of the similarities and differences between the different foods. A bar chart, on the other hand, is a good way to show this.

Food (value per 100 g)	Energy (kJ)	Water (g)	Carbohydrates (g)	Fats (g)	Proteins (g)
Peanuts	2374	6.5	16.13	49.24	25.80
Chips[1]	636	64.85	26.98	3,76	2,57
Milk (3.25% fat)	252	88.32	4.52	3.25	3.22
Hard-boiled egg	647	74.62	1.12	10.61	12.58
Apple	218	85.56	13.81	0.17	0.26
Salmon	761	59.62	0.00	8.13	25.44

Table 3 **Energy values for and nutrients in different foods** *From: www.nal.usda.gov/fnic/foodcomp/search/*

Follow these easy steps:

1 Draw a horizontal x-axis and a vertical y-axis on your page.
2 Find the highest number you have to plot (e. g. for nutrients in Table 3 it is 88.32 g) and round[2] this figure off to a higher number (e. g. a figure that you can divide[3] by 10). Make this the highest point on your y-axis. So for our chart, the top of the y-axis can be 90 g.
3 Next, mark the y-axis with logical divisions (divisions of ten work best in this case).
4 Label the x-axis with the four different nutrients, and draw a bar above each label. (The bars should not touch each other.) Find the correct height for each bar by aligning your ruler horizontally with the correct number on the y-axis for each nutrient.
5 Use the same scale and the same order of information for each bar chart that you draw.

1 chips (BE) = (AE) French fries *Pommes frites*
2 round off *aufrunden*
3 divide sth. by sth. *etwas teilen durch etwas*

1 **a** Make bar charts for hard boiled eggs, apples and salmon.
b Describe your charts to a partner using phrases from **Activate your English**, 'Describing and comparing bar charts', on the inside cover.

2 **a** Look at the bar charts you have made and compare them. For language help see **Activate your English**, 'Comparing and contrasting', on the inside cover. These phrases may also help you: *Peanuts/Milk/... contain(s) about ... grams of fat/...; ... consists / is made up mainly of ...; ... contains very little / only a small amount of ...; ... is rich in ... but poor in ...*
b Decide the role that each of these foods should play in a balanced diet.

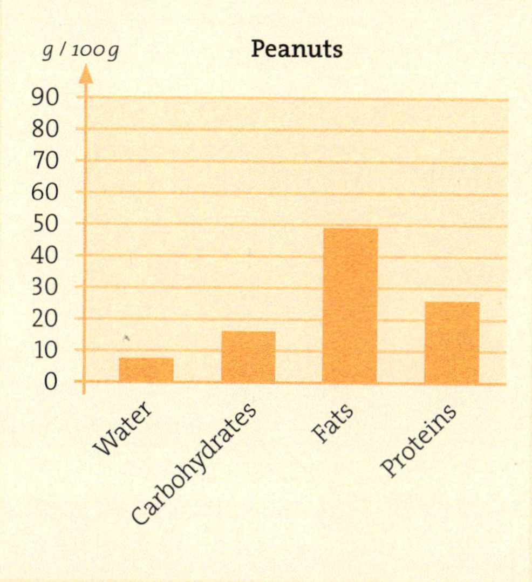

Figure 17 **A bar chart showing the amounts of nutrients in a 100 g serving of peanuts**

Conducting[1] food tests: Learning at stations[2]

Do apples contain protein? Does bread contain fat? Do peanuts contain carbohydrates? To answer these questions, there are a few simple tests you can do.
Learning at stations allows every pupil in your class to do a number of experiments, because several groups can work at the same time. Also, everybody can work at their own pace[3]. This will help you to develop your skills in conducting tests and recording results, and you will also learn more about the food you eat.

Key terms

Food tests
- blank sample *Blindprobe*
- evaporate *verdampfen*
- indicate sth. *etwas nachweisen*
- indicator *Indikator*
- log sheet *Laufzettel*
- Petri dish *Petrischale*
- pipette *Pipette*
- precipitation *Niederschlag*
- safety goggles *Schutzbrille*
- sample *Probe*
- test tube *Reagenzglass*

Here are some rules for learning at stations:
- Study the instructions[4] thoroughly[5] before you start each test.
- Stick to the rules, especially the safety warnings (see 'Safety in the laboratory' below).
- If you have any problems at one station, first discuss them with your group and try to find a solution. Only call the teacher if you cannot solve the problem yourselves.
- Handle all materials carefully – you want your next station to work properly, so you need to leave each station ready for the next team to use.
- Record everything you do on a **log sheet** (see Table 4 below).

In your exercise book make a log sheet like the one below. By the end of this sequence, you should have completed all the sections of your log sheet.

Station	***Tested substance***	***Result of test***	***Comment/Conclusion***
1. Testing for glucose	*Table sugar* *Sunflower oil*	*Solution turns brick-red[6].* ...	*Sample contains pure glucose.* ...
2. ...	...	...	...
3. ...	...	...	...

Table 4 **Example of a log sheet**

Safety in the laboratory

For some of these tests you will be using chemicals. Whenever you handle chemicals, there are safety rules you have to follow:
- Wear **safety goggles** at all times to protect your eyes.
- When you heat substances in a **test tube**, point the opening of the tube away from yourself and others.
- Wash your hands thoroughly when you have finished each experiment.
- Never let any chemical touch your eyes, nose, lips or tongue.
- When working with two or more test tubes, be careful not to put any substances into the wrong test tubes. For example, do not use the same pipette for different solutions without rinsing[7] it thoroughly first.

1 conduct sth. *etwas durchführen*
2 learning at stations *Lernen an Stationen*
3 pace *Schritt, Geschwindigkeit*
4 instruction *Anweisung*
5 thoroughly *gründlich*
6 brick-red *ziegelrot*
7 rinse sth. *etwas ausspülen*

Station 1 Use an indicator to test for glucose

Figure 18 Testing for glucose

Materials
- several test tubes
- **pipette**
- an **indicator**: Fehling's solution A and B or Benedict's solution
- hot water bath
- **sample** substances in solution (e.g. sunflower oil, orange juice, juice from onions, glucose)

Procedure
1. Mix 3 ml of Fehling's solution A with the same amount of solution B in a test tube.
2. Add about 5 ml of the sample solution and put it in the hot water bath for two to three minutes.
3. Observe any changes in the solution. Any red, orange or yellow **precipitation indicates** the presence of glucose in decreasing concentration. A brick-red colour shows that the sample contains pure glucose.
4. Carefully record on your log sheet the results for all the substances you test.

(Note: With Benedict's solution the test can be carried out more easily, because you do not need to prepare the indicator. Just add 5 ml of Benedict's solution to 5 ml of the sample solution and put the test tube in the water bath for five minutes.)

Station 2 The iodine test for starch

Figure 19 Testing for starch

Materials
- several test tubes
- pipette
- iodine solution (i.e. iodine dissolved in an aqueous solution of potassium iodide)
- sample substances (e.g. potatoes, pure starch, bananas, apples)

Procedure
1. Put a small amount of one of the sample substances into a test tube, into the lid of a **Petri dish** or onto a white tile.
2. Add a few drops of iodine solution.
3. Observe any changes in colour. If the colour changes to a deep blue or black, it is an indication of the presence of starch.
4. Test all the different substances and record the results on your log sheet, even if there is no change in colour.

Station 3 The filter test for fats

Materials
- filter paper
- lamp
- sample substances (e.g. margarine, bacon, bread, peanuts, oil, mashed banana)

Procedure
1. Put a sample of your substance on the filter paper and circle it with a red pen.
2. To make a **blank sample**, put a drop of water on the filter paper some distance from the red circle and circle it with a blue pen.
3. Hold the filter paper in front of a light to make the liquids **evaporate**. This can take up to ten minutes.
4. Observe the results. If your sample contains fat, a stain will remain on the filter paper inside the red circle. In the blue circle, you should see nothing.
5. Record your findings on your log sheet.

Figure 20 Testing for fats

(Note: You can also put several samples in different spots on the same piece of filter paper; just make sure to label each clearly.)

Station 4 Use test strips to test for proteins and vitamin C

Materials
- test strips for proteins and vitamin C
- samples (e.g. lemon juice, milk, nuts, eggs, apples)

Procedure
1. There are specially made test strips for many different substances. In this case, we want to test for proteins and vitamin C. Simply follow the instructions on the back of the test-strip packaging.
2. Carefully record all your findings.

Figure 21 Testing for proteins

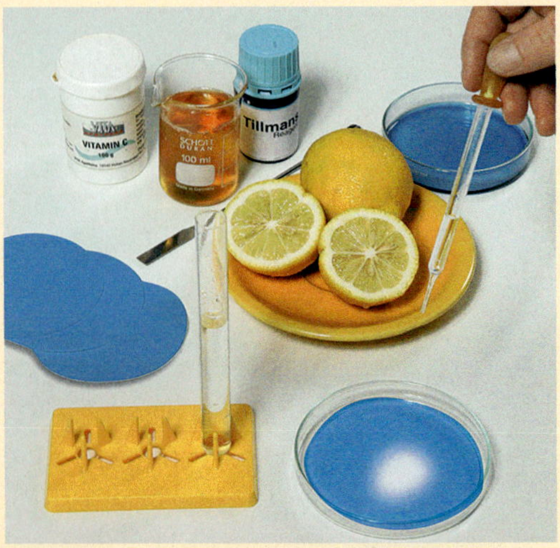

Figure 22 Testing for vitamin C

Summary

Food has three main functions: supplying energy, providing materials for growth and repair, and keeping the body healthy. Energy is mostly supplied by fats and carbohydrates. Proteins are mainly used for growth and repair. Vitamins, minerals, fibre and water are also necessary to stay healthy.
Foods can be classified in groups depending on how much of each nutrient they contain. A third of the food you eat should be either potatoes or wholegrain products. You should also eat at least five servings of fruit or vegetables a day and only a few snacks, as these are often high in unhealthy fat.
If you eat too much, you take in more energy than you need (an average teenage boy needs about 9270 kJ, a girl about 7720 kJ). Over time this can lead to obesity, which has many negative effects on your health. Other serious eating disorders are bulimia and anorexia. A lot of teenagers struggle with these disorders and should talk to somebody about it to get help.

Key terms

See the boxes:

Aspects of food .... p. 58
Nutrition 1, 2 .... pp. 59, 60
Food and lifestyle .... p. 61
Energy needs .... p. 65
Body weight and eating habits ...... p. 66
Food tests .... p. 70

A Study tasks

1. Create a two-day plan with the right amount of foods from all the main groups for a:
   a) non-vegetarian
   b) vegetarian.
2. Section 4.3 tells you how much from each food group you should be eating regularly. Draw a diagram arranging these amounts in a pyramid, placing the food group you should eat most of at the bottom.
3. Use the information in Table 1 (p. 64) to make a bar chart that shows the different energy requirements of males and females between the ages of 1 and 18.
4. What does the body need energy for when it is at rest?

B Practise your English

Complete these sentences with words from the chapter.
1. Salmon ... more fats than carbohydrates.
2. The energy in fats can be ... into a form that the body can use.
3. Fruits and vegetables are good for your digestion because of the ... in them.
4. Fats, carbohydrates and proteins are ... that give you energy.
5. ... are foods that are made from milk.
6. The ... tells you how much energy your body needs when it is not doing any work.

C Say it in German

Together with a partner, look again at p. 66 and quiz each other in German on the body mass index (BMI) and how it works. Then discuss the results of your own BMI calculations and what they might mean for your diet.

D Critical thinking

1 Use the Internet and other sources to prepare an in-depth[1] presentation on eating disorders.
2 Go to the website of a fast-food restaurant chain. Look at the information they offer on the ingredients in and the nutritional value[2] of their food. Decide whether 'healthy fast food' exists.
3 This chapter has focused on the recommendations[3] of the British health services. Using the Internet, find out what the US and German governments suggest.

1 in-depth *detailliert*
2 nutritional value *Nährwert*
3 recommendation *Empfehlung*

Chapter 5

Digestion and Absorption

Chapter objectives

- To **name** and **locate** the main organs of digestion
- To **apply**[1] the rules for healthy teeth and gums
- To **describe** how food is changed on its way through the alimentary canal
- To **explain** how nutrients are absorbed into the body
- To **plan** an experiment to find out about the action of digestive enzymes

Key terms

The digestive organs

- alimentary canal *Verdauungskanal*
- anus *Darmausgang*
- appendix *Wurmfortsatz (des Blinddarms)*
- caecum *Blinddarm*
- digestion *Verdauung*
- duodenum *Zwölffingerdarm*
- gall bladder *Gallenblase*
- gullet *Speiseröhre*
- large intestine *Dickdarm*
- liver *Leber*
- mouth cavity *Mundhöhle*
- pancreas *Bauchspeicheldrüse*
- rectum *Mastdarm*
- salivary gland *Speicheldrüse*
- small intestine *Dünndarm*
- stomach *Magen*

Each of the billions of cells that make up your body needs a constant supply of water and nutrients to stay alive and to function properly. The process that changes the food you eat, so that the cells are able to use it, is called **digestion**. This process happens in several steps while the food moves slowly from the mouth through the different organs of the **alimentary canal**, which is between five and six metres long. The food is broken down into its simplest building blocks by mechanical and chemical processes, and these tiny particles[2] pass into your bloodstream[3] and lymph[4], to be carried to all body cells. Finally, the waste material[5] passes out of your body when you go to the toilet.

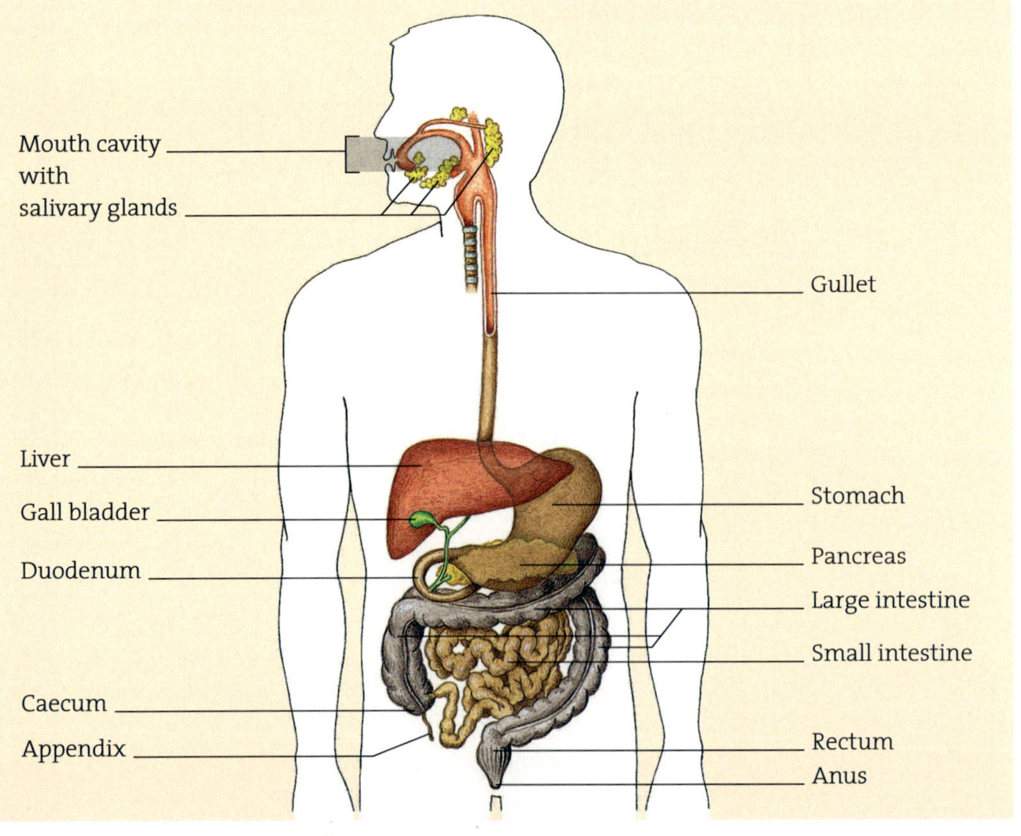

Figure 1 The human digestive[6] organs: the alimentary canal

1 apply *anwenden*
2 particle *Teilchen*
3 bloodstream *Blutbahn*
4 lymph *Gewebsflüssigkeit, Lymphe*
5 waste material *Abfallstoff, Reststoff*
6 digestive *die Verdauung betreffend*

5.1 Digestion starts in the mouth

The first stage[1] of digestion takes place when you put food into your mouth and eat it. While you are chewing[2] the food with your teeth, breaking it into smaller pieces, the salivary glands release a special fluid called **saliva** into your mouth cavity. Saliva consists mainly of water and makes the food moist and slippery[3] and easier to swallow[4]. But saliva also contains a special digestive **enzyme** that breaks starch, a carbohydrate, into sugar molecules. Enzymes work best when the food particles are small – this is why it is important to chew your food well before you swallow it. It is also a reason why you need healthy teeth, as they play a key role in this first stage of digestion.

The tongue is also important in the processes of chewing and digestion. It is fastened at the back of the mouth and consists of many bundles of voluntary muscles that run in different directions. This enables the tongue to carry out quick movements in different directions. By moving from side to side, it pushes the food between the chewing surfaces of the teeth, where it is reduced to smaller pieces. It also moves the food from one area of the mouth to another, thus mixing it with saliva. By curling[5] upwards it presses the mashed[6] food against the top of the mouth cavity and forms the food into a ball, which can be swallowed more easily. As a last function connected with digestion, the tongue cleans the mouth of food remains[7].

Key terms

Digestive juices
- enzyme *Enzym, Ferment*
- juice *Saft*
- saliva *Speichel*

Did you know?

You have three pairs of salivary glands, which empty about two litres of saliva into your mouth every day. The smell, the sight or even the thought of something tasty stimulates these glands and 'makes your mouth water'.

1 Think about the processes in your mouth that break the food into smaller pieces. Which are part of mechanical[8] digestion, and which are part of chemical[9] digestion? Explain.
2 Why is it important that you chew your food well?
3 Explain how the teeth and tongue work together in the digestion of food in the mouth.

Further facts

Focus on enzymes

Enzymes are substances that cause chemical reactions. For example, they can make complicated molecules split into smaller and simpler ones, or make simple molecules combine into more complicated ones. Without enzymes these reactions would need a much longer time or a temperature high enough to destroy the proteins in living cells.

Enzymes mainly consist of protein, which means that they work well at a normal body temperature of 37 °C but are destroyed at temperatures above 42 °C. On the other hand, when the temperature gets lower, the activity of the enzymes is reduced.

The level of acidity[10] in the surrounding environment[11] also affects how well enzymes work. Some enzymes only work in a neutral[12] medium[13], others in acidic[14] surroundings, and a third group needs a basic[15] medium in order to activate a reaction.

Enzymes are not destroyed or changed by the processes they cause and each enzyme can only control a specific chemical process.

For more on enzymes, see Section 5.7, 'Enzymes are important for chemical digestion', p. 86.

1 stage *Stadium*
2 chew *kauen*
3 slippery *gleitfähig, schlüpfrig*
4 swallow *schlucken*
5 curl *rollen*
6 mashed *zerdrückt, zerstampft*
7 food remains (pl.) *Nahrungsreste*
8 mechanical *mechanisch*
9 chemical *chemisch*
10 acidity *Säure, Säuregehalt*
11 environment *Umgebung*
12 neutral *neutral*
13 medium *Milieu, Nährboden*
14 acidic *sauer, säurehaltig*
15 basic *basisch, alkalisch*

5.2 Teeth need proper care

Our teeth allow us to chew food, making it smaller and thus beginning the process of digestion. Therefore it is very important that we look after them and keep them healthy for as long as possible.

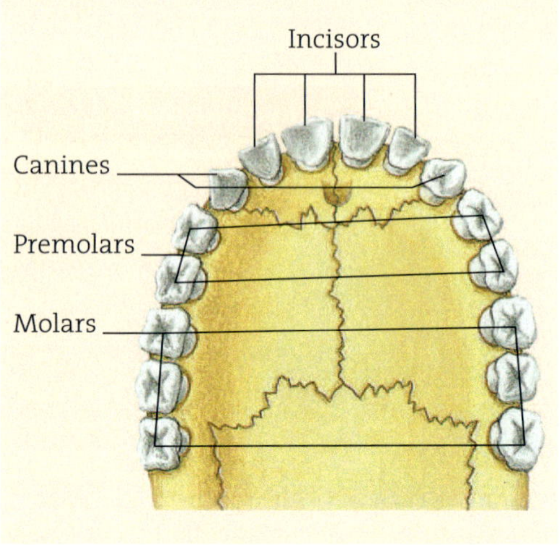

Figure 2 **Set of teeth in the upper jaw of an adult**

Key terms

Teeth

- canine *Eckzahn*
- cement *Zahnzement*
- crown *(hier) Zahnkrone*
- dentine *Zahnbein*
- enamel *Zahnschmelz*
- fluoride *Fluor*
- gum(s) *Zahnfleisch*
- incisor *Schneidezahn*
- molar *hinterer Backenzahn*
- permanent teeth *bleibende Zähne*
- premolar *vorderer Backenzahn*
- primary or milk teeth *Milchzähne*
- pulp *Zahnmark*
- pulp cavity *Zahnhöhle*
- root *(hier) Zahnwurzel*
- set of teeth *Gebiss*
- socket *Hohlraum für den Zahn im Kiefer*

Newborn babies have no teeth, but by the time children are about three years old, they normally have twenty teeth. These **primary teeth** or **milk teeth** fall out and are gradually replaced[1] by **permanent teeth** from the age of five or six onwards. In this process the number of teeth grows to 32. An adult person has four **incisors**, two **canines**, four **premolars** and six **molars** each in the upper and lower jaw[2].

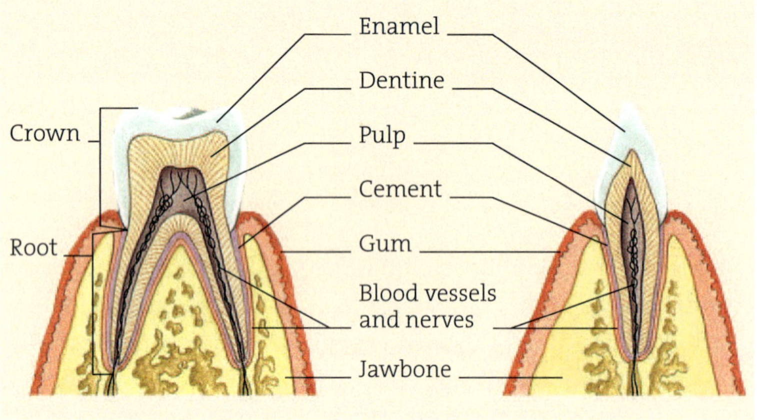

Figure 3 **Longitudinal section[3] of a molar and a canine**

Our teeth are different shapes and sizes, but they all have the same basic plan. The **crown**, the visible part of the tooth, is covered by a hard layer of **enamel** on the outside. Enamel is the hardest substance in the human body. It protects the main part of the tooth, the **dentine**, which is a much softer, bony material. The **root** of the tooth is covered with a thin layer of **cement**, another bony substance, which holds the tooth in its **socket** in the jaw. The most sensitive part of the tooth is the area where the **gums** touch the tooth. The **pulp cavity** is a space in the centre of the tooth filled with a soft tissue, the **pulp**. The pulp contains nerves and blood vessels, which keep the tooth alive.

The different types of teeth have different jobs to do. The front teeth, the incisors, are like chisels[4]. They are for cutting the food into smaller pieces. The canines, which are sharp and pointed, hold and tear the food. The premolars and molars are broader. They have a bumpy[5] surface for chewing and grinding up[6] the food to make it easier to swallow. Molar teeth are larger than premolars and always have more than one root to anchor them safely in the jaw.

1 replace sth. by sth. *etwas durch etwas ersetzen*
2 jaw *Kiefer*
3 longitudinal section *Längsschnitt*
4 chisel *Meißel*
5 bumpy *uneben*
6 grind sth. up *etwas zermahlen*

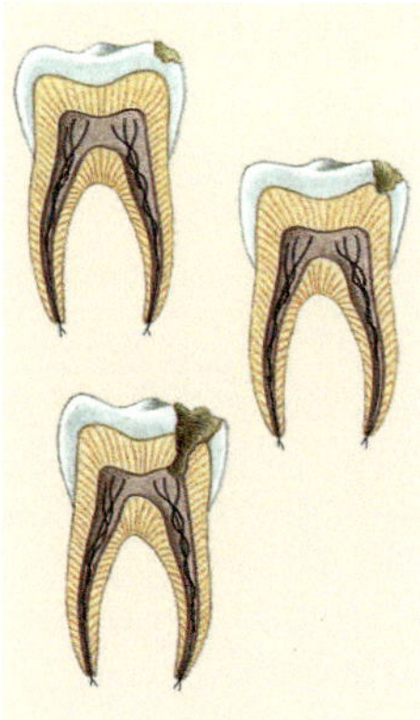

Figure 4 **Development of tooth decay**

From the moment of birth, bacteria invade[1] the human body. Most of them are harmless or even help in digestion, but some bacteria are harmful[2], as they can cause **tooth decay** and **gum disease**. When you eat sweet foods containing sugar, bacteria form a sticky layer on the surface of your teeth called **plaque**, especially next to the gums and between the teeth. Plaque bacteria break down sugar and produce an acid which dissolves enamel and makes tiny cracks[3] and holes in your teeth. When bacteria reach the dentine, they can spread rapidly to the pulp and the roots of the tooth. When the tooth starts to hurt, your dentist has to use a drill[4] to remove[5] the decayed[6] tissue and fill the hole with a special material. If you wait too long before seeing your dentist, the bad tooth may have to be pulled out.

If plaque is left on the teeth for a few days, it hardens[7] and cannot be brushed off anymore. This hard layer can lead to an **inflammation** of the gums, which swell[8] and hurt. At a later stage they start to bleed when they come into contact with a toothbrush. Gradually the gums shrink[9] and pull away from the teeth, which become loose[10]. The teeth may even fall out if the plaque is not removed and the infection is not treated[11] in time[12].

Key terms

Teeth-related problems

- gum disease *Zahnfleischerkrankung*
- inflammation *Entzündung*
- plaque *Zahnbelag*
- tooth decay (caries) *Zahnfäule, Karies*

But gum disease and tooth decay can be prevented if you follow a few simple steps for healthy gums and teeth:

- Eat a balanced diet including a lot of fresh fruit, vegetables and milk products.
- Don't eat too many sweets, and when you do eat sweet things, don't chew them for too long.
- Brush your teeth in the morning, at night and after meals, if possible.
- Use a **fluoride** toothpaste to protect the enamel.
- Use dental floss[13] once a day to clean the areas between the teeth that your toothbrush cannot reach.
- See your dentist twice a year for a check-up[14].

1 invade sth. *in etwas eindringen*
2 harmful *schädlich*
3 crack *Riss*
4 drill *Bohrer*
5 remove sth. *etwas entfernen*
6 decay *faulen, zerfallen*
7 harden *härten*
8 swell *anschwellen*
9 shrink *schrumpfen*
10 become loose *sich lockern*
11 treat sth. *etwas behandeln*
12 in time *rechtzeitig*
13 dental floss *Zahnseide*
14 check-up *Kontrolluntersuchung*

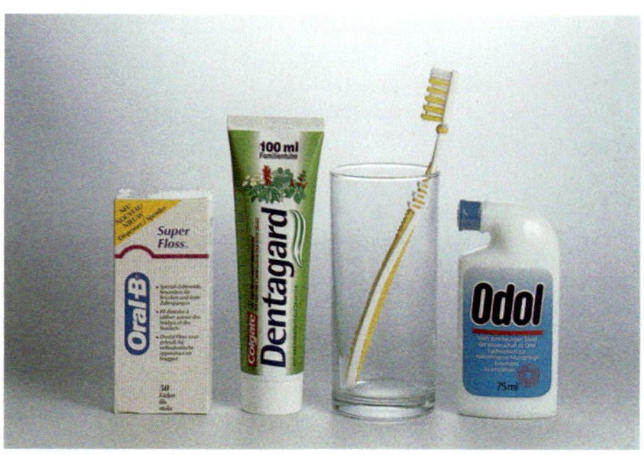

Figure 5 **Articles for dental care**

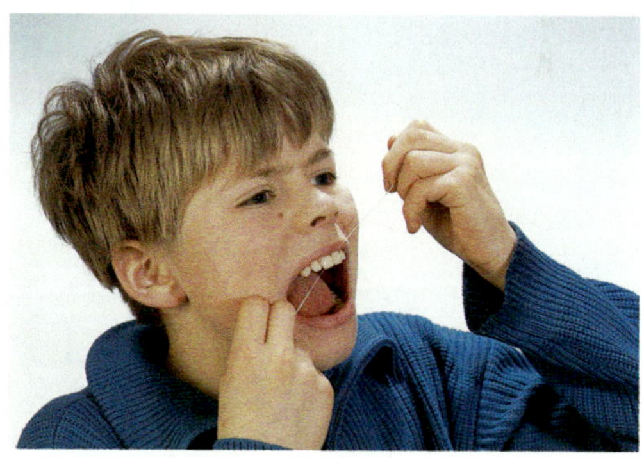

Figure 6 **Using dental floss**

1 Draw a diagram showing the structure of a tooth. Label as many different parts as you can using the Key terms box above.

2 In what ways are an adult's teeth different from those of a three-year-old child? Discuss this question with a partner.

3 Explain why it is best not to eat too many sweet foods.

4 Write a short report describing the proper technique for brushing your teeth. Don't forget to say how long you should brush to remove all food remains and plaque.

5 Working with a partner, make a poster showing how people can keep their teeth healthy. Display your poster in the classroom.

5.3 The gullet transports food from the mouth to the stomach

When your tongue pushes a small ball of moist food (bolus[1]) to the back of your mouth, it touches the **soft palate**. This makes you swallow the food automatically[2]. There are two tubes at the base of the throat: the windpipe[3] at the front and the gullet at the back. When you swallow, a flap of tissue[4] called the **epiglottis** automatically covers the opening of the windpipe. This mechanism[5] stops food from going down the wrong passage[6]. After swallowing, the food enters the gullet, which is between 25 and 30 cm long. The **ring muscles** of the gullet push the food downward[7] to the stomach in a wavelike motion called **peristalsis**.

Key terms

From the mouth to the stomach
- epiglottis *Kehldeckel*
- hard palate *harter Gaumen*
- peristalsis *peristaltische Bewegung*
- ring muscle *Ringmuskel*
- soft palate *weicher Gaumen*

1 Bolus *rundlicher Nahrungsbrocken*
2 automatically *automatisch*
3 windpipe *Luftröhre*
4 flap of tissue *Klappe oder Lasche aus Gewebe*
5 mechanism *Mechanismus*
6 passage *Durchgang*
7 downward *abwärts*
8 pharynx *Rachen*
9 upside-down *umgedreht*

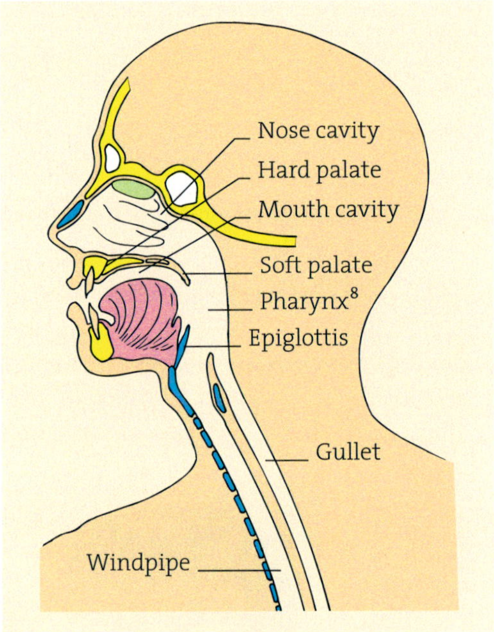

Figure 7 Position of the windpipe and gullet

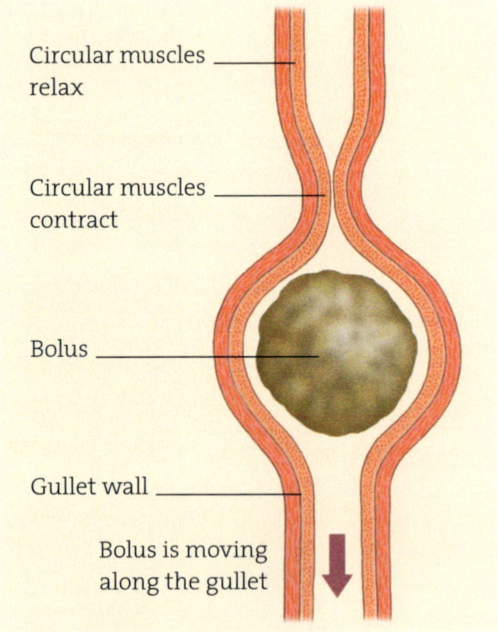

Figure 8 Longitudinal section of the gullet showing peristalsis

1 Make a flow chart which shows how a piece of food moves from the mouth to the stomach. For help see 'Working with flow charts', p. 84.

2 Have you ever eaten something and then started to cough because a small piece of food got into your windpipe? Why is this unusual? How does it happen? (See also Figure 12, p. 39) Discuss these questions with a partner and try to think of what you can do to help someone when their food 'goes down the wrong way'.

3 Why do you think it is possible to drink with a straw while you are in an upside-down[9] position, for example in a handstand or headstand? Discuss the question in class. You may want to try it yourself!

5.4 Protein digestion starts in the stomach

When food reaches the stomach, protein digestion begins. The stomach is shaped like a bag and holds up to two litres of food. The **stomach lining** has millions of tiny **glands**, which release **pepsinogen**, **hydrochloric acid** and **mucus**. As the thick muscular walls of the stomach contract and relax, food and **gastric juice** mix together. When the pepsinogen comes into contact with the hydrochloric acid, it is activated and becomes the enzyme **pepsin**. This enzyme starts the digestion of the long chains of proteins, which are broken up into smaller pieces.

Key terms

The stomach
- gastric juice *Magensaft*
- gland *Drüse*
- hydrochloric acid *Salzsäure*
- longitudinal muscle *Längsmuskel*
- mucus *Schleim*
- pepsin *Pepsin*
- pepsinogen *Pepsinogen*
- stomach lining *Magenschleimhaut*

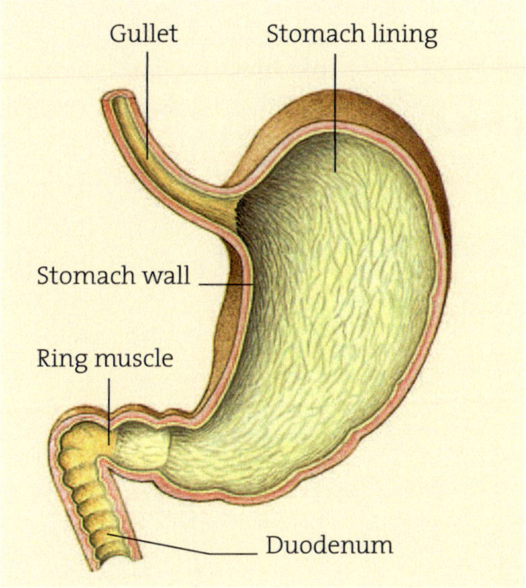

Figure 9 Cross-section of the stomach

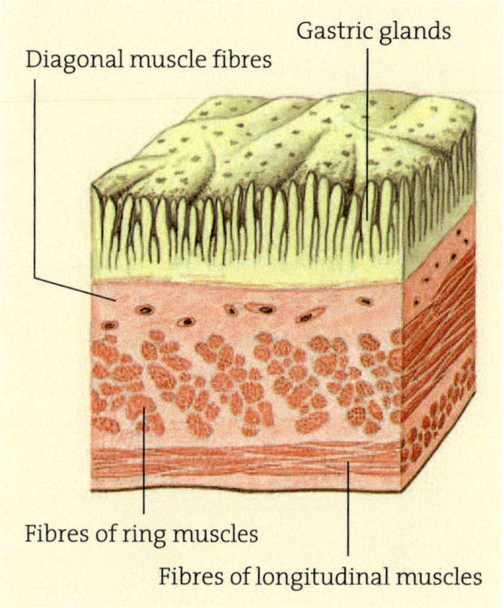

Figure 10 The stomach wall in detail

Stomach acid[1] has two important functions: it activates the enzymes for protein digestion, which are produced in the stomach lining in an inactive state[2], and it kills bacteria which may have entered the stomach together with the food. The mucus protects the stomach lining against the acid as well as pepsin.

The length of time that food remains in the stomach depends on the type of food you have eaten. Water may pass through in a few minutes, whereas milk and a soup with rice or oats will remain in the stomach for between one and two hours. Fat slows down the digestive processes in the stomach. A dish[3] of roast beef or roast pork will keep your stomach busy for up to seven hours. After that time, the ring muscle at the lower end of the stomach opens (in regular intervals[4]) and small amounts of the partly digested food flow into the small intestine.

1 stomach acid *Magensäure*
2 state *Zustand*
3 dish *Gericht*
4 interval *Zeitabstand*

1 Describe the structure of the stomach wall (see Figure 10) and its function in the process of digestion.
2 Which of the processes in your stomach can be described as mechanical digestion and which as chemical digestion? Explain.
3 The cells of your stomach consist of protein. What prevents the stomach from digesting itself?

5.5 Food is further digested and absorbed in the small intestine

After leaving the stomach, the partly digested food enters the small intestine, which is about 7 metres long and 2.5 cm wide. It is the longest part of your **gut** and is twisted[1] and folded[2] in order to fit into the space in the abdomen. The first 30 cm of the small intestine are called the duodenum. In the duodenum, two important glands add their digestive liquids to the food. Firstly, the liver produces **bile**, a bitter-tasting liquid which is stored in the **gall bladder** and flows through a small tube into the duodenum. Special salts in the bile help to digest fat by breaking it down[3] into small droplets[4]. Secondly, the pancreas, which is located between the stomach and small intestine, adds **pancreatic juice** to the food. This juice contains a variety[5] of enzymes which break down carbohydrates, fats and proteins.

Key terms

The intestines
- absorption *Resorption*
- bile *Gallenflüssigkeit*
- gall bladder *Gallenblase*
- gut *Darm*
- microvillus (pl. microvilli) *Microvillus*
- pancreatic juice *Bauchspeicheldrüsensaft*
- villus (pl. villi) *Darmzotte*

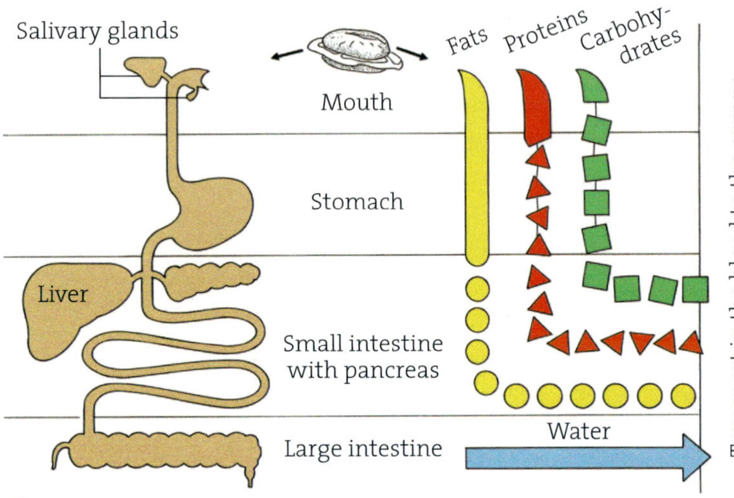

Figure 11 Stages of digestion and the uptake of digested nutrients into the blood and the lymph

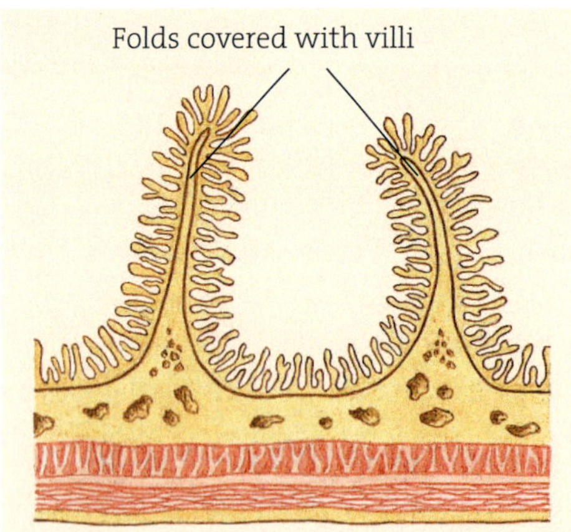

Figure 12 Structure of the intestinal wall

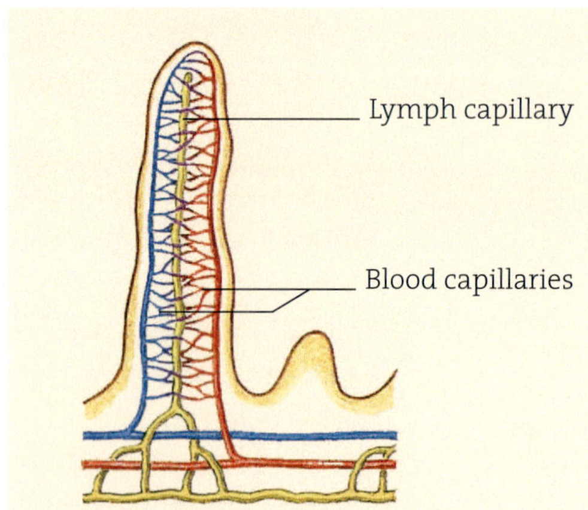

Figure 13 Inner structure of a villus in detail

Enzymes produced by special glands in the wall of the small intestine complete the process of digestion. By this stage the food has broken down into molecules, which are so small that they can pass through the lining of the intestine into the blood and the lymph capillaries[6]. This process is called **absorption**. The surface area of the inner lining of the small intestine is greatly increased by millions of tiny, finger-like structures called villi. Each **villus**,

Did you know?

The folds in the lining of your small intestine, together with the villi and microvilli, increase the surface area of your small intestine to about 200 square metres. If you compare this to the size of your classroom, you will get an idea how big the inner surface of your intestine is.

1 twisted *gebogen, gedreht*
2 folded *gefaltet*
3 break sth. down *etwas abbauen*
4 droplet *Tröpfchen*
5 variety *Vielfalt*
6 lymph capillaries *Lymphkapillaren*

whose lining is only one cell thick, is about 1 mm long and contains blood and lymph capillaries. The surface of a villus is covered by a fur[1] of hair-like structures, the **microvilli**, which can only be seen under the microscope.

Digestive glands or organs	Digestive liquid	Digested nutrient
Salivary glands	Saliva	Carbohydrates
Stomach	Gastric juice	Proteins
Liver	Bile	Fats and oils
Pancreas	Pancreatic juice	Carbohydrates, fats, proteins
Small intestine	Intestinal juice	Carbohydrates, proteins

Table 1 Digestive processes in the human body

5.6 The large intestine absorbs water and stores undigested food

In the final stage of the digestive process plant fibres and other parts of your food that cannot be digested, pass from the small intestine to the large intestine. These substances, also called **roughage**, are indigestible because our digestive organs do not produce the enzymes needed to break them down into molecules. However, roughage stimulates the muscles of the small intestine and supports the digestive process. The large intestine is about 1.4 metres long and twice as wide as the small intestine. The lining of the large intestine absorbs water, together with some minerals and vitamins, into the blood. The remaining[2] **wastes**, which are now more solid[3], move into the last part of the large intestine, the rectum. Here they are stored for some time. Finally they leave the body through the anus as **faeces** when you go to the toilet.

Key terms

The final stage
- faeces *Kot*
- roughage *Ballaststoffe*
- waste *Abfallstoff*

Did you know?

The appendix was considered to have no meaningful function in the human body, until scientists found out that it plays a role for the immune system. However, in many plant-eating animals such as guinea pigs[4] it is the largest organ in the abdomen and is needed for digesting plant material.

1 If nothing is digested in the liver and pancreas, why do they belong to the digestive organs?
2 Working with a partner, one of you is Student A and one is Student B. Student A: Describe the inner surface of the small intestine. Student B: Ask your partner questions about the purpose of this structure. Then swap roles[5] and repeat the activity. The words and phrases in **Activate your English** may help you.
3 Why are plant fibres, which your body cannot digest, part of a healthy diet? What would happen if you did not eat enough fruit and vegetables?
4 Research what happens in your alimentary canal when you suffer from diarrhoea[6].

Activate your English

Describing an organ and its function
The surface of the ... is covered with ...
The surface area of the ... is increased by ...
Each ... contains ...
These look like ...
These carry ... to the ...
The purpose of the ... is to ...

1 fur *Pelz*
2 remaining *übrig geblieben*
3 solid *fest*
4 guinea pig *Meerschweinchen*
5 swap roles *die Rollen tauschen*
6 diarrhoea *Durchfall*

1. Planning and carrying out a scientific experiment

In your biology class you are often asked to explain why something happens. In order to prove whether or not your explanation is correct, you can do an experiment.

How scientists work

An **investigation** in biology or any other science always starts with a problem or a question about the world around us or about our own body.
After first asking a question, scientists develop ideas about a possible solution to the problem by using all the knowledge they have from books or previous experience. In other words: they develop a **hypothesis**. Now they need to prove whether their hypothesis is correct by planning and performing an experiment. They make a **prediction** about the **outcome** of the experiment on the basis of this hypothesis.

> **Key terms**
>
> **Scientific experiments**
> - control (experiment) *Kontrolluntersuchung, Kontrollgruppe*
> - follow-up experiment *Anschlussexperiment*
> - hypothesis (pl. hypotheses) *Hypothese*
> - investigation *Untersuchung*
> - outcome *Ergebnis*
> - prediction *Voraussage*

When you plan an experiment, you should know exactly what you want to find out and which materials you need for the experiment.
Be sure to note down everything you do and everything that happens during the experiment so that you or somebody else can repeat it and check the results.
When experimenting with an unknown[1] substance under special conditions, it is best to have a parallel[2] experiment as a **control** with a substance whose reaction or effect you know under identical conditions (water, for instance).
The results of the experiment will show if the hypothesis you are trying to prove is correct or wrong. Often the results of an experiment lead to new questions and further experiments.

Experiment with saliva

Aim: To find out if saliva contains enzymes which split[3] starch into sugar molecules

Materials:
- four test tubes[4] labelled A, B, C and D
- starch solution
- medicine dropper or pipette
- Lugol's iodine solution
- saliva
- water bath
- test-tube holder[5]

Procedure:
1. Pour about 2 ml of starch solution into test tubes A and B. Add a drop of iodine solution to both tubes with the medicine dropper or the pipette and shake them. Observe the change in colour.
2. Collect about 2 ml of saliva in the third test tube (C) and put the same amount of water into the fourth one (D). Then empty the contents of both tubes separately into the tubes containing the starch solution from step 1. Shake the mixture in both test tubes carefully and place them in a water bath of 37 °C. Observe the test tubes A and B to see if you can detect a colour change.

1 unknown *unbekannt*
2 parallel *parallel*
3 split sth. *etwas teilen, aufspalten*
4 test tube *Reagenzglas*
5 test-tube holder *Reagenzglashalter*

1 What happens when the iodine solution is added to the two test tubes with starch solution? Explain.
2 Explain the effect of adding saliva to the two solutions.

3 Why do the test tubes have to be placed in a water bath of 37 °C?
4 Why are two test tubes used in both experiments?
5 Does the result fulfil the original aim of the experiment? Explain.

A follow-up experiment

If the research question has not been completely answered, it is your turn to plan and carry out an experiment that can provide the decisive[1] answer. Discuss this project with your classmates, then choose a partner to carry out the experiment with. You may want to use the information gained in the experiments in Chapter 4. The words and phrases in **Activate your English** below will help you.

Note: Always wear safety goggles and use a test-tube holder when working with chemicals.

Activate your English

Talking and writing about a laboratory experiment

Pointing out the aim of the experiment

I want	*to find out if*	*by ...ing ...*
My aim is	*to prove that*	*by looking at ...*

In order to achieve this, I will ...
The solution x will be used as an ***indicator*** *for the presence of ...*

I expect that	*the iodine solution will make the starch solution change colour.*
	the test sample will react differently to the control sample.
	... .

Describing the procedure for testing a liquid and the results

The test tubes used in the experiment are labelled ...
First you pour ... ml of ... into test tube ...
... solution is used as a test reagent for ...
The solution / content of the test tube is mixed by ...ing ...

The test tube is placed in	*a hot water bath*	*for ... minutes.*
	a water bath of ...°C	
	...	

Observe/Watch the test tube to see if ...

Explaining the results

If the colour of the solution changes from ... to ..., the test for ... is positive.

Drawing conclusions

The result of the experiment	*indicates that ...* *proves that ...* *does not prove that ...*
My predictions were	*correct.* *wrong.*
My expectations were	*fulfilled.* *partly fulfilled.* *not fulfilled.*
The results	*correspond*[2] *with my predictions.* *do not correspond with my predictions.*

As the results do not agree with my predictions, I have to change/modify[3] *the experiment.*

1 decisive *entscheidend*
2 correspond with sth. *mit etwas übereinstimmen*
3 modify sth. *etwas abwandeln, modifizieren*

2. Working with flow charts

Flow charts[1] help you to understand and memorize texts that describe complex activities. As they visualize[2] the steps in a process, flow charts are also useful when giving a presentation.

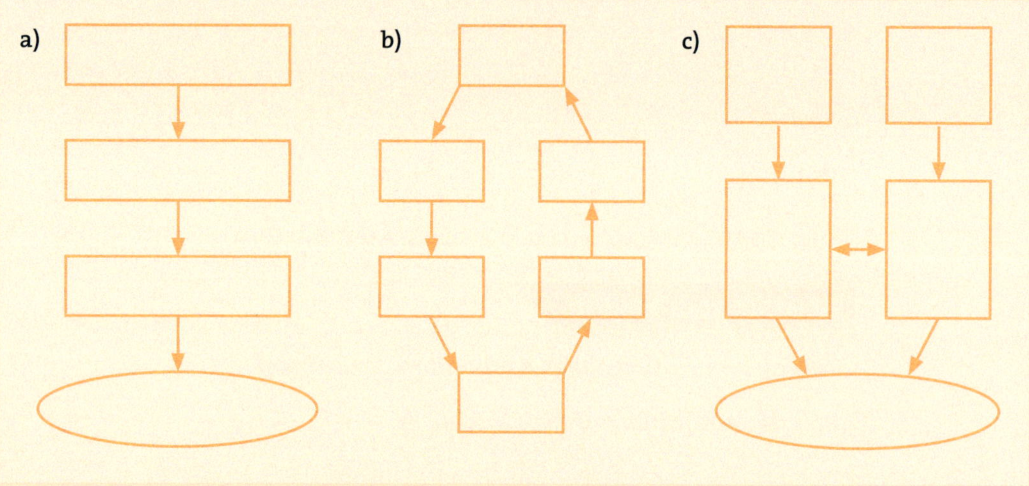

Figure 14 **A flow chart can say more than a thousand words**

Flow chart type **a** in Figure 14 shows the steps in a process leading to a result; type **b** demonstrates how the result of a process has an influence on its beginning; type **c** shows connections or conflicts between two or more processes.

1. How to make a flow chart from a text

First study the text carefully and write down the most important words and phrases[3]. Make sure you understand how they are related. Then arrange these key words and phrases in boxes and/or circles, using arrows to show the connections between them.

Example: As digestion consists of a succession[4] of events[5] taking place in the human body, you can practice making a flow chart from a text by choosing any of the texts in this chapter, from Sections 5.1 to 5.6.

2. How to develop a flow chart to visualize a complex process

First divide the process into its main stages. Then try to understand what happens during the different stages of the process and write it down in note form[6].
This will help you to decide which type of flow chart is suitable for visualizing the process.

Example: Use Table 1 (p. 81) to draw a flow chart showing how a sandwich with butter and cheese is digested. Your drawing should contain the three main nutrients, the places of digestion, the names of the digestive organs, the digestive juices and the products of their action. In small groups, use your flow chart to explain the process of digesting the sandwich.

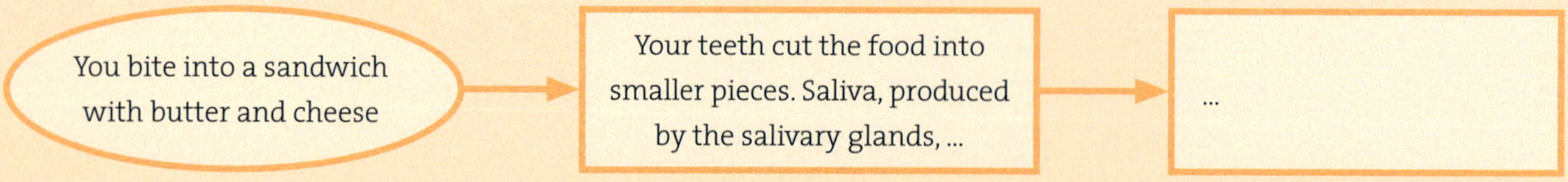

1 flow chart *Fließdiagramm*
2 visualize sth. *etwas veranschaulichen, bildlich darstellen*
3 phrase *Satz*
4 succession *Abfolge*
5 event *Ereignis*
6 in note form *in Stichworten*

Activity 1

Explaining changes in taste

Materials:
- water
- white bread

Procedure:
Rinse[1] your mouth with water, then chew a piece of white bread (ideally from a baker's shop or bread that somebody has just baked) for about two minutes.

1. What changes in taste do you notice?
2. What possible explanations can you think of for these changes?

For language help with describing a process, describing physical sensations, and explaining why things happen, see **Activate your English** on the inside cover.

Activity 2

Simulating[2] peristalsis

Materials:
- golf ball
- nylon stocking[3]

Procedure:
1. Put the golf ball into the open end of the nylon stocking.
2. Squeeze the stocking together with your hand directly behind the ball, pushing the ball along slightly. Open your hand again and repeat the action.
3. The ball is moved along the inside of the stocking in much the same way as food is moved along the gullet and the intestine by the wave-like action of peristalsis.

Activity 3

Investigating the action of bile

Materials:
- 2 test tubes
- test-tube rack
- measuring pipette[4]
- pen for labelling the tubes
- distilled water
- ox bile[5] (from a shop that sells materials for drawing and painting)
- cooking oil

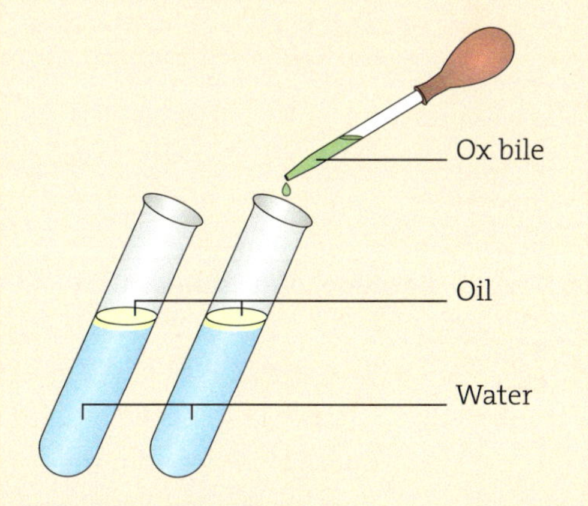

Figure 15 Effect of bile

Procedure:
1. Label[6] two test tubes with the numbers 1 and 2.
2. Pour 2 ml of oil and 5 ml of distilled water into test tube 1.
3. Put 2 ml of oil and 5 ml of ox bile into test tube 2.
4. Shake both test tubes carefully to mix the substances.
5. Then put both tubes into a test-tube rack[7] and leave them there for about 1 minute.

1. Describe the appearance[8] of the substance in each test tube.
2. Explain the differences.

1 rinse sth. *etwas spülen*
2 simulate sth. *etwas simulieren*
3 nylon stocking *Nylonstrumpf*
4 measuring pipette *Messpipette*
5 ox bile *Ochsengalle*
6 label sth. *etwas beschriften*
7 test-tube rack *Reagenzglasständer*
8 appearance *Aussehen*

5.7 Enzymes are important for chemical digestion

When food enters the mouth, three pairs of salivary glands release saliva into the mouth cavity: one pair in your cheeks, the second under your tongue and the third on both sides of your lower jaw. Saliva contains the enzyme **amylase**, which splits part of the starch into maltose[1], a double sugar[2]. The longer you chew on food containing carbohydrates, the more of the starch is converted[3] into sugar molecules. In order to function, amylase needs a neutral or slightly basic medium.

Key terms

Digestive enzymes
- amylase *stärkespaltendes Enzym*
- lipase *fettspaltendes Enzym*
- maltase *malzzuckerspaltendes Enzym*
- pepsin *eiweißspaltendes Enzym*
- peptidase *peptidspaltendes Enzym*
- protease *eiweißspaltendes Enzym*

In the stomach the enzyme **pepsin** acts on proteins and splits them into polypeptides[4], which are smaller strings of amino acids[5]. Hydrochloric acid, which is released from cells in the stomach lining, provides an acidic environment, which pepsin needs in order to work.

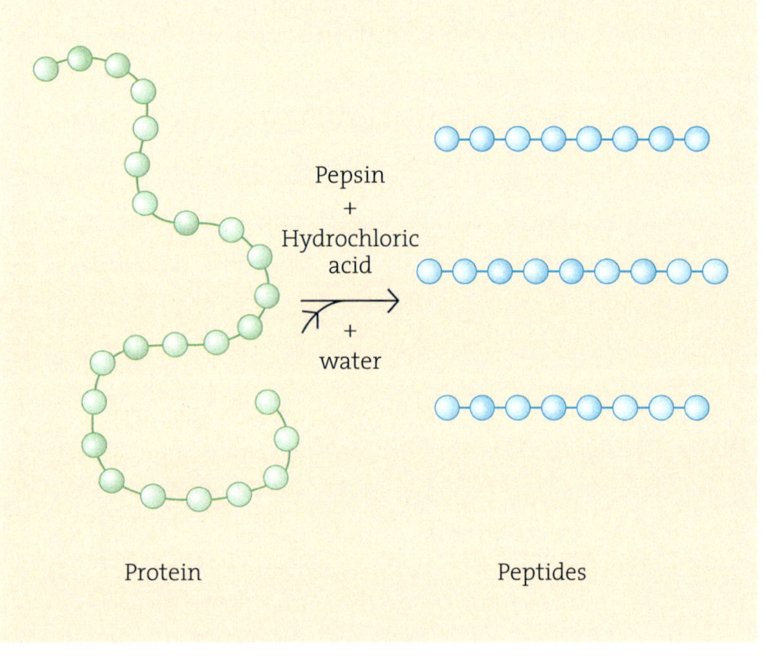

Figure 16 Protein digestion in the stomach

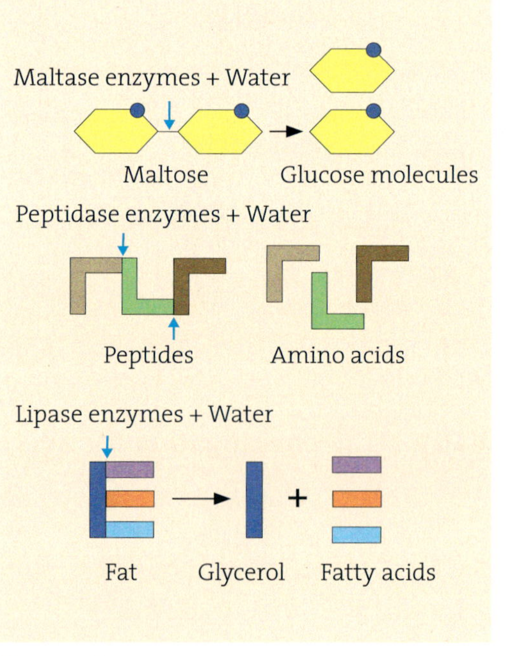

Figure 17 Digestion in the small intestine

The digestion of fats and oils starts in the duodenum (the first 30 cm of the small intestine), where bile splits the fat into tiny droplets, providing a larger surface area for the attack of the **lipase** enzymes from the pancreas. In the basic medium of the small intestine, fats are split into their building blocks: fatty acids[6] and glycerol[7]. Enzymes from the pancreas together with those from glands in the intestinal wall (amylase, **maltase**, **protease** and **peptidase**) convert starch and double sugars into single sugars[8]. They also break down proteins and polypeptides to single amino acids.

The basic building blocks of carbohydrates, proteins and fats leave the small intestine on two separate routes. Single sugars, amino acids, some glycerol and smaller fatty acid molecules are taken up by the capillaries inside the villi and are carried away by the blood. Most of the glycerol and fatty acid molecules have been built up into fats again in the cells of the intestinal lining. From there they enter the lacteals[9] in the villi and are carried away in the lymph.

1 maltose *Maltose*
2 double sugar *Doppelzucker*
3 convert sth. *etwas umwandeln, umbauen*
4 polypeptide *Polypeptid*
5 amino acid *Aminosäure*
6 fatty acid *Fettsäure*
7 glycerol *Glyzerin*
8 single sugar *Einfachzucker*
9 lacteal *Lymphkapillare*

5.8 Your liver is a chemical factory

The liver is the largest of the inner organs, weighing about 1500 grams. It is positioned just under the diaphragm[1], covering part of the stomach (see Figure 1, p. 74). It has a red-brown colour as it receives not only the oxygenated[2] blood from the heart, but also blood from the small intestine, which carries the absorbed nutrients. The bigger of the two parts of the liver is connected to the gall bladder, which stores about 50 ml of concentrated bile. Every day the liver produces about 600 ml of this greenish[3] liquid.

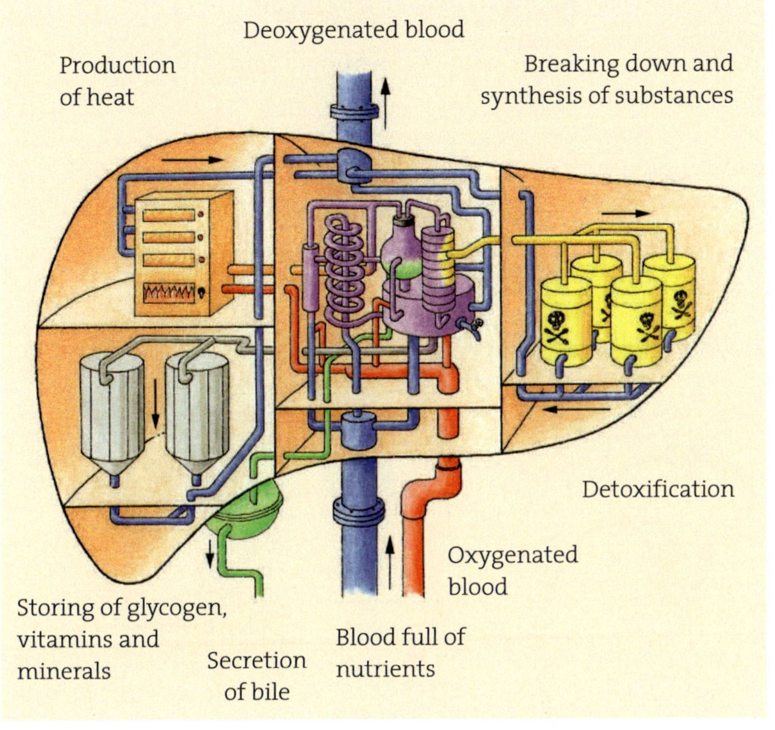

Figure 18 **Schematic picture showing some of the main functions of the liver**

Key terms

The liver
- ammonia *Ammoniak*
- detoxification *Entgiftung*
- excrete sth. *etwas ausscheiden*
- glycogen *Glykogen*
- metabolism *Stoffwechsel*
- urea *Harnstoff*

Besides its role in fat digestion, the liver has many other important functions. It is the centre of **metabolism** in the human body, in other words the place where substances are changed, broken down or stored. Body-specific[4] blood proteins like fibrinogen[5] for the blood-clotting process are formed here out of amino acids. Glucose, amino acids and other substances are converted into **glycogen**, a carbohydrate that is insoluble in water. About 100 grams of glycogen can be stored in the liver. When energy is needed, the glycogen is split into glucose and the glucose level in the blood rises. The liver can also store the vitamins A, B_{12}, D, K and some minerals. It contains about 20 per cent of the body's blood. It breaks down old red blood cells and it stores the iron of the haemoglobin[6] in its cells. The liver produces about 80 % of the cholesterol[7] in your body, a substance which is important for making and repairing cell membranes[8].

The liver is also an important organ of **detoxification**. It changes poisonous substances in medicines[9] or drugs like alcohol into harmless substances. Poisonous **ammonia**, produced when bacteria break down amino acids, is changed in the liver into non-toxic **urea**. As urea is water-soluble, it is taken up by the blood and transported to the kidneys, where it is **excreted**. These processes actually destroy liver cells, but compared to other organs the liver has a particularly high rate of regeneration[10].

Many of the chemical changes that take place in the liver release energy in the form of heat. This heat is spread through the whole body by the circulation of the blood. It helps us to keep up a constant body temperature.

1 diaphragm *Zwerchfell*
2 oxygenated *mit Sauerstoff angereichert*
3 greenish *grünlich*
4 body-specific *körpereigen*
5 fibrinogen *Fibrinogen*
6 haemoglobin *Hämoglobin, roter Blutfarbstoff*
7 cholesterol *Cholesterin*
8 cell membrane *Zellmembran*
9 medicine *Medikament, Medizin*
10 regeneration *Erneuerung, Regeneration*

1 Working with a partner, locate the position of your liver and describe its position.

2 How does the liver help to provide the body with energy when it is needed? Describe a situation in which this function might be helpful to you.

3 Describe how the liver stores useful substances and destroys harmful ones.

5.9 The human body is a complex biological system

Every minute of your life, when you are busy and when you are asleep, your organs are working. Muscles keep your lungs moving, the lungs load blood with oxygen, your heart pumps the blood through your body, food is digested and absorbed, and waste substances are excreted. Human life becomes possible because the different organs work together in a coordinated way and can adjust[1] to changing conditions inside and outside the body. The teamwork among the organs characterizes a **biological system**.

The demands made on the body, for instance by changes in the outer temperature or by increased muscle action during sport, determine the speed in which food substances and oxygen are taken up and transported to the body cells. Such changes also determine how quickly these are transformed into specific substances to be used for building new cells or providing the organs with energy.

Key terms

Processes in the human body
- absorption *Stoffaufnahme*
- biological system *biologisches System*
- cell respiration *Zellatmung*
- metabolic process *Stoffwechselprozess*
- excretion *Ausscheidung*

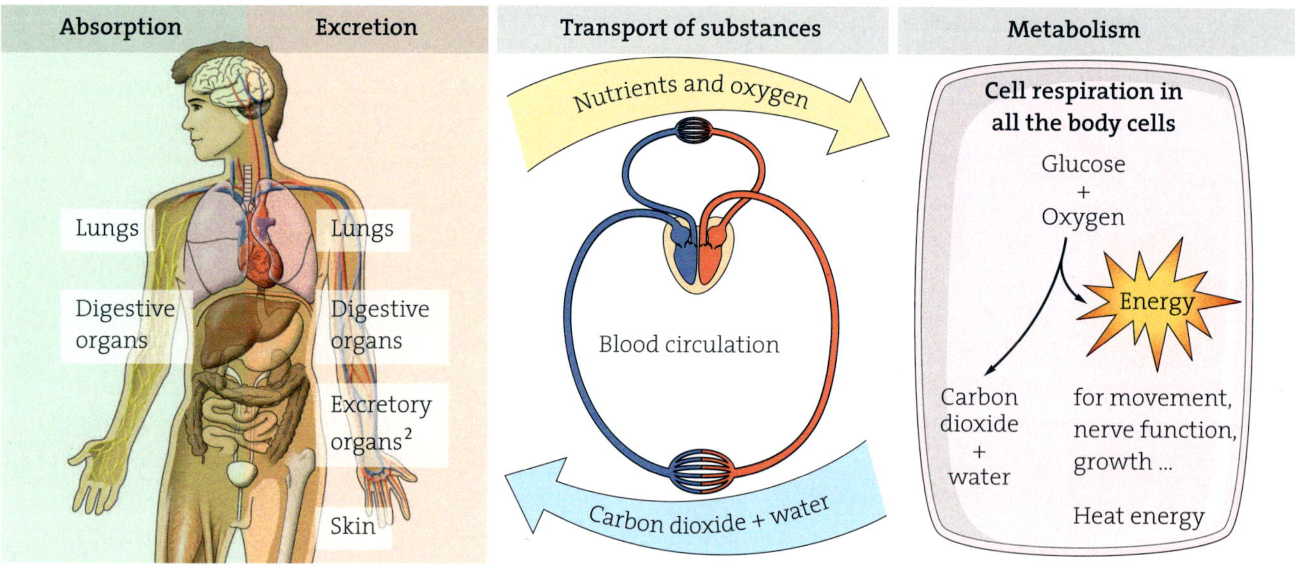

Figure 19 **Absorption, transport, metabolism and excretion in the human body**

In the complex[3] metabolism of body cells oxygen is used to split glucose into carbon dioxide and water. Part of the chemical energy which is stored in glucose is released and used to build up ATP. The rest is set free in the form of heat. This **metabolic process** is called **cell respiration**.

Glucose + Oxygen ➡ Carbon dioxide + Water + Energy (ATP + Heat)

ATP is used for all processes in the body which need energy, i. e. movement, transfer of information, growth and healing of wounds. (see also Chapter 4, p. 58)

1 adjust to sth. *(sich) an etwas anpassen*
2 Excretory organs *Ausscheidungsorgan*
3 complex *komplex*

1 Use Figure 19 to explain how the processes of absorption, transport of substances and excretion are connected.
2 Explain the formula for cell respiration.
3 How are your lungs, your blood circulation, your cell respiration and your skin affected by a long-distance run?

Summary

The process of digestion breaks down the food you eat by mechanical and chemical processes. In the different parts of the alimentary canal the nutrients are mixed with digestive juices. These juices contain enzymes that split carbohydrates, fats and proteins into their smallest building blocks. These can then pass through the lining of the small intestine into the blood and the lymph system. Those parts of the food which cannot be digested become solid in the large intestine, where water is absorbed into the blood. The remaining wastes pass out of the body when you go to the toilet.

A Language task

Complete each sentence on the left using the appropriate sentence part on the right.

1. Water is removed from digested food ...
2. Glands in the stomach wall produce mucus ...
3. Only small molecules can pass through the intestinal lining, ...
4. When food is swallowed, ...
5. Bile helps to digest fat ...
6. Having entered the gullet, ...
7. If plaque is left on the teeth, ...

a) so food must be broken down to its smallest building blocks before we can benefit from it.
b) the food is pushed towards the stomach by peristalsis.
c) by breaking it down into small droplets.
d) because it would be a waste to excrete it.
e) bacteria can cause gum disease.
f) the epiglottis closes the opening of the windpipe.
g) so that this organ does not digest itself.

B Bio quiz

True or false? Write down the correct information in a complete sentence.

1. The normal number of teeth for adults is 20.
2. The pulp cavity contains blood vessels and nerves.
3. A hard layer of cement keeps the crown of the teeth safe from bacteria.
4. Canines are for crushing and grinding the food.
5. Caries is an inflammation of your gums.
6. Bile helps the body digest fats.
7. The digestion of proteins starts in your mouth.
8. Enzymes belong to the three main nutrients.
9. Water is removed from undigested food in the large intestine.
10. The liver produces enzymes for the digestion of carbohydrates, fats and proteins.

Key terms

See the boxes:

The digestive organs .... p. 74
Digestive juices .... p. 75
Teeth .... p. 76
Teeth-related problems .... p. 77
From the mouth to the stomach .... p. 78
The stomach .... p. 79
The intestines .... p. 80
The final stage .... p. 81
Scientific experiments .... p. 82
Digestive enzymes .... p. 86
The liver .... p. 87
Processes in the human body .... p. 88

C Critical thinking

1 Explain why you should see your dentist twice a year, whether you have problems with your teeth or not.

2 Explain why the digestion of food substances happens in several steps in different organs and not in one place at the same time.

3 Some people get sick when they drink milk because they cannot digest lactose, the sugar in milk. Which areas of the digestive system may be involved in this inability?

Chapter 6 **Reproduction and Development**

Chapter objectives

- To **describe** the physical, emotional and mental changes that take place during puberty
- To **describe** the formation of sperms
- To **outline** what teenagers should do for their personal hygiene
- To **explain** the development of a human life from the fertilized egg cell to birth
- To **analyse** the advantages and disadvantages of different contraceptives
- To **know how** to protect yourself against sexual abuse and sexually transmitted diseases

If you look at a picture of yourself taken three years ago and compare it to what you see in the mirror, you will notice many changes. Between the ages of 9 and 17 you slowly develop from a child into an adult. This time is called adolescence, and it brings many changes, not only to the way your body looks from the outside, but also inside your body and your mind. This chapter will help you to understand some of these changes and make suggestions for how to adjust[1] to them.

6.1 Puberty is a time of physical, mental and emotional change

Figure 1 The peer group grows in importance.

Key terms

Reaching puberty (1)

- adolescence *Jugendzeit*
- ejaculation *Samenerguss*
- hormone *Hormon (Botenstoff)*
- menstrual cycle *der Menstruationszyklus*
- menstruation *Menstruation (monatliche Blutung)*
- ovary (pl. ovaries) *Eierstock*
- penis *Penis*
- period *Periode, Menstruation*
- puberty *Pubertät*
- reproduce *sich fortpflanzen*
- sperm (pl. sperm or sperms) *Spermium, Sperma; (Spermien)*
- testis (pl. testes) *Hoden*

While **adolescence** is the complete period of change from a child into an adult, **puberty** is a biological term describing the stage an organism reaches when it becomes able to **reproduce**, or produce offspring[2]. During puberty girls' **ovaries** become bigger and start producing egg cells. This becomes evident[3] when the **menstrual cycle** with its monthly flows of blood, or **periods**, starts. As part of puberty, boys' **testes** increase in size and begin to produce **sperms**. This is the time when a boy has his first **ejaculation**, in which sperm is released from the **penis**. These physical changes are covered in more detail in Sections 6.3 and 6.4. Both **menstruation** and ejaculations are a normal part of being an adult and are nothing to be ashamed of.
But puberty is much more than just a series of changes in your reproductive system. When puberty begins (at some point between the ages of eight and twelve) **hormones** in your brain and your sexual organs cause all kinds of changes in your whole body.

1 adjust to sth. *sich auf etwas einstellen, sich an etwas gewöhnen*
2 offspring *Nachwuchs*
3 evident *deutlich, offensichtlich*

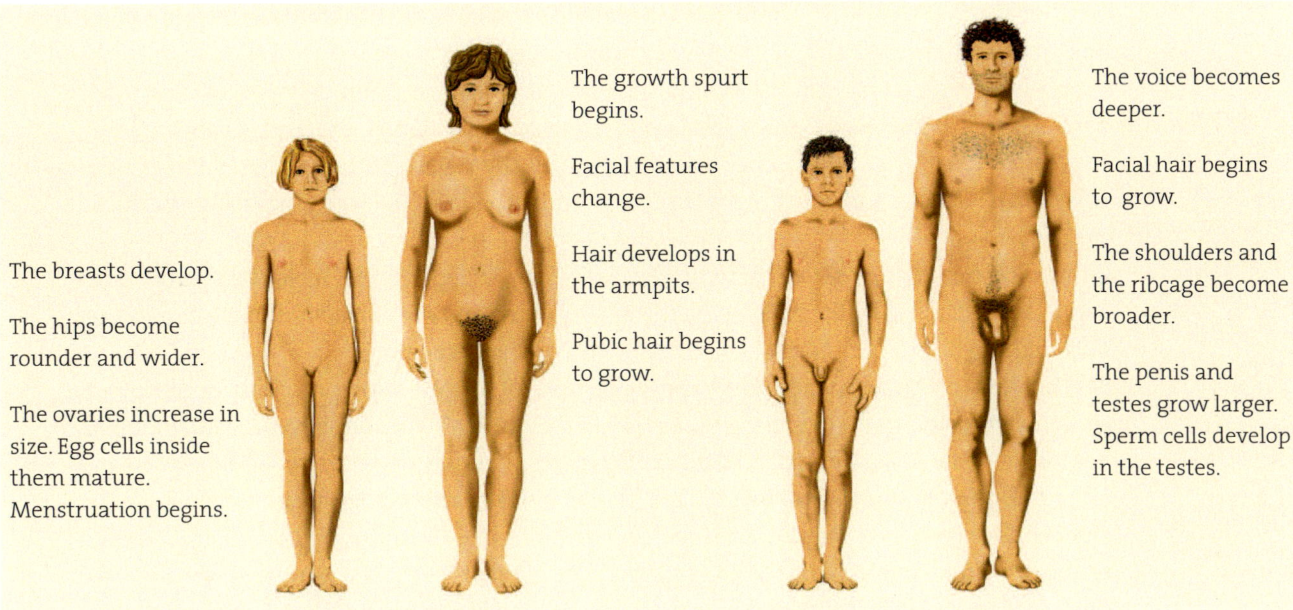

Figure 2 Physical changes during puberty

Both boys and girls start to grow hair on their arms and legs, in their armpits[1] and around their **genitals** (where it is called **pubic hair**). Boys also grow **facial hair** and sometimes **chest hair**. Girls' **breasts** start to develop and become bigger, while boys experience penis growth.
Both sexes undergo a time of rapid bodily growth, called a **growth spurt**. When you look around your class, you might see that most of the girls are taller than most of the boys. Figure 3 shows that this is quite normal: girls are younger when they have their first growth spurts, so the boys remain smaller for a certain time. By the age of about 15, however, these differences have mostly levelled out[2].

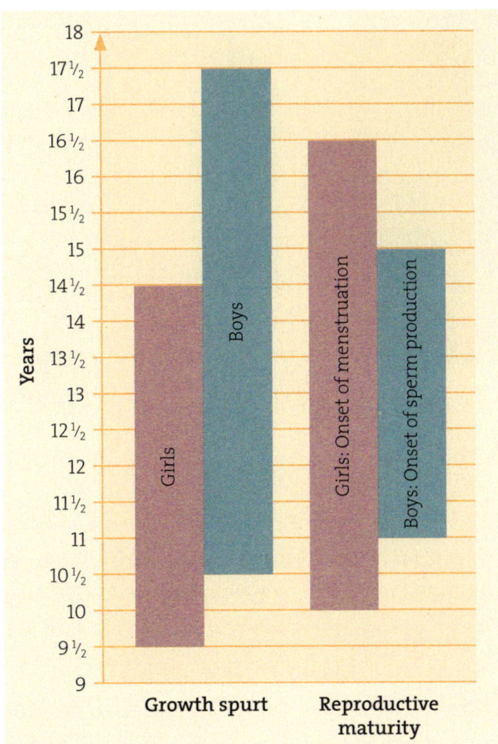

Figure 3 Average age of physical changes during puberty

Not only the height but also the shape of the body changes during puberty. In girls' bodies the **pelvis** widens in order to make room for a future pregnancy. In boys' bodies the shoulders and the chest become broader while the

Key terms

Reaching puberty (2)
- breast *Brust*
- chest hair *Brusthaare*
- facial hair *Gesichtsbehaarung*
- genitals (pl.) *Genitalien*
- growth spurt *Wachstumsschub*
- peer group *Gruppe von Gleichaltrigen*
- pelvis *Becken*
- pubic hair *Schambehaarung*
- sanitary towel (or sanitary pad) *Monatsbinde*
- tampon *Tampon*
- vocal cord *Stimmband*
- voice box *Kehlkopf*

1 armpit *Achselhöhle*
2 level out *sich ausgleichen*

pelvis only grows slightly. The hormones that are released in puberty also make the male **voice box** grow and change shape, with the **vocal cords** becoming larger and thicker. For some boys the change seems to happen slowly, and the voice gradually becomes deeper. However, many boys experience a sudden change over just a few weeks or months. When this happens, their voice often 'breaks', or is high one moment and low the next. Controlling the pitch[1] of your voice can be difficult and you may feel embarrassed. But before long, your voice will be balanced again.

Girls' voices also change during puberty. Their voices may deepen by a couple of tones (whereas a boy's voice can fall by an octave). At this time, girls often have difficulty singing at an exact pitch, although some girls may not notice this change happening.

Besides these physical changes, adolescents change a lot in their minds, too. First of all, boys or girls need to come to terms with[2] the changes in their bodies. Very often this is harder for girls, because it seems as if their bodies change much more, and they have to start wearing bras[3] and using **tampons** or **sanitary towels**. Young people of both sexes struggle with becoming more independent from their parents. They question the things that authority figures, such as parents, teachers, trainers, and other adults, tell them. They start to reflect on[4] who they are and who they want to be. While trying to find new points of orientation – which can be very confusing and frustrating at times – the **peer group** grows in importance. Finding the answers to all the questions that arise is very important, but it is not easy and it takes time, often much longer than puberty itself.

During adolescence young people leave behind the things they were interested in as children and try out new interests. Very often adolescents radically change their taste in music, films, books or fashion at this time of their lives and many start to develop an awareness[5] of political and environmental[6] issues. This is also the time when many young adults become interested in the opposite sex and have their first experiences of romantic relationships.

It is important to remember that all aspects of puberty – physical, mental and emotional – can differ very much from person to person. One girl might have her first period at the age of nine, while a classmate might only start at 15. One boy does not have his growth spurt until 16, while another had started to grow chest hair five years earlier. It does not matter when you start puberty. There is no reason to worry because your body will develop at its own pace[7].

1 pitch *Stimmlage*
2 come to terms with sth. *mit etwas klar kommen*
3 bra *BH*
4 reflect on sth. *über etwas nachdenken*
5 awareness *(hier) Bewusstsein*
6 environment *Umwelt*
7 pace *Tempo*
8 curvy *kurvenreich*

Figure 4 **Puperty also means emotional change**

1 What changes take place in boys and girls during puberty? Make a table.
2 Explain why women's voices do not become as low as men's voices.
3 Many adolescents say that they do not feel 'comfortable in their skin'. Try to find reasons for this and make suggestions about how they could feel better about themselves. For language help, see **Activate your English**, 'Making suggestions', on the inside cover.
4 In puberty, a girl's pelvis widens and her body becomes more 'curvy[8]'. This has nothing to do with being overweight. What problems might arise from this and how can they be overcome?
5 Ask your parents and/or grandparents about the experiences they had during puberty. What kinds of problems did they have with their parents, and how did they solve them? What do you think is different today in the way young people and their parents deal with the problems that arise during puberty? Discuss in class.

6.2 'Love don't come easy'

Figure 5 **Relationships can develop in different ways.**

As you change from a child into an adult, your relationships change too. Friendships with your peers[1] become more important while the process of becoming more independent from your parents sometimes puts a strain on[2] your relationship with them. The teenage years are also the time when most people begin to experience feelings of physical attraction.

You might have experienced the following: a certain person enters the room and your heart starts thumping[3]. Or your mouth goes dry and your hands go moist as soon as this special girl or boy asks you something. Or whenever you see a certain actor, actress or athlete, intense emotions flood your body.

These sometimes overwhelming[4] feelings of attraction to another person are called **infatuation**. They are normal and healthy and can help you learn to form strong relationships later. A lot of teenagers wonder if these feelings are the same as love. In reality, they are not, although they are an important step for love to develop. If you are infatuated with someone for a longer period of time you can say that you are 'in love[5]'. In contrast to being in love, 'real' love is much harder to describe and much more complex. The most important difference is probably that love is not dependent on butterflies in your stomach, and you still want to be with the person even when he or she is in a bad mood or is sick, and even though you know that he or she sometimes makes mistakes. Real love lasts after the first period of infatuation has passed.

When you have been going out with a person you are in love with, the desire for **physical intimacy** may arise. This brings with it a lot of questions: How can I show that I like my partner without going too far? Are these feelings normal? Am I ready for a sexual relationship? These are important questions without easy answers. You need to think about the consequences your decisions might have for yourself and for your relationship with your friend. A lot of factors can influence your decisions, such as your family, friends, your religion or cultural background[6], and of course the media (TV, radio, magazines, the Internet). All of these shape your values[7], and going against them can feel deeply dissatisfying[8]. Sticking[9] to your values, on the other hand, can sometimes be hard but can make you feel strong and secure about yourself.

Key terms

Love and sexual relationships

- abstinence *Enthaltsamkeit (d.h. nicht mit jemandem schlafen)*
- infatuation *Verliebtheit*
- masturbation *Masturbation, Selbstbefriedigung*
- petting *Petting (Streicheln der Sexualorgane)*
- physical intimacy *körperliche Intimität*
- pregnancy *Schwangerschaft*
- sexual intercourse *Geschlechtsverkehr*
- sexually transmitted disease (STD) *Geschlechtskrankheit*
- stimulate sb./sth. *jdn./etwas stimulieren, jdn. erregen*

1 peer (usu. pl.) *Gleichaltrige/r*
2 put a strain on sth. *etwas belasten*
3 thumping *pochen*
4 overwhelming *überwältigend*
5 in love *verliebt*
6 cultural background *kulturelle Herkunft*
7 values (pl.) *Werte, Überzeugungen*
8 dissatisfying *unbefriedigend*
9 stick to sth. *an etwas festhalten*

Figure 6 Different aspects of relationships

The decision to become physically intimate can have a big effect on all areas of a relationship. It can create a lot of pressure, for example, if one person wants to have sex and the other does not. Preparing yourself for these issues[1] can take a lot of time, especially since many couples say that once you have become physically intimate with someone, it is almost impossible to return to the old state of the relationship. You also have to be aware of the possibility of **pregnancy** and **sexually transmitted diseases** (STDs) when you enter a sexual relationship with someone (more on these topics in Sections 6.6–6.9).

Becoming physically intimate with someone does not necessarily mean having **sexual intercourse** with him or her. It can also mean kissing or **petting**. Some teenagers (and some adults, too) choose not to go further than petting during long phases of a relationship. Choosing not to sleep with someone is called **abstinence**, and is a good way of being physically satisfied without having to worry about pregnancy and STDs. When you need to decide how intimate to be with someone, it's a good idea to talk it over[2] with people whose opinions you respect, including, of course, your partner.

As sexual feelings develop during puberty, many teenagers have to address another question: **masturbation**. Masturbation means **stimulating** your own genitals and other parts of the body to give yourself sexual pleasure. There are no medical arguments against it, despite what some people say. A lot of people feel it is an important part of their sexuality because it allows them to get to know their body more intimately. However, people in different cultures and at different times have had varying views about masturbation, and you have to find out for yourself how you feel about it.

While most people develop feelings for members of the opposite sex (which is described as being heterosexual), others feel attracted to people of the same sex (which is called being homosexual). Being attracted to or infatuated with a member of your own sex at some stage during puberty does not automatically mean that you are homosexual: many people experience feelings like this and then go on to follow a heterosexual lifestyle. This exploration of your sexual identity is part of the process of becoming an adult. If you are not sure about your feelings, talk about them with a person you trust.

1 issue *(Streit-)Frage, Thema*

2 talk sth. over with sb. *etwas mit jdm. besprechen*

This whole chapter is about sex, sexual organs and how they work. Do not forget, though, that being physically intimate with someone, whether it is petting or sexual intercourse, is only one aspect of a relationship. Being emotionally intimate, sharing time with each other and with friends, and finding out who the other person really is, is another very important and exciting part of a relationship.

Did you know?

People in different cultures go about relationships in very different ways. In America, for example, it is quite usual for teenagers to go out on a date with a different person each weekend.
When someone has been on a few dates with the same person, they might want to say that they are 'exclusive', which means that they do not want to date any other people at the moment. At this stage, they might say that they are 'in a relationship' or 'going steady'.
But of course there are huge differences between people, and there are teens who rarely go on dates but nevertheless have strong relationships.

1 It can be difficult to tell the difference between infatuation and 'real love'. Using your own experiences and your knowledge from films, books and real life, make a table with two columns containing words that describe these two emotions/feelings.
2 Figure 6 shows different aspects of relationships. Make a list of these and other aspects and discuss how important each one is for a 'good' relationship.
3 Interview different couples about how they met each other. Find out how they knew that what they felt was more than just infatuation.

6.3 The male reproductive system produces hormones and sperm

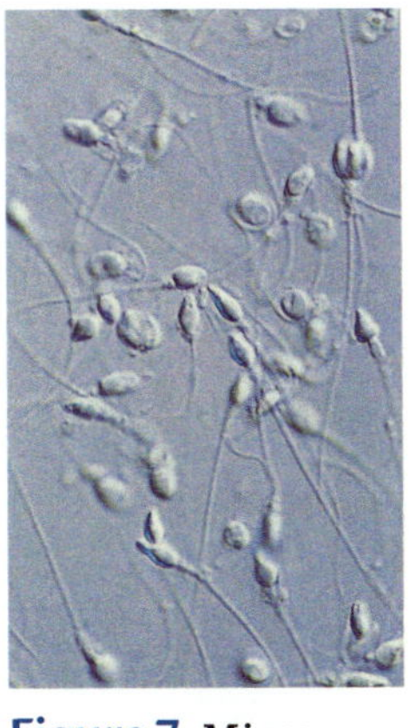

Figure 7 Microscopic picture of sperms

Both males and females produce sex cells, called **gametes**, each of which contains half the genetic information needed to make a new human being. The male sexual organs which can be seen on the outside of the body are the penis and the **scrotum**. The head of the penis, or **glans penis**, which is very sensitive, is usually covered by the **foreskin** for protection.
The male sex cells, or sperms, are produced in the **testes**. These are two egg-shaped organs which lie inside the scrotum (see Figure 8). After the sperms are produced, they travel from the testes to the **epididymis**, a gland which lies on top of each testis, where they mature[1] and are stored until ejaculation.
During **sexual arousal** the penis becomes stiff and erect[2], because the blood that enters the **erectile tissue** is prevented from flowing out again. If arousal continues, the point of

1 mature *heranreifen*
2 erect *aufgerichtet, aufrecht*

Key terms

The male reproductive system
- Adam's apple *Adamsapfel*
- epididymis (pl. epididymides) *Nebenhoden*
- erectile tissue *Schwellkörper*
- foreskin *Vorhaut*
- gamete *Keimzelle*
- glans penis *Eichel*
- hormone *Hormon*
- orgasm *Orgasmus*
- prostate gland *Prostata*
- scrotum *Hodensack*
- semen *Sperma*
- seminal vesicles *Bläschendrüsen*
- sexual arousal *sexuelle Erregung*
- sperm duct *Spermienleiter*
- testis (pl. testes) *Hoden*
- testosterone *Testosteron*
- urethra *Harnspermienröhre*

orgasm is reached when several muscles in the abdomen contract simultaneously. This is when ejaculation occurs.
During ejaculation the sperms travel from each epididymis through a tube called the **sperm duct** to the beginning of the **urethra**, just under the urinary bladder[1] (see Figure 8). There the tubes from the two epididymides join and liquids are added to the sperm, making **semen**. The liquids come from different glands, including the **seminal vesicles** and the **prostate gland**. Without the fluid and nutrients contained in these liquids, the sperms could not move and thus could not travel through the female reproductive system (see Figure 9).
The sudden contraction of the abdominal muscles during ejaculation propels[2] the semen through the urethra and out of the penis. If the sperms come into contact with an egg cell in the female reproductive tract, pregnancy can take place.
The testes also produce **hormones**, most importantly **testosterone**. Hormones function as messengers in the body. When puberty sets in, testosterone informs the larynx[3] to grow into the larger **Adam's apple** (which gives boys a much lower voice). Testosterone also plays an important part in making sperm and, like other hormones, has a big influence on the emotions.

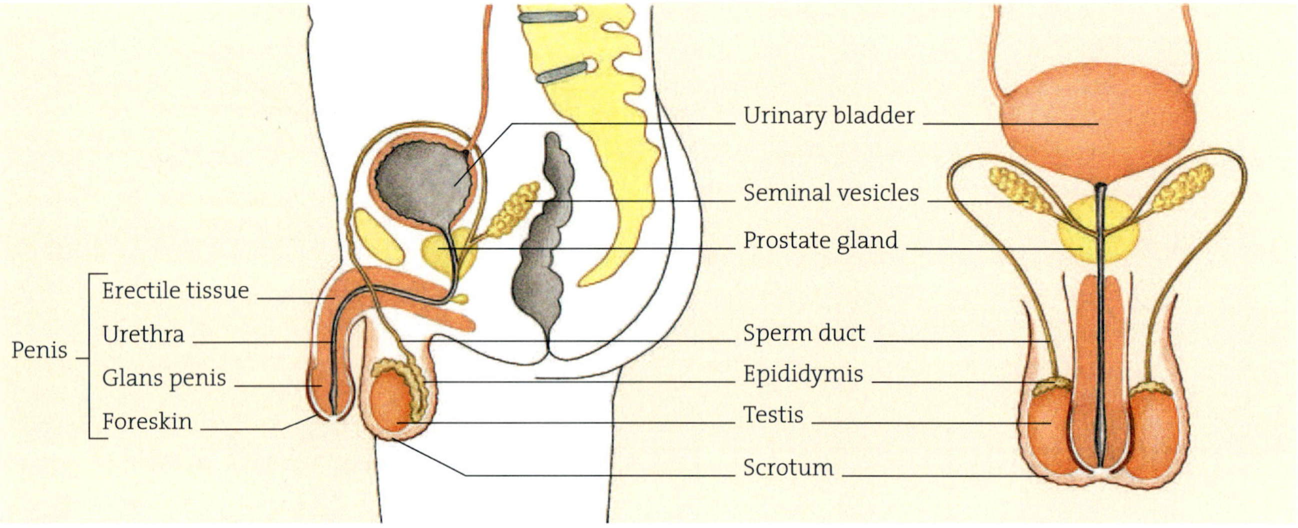

Figure 8 The male reproductive system

1 What is semen and what is its function?
2 Why and how does the penis become stiff?
3 Make a flow chart illustrating the path of a single sperm from production to ejaculation. For help see 'Working with Flow charts', p. 84.
4 A male starts producing sperm from about the age of twelve and sperm production usually continues until his death. Ignoring the fact that the production rate slows down in old age, calculate how many sperm cells have been made by the time he dies at the age of 70. The information in the 'Did you know?' box on the right will help you.
5 One of the treatments for prostate cancer is to remove the prostate and seminal vesicles surgically. Find out what the consequences of this are for orgasms and ejaculation. For language help, see **Activate your English**, 'Describing causes and effects', on the inside cover.

Did you know?

- The testes produce about 1000 sperms per second. It takes them 90 days to reach full maturity.
- The best temperature for sperms to grow is about 33 °C. That is why the testes lie in the scrotum: the temperature is 2–3 °C lower than in the rest of the body.

1 urinary bladder *Harnblase*
2 propel sth. *etwas vorwärts treiben*
3 larynx *Kehlkopf*

6.4 The female reproductive system is prepared for pregnancy

At some stage during puberty, a girl will have her first **period**, or monthly flow of blood from the **vagina**. Some girls cannot wait for the onset[1] of **menstruation**, while others might feel a little afraid. It is important for both boys and girls to understand how this cycle works and why menstruation takes place, because it is a natural part of growing up.

As in males, the female reproductive system has an external and an internal part. The external part, or **vulva**, is made up of six organs: the **clitoris**, two larger and two smaller folds of skin (the **labia majora** and **labia minora**), the opening of the **urethra** and the opening of the vagina with the **hymen** (see Figure 9).

The clitoris is a small organ at the front of the vulva and has no other function than to create sexual arousal when stimulated. The clitoris also is the place where the two labia majora meet. These are covered with pubic hair and surround the labia minora, which do not have any hair. Both pairs of skin folds have glands that moisten the opening to the vagina during sexual arousal so that the penetration of the penis does not hurt the tissue during intercourse.

Key terms

The female reproductive system (1)
- clitoris *Klitoris (Kitzler)*
- cervix *Muttermund*
- hymen *Jungfernhäutchen*
- labia majora *große Schamlippe*
- labia minora *kleine Schamlippe*
- menstruation *Regelblutung, Menstruation*
- urethra *Harnröhre*
- vagina *Vagina, Scheide*
- vulva *weibliche Scham*

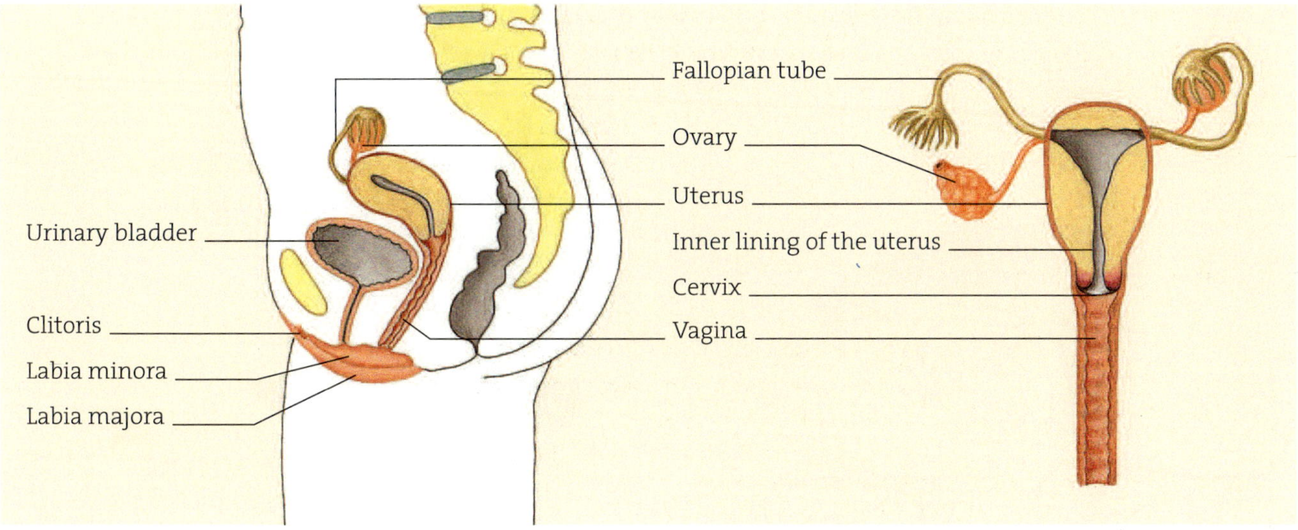

Figure 9 The female reproductive system

In contrast to the male reproductive system, where semen and urine flow through the same tube, women have a separate tube for urine, called the urethra. The opening to the vagina is partly covered by a thin mucous membrane[2], the hymen. It can sometimes tear[3] during sports activities or when a tampon is used, but usually only breaks during the first sexual intercourse.

The internal part of the female reproductive system consists of four organs: two **ovaries**, two **Fallopian tubes**, the **uterus** (also sometimes called the **womb**) with the cervix, and the vagina. These organs play an important role in the **menstrual cycle**.

- During the first two weeks of this cycle, one of the egg cells in the ovaries starts to develop and reach maturity. At almost the same time, the **inner lining** begins to build up in the uterus (see Figure 10).

1 onset *Anfang, Beginn*
2 mucous membrane *Schleimhaut*
3 tear *reißen*

• Once the egg cell in the ovary is fully mature, it is released into the Fallopian tube, a process called **ovulation**. This occurs around the middle of the menstrual cycle (about day 14). Since the egg is now ready to be **fertilized**, the likelihood of becoming pregnant during sexual intercourse is highest on the three to four days around ovulation. This, of course, does not mean that you cannot become pregnant on other days, too, because the menstrual cycle can be irregular. One can say in general that a woman is fertile for about one third of her menstrual cycle.

• After the egg has entered the Fallopian tube, tiny hairs and muscle movements transport it down the tube towards the uterus. During the three-to-five-day journey, the lining of the uterus continues to grow and store lot of nutrients. If the egg is fertilized and a pregnancy occurs, the developing egg then sinks into the lining of the uterus (a process called implantation) and develops further.

• Usually, however, the egg is not fertilized and the lining of the uterus is not needed for pregnancy. When this happens, the lining is then discarded[1]. Changes in hormone levels cause the lining to be shed[2], and this is when menstruation begins, marking the first day of the new cycle. The **menstrual flow** consists of blood and cells from the lining. Once the lining has been completely shed, the uterus begins building a new one between days four and six of the new cycle.

1 discard sth. *etwas abwerfen*
2 shed sth. *(hier) etwas ausscheiden*

Key terms

The female reproductive system (2)
- Fallopian tube *Eileiter*
- fertilization *Befruchtung*
- fertilize *befruchten*
- inner lining of the uterus *Gebärmutterschleimhaut*
- menopause *Wechseljahre*
- menstrual cycle *Menstruationszyklus*
- menstrual flow *Menstruationsblut*
- ovary (pl. ovaries) *Eierstock*
- ovulation *Eisprung*
- uterus *Gebärmutter*
- womb *Gebärmutter (see also uterus)*

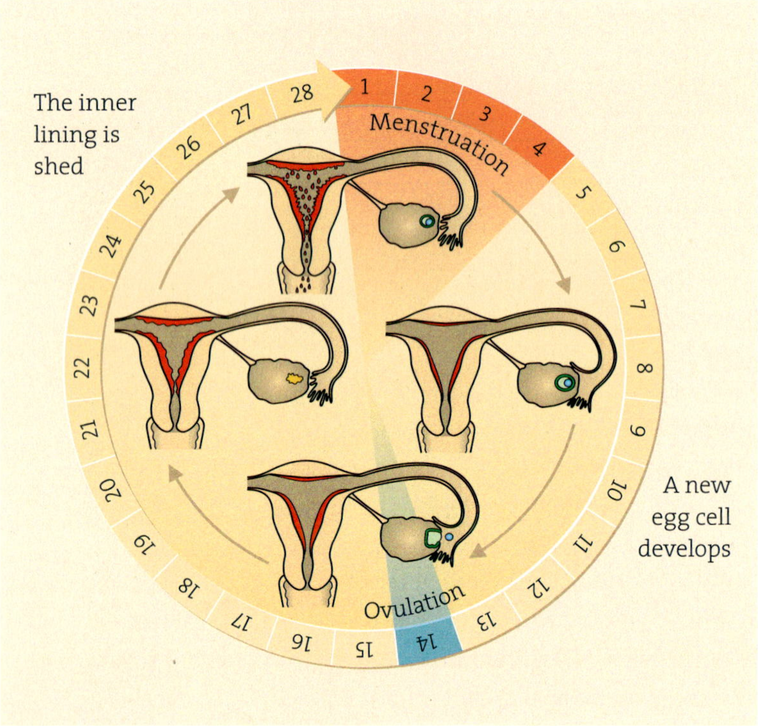

Figure 10 The menstrual cycle

A cycle takes between 28 and 33 days, the length varying from woman to woman. Each time a new egg cell matures, the woman's whole body prepares for the possibility of a pregnancy.

Did you know?

- A girl has up to 400,000 immature egg cells in her ovaries. They all formed before she was born.
- The menstrual cycle starts during puberty and ends during menopause, which marks a time of hormonal changes after which no more egg cells are released. Most women are between 45 and 55 when menopause sets in, but it can come earlier or later. During and after menopause it is still possible to have a satisfying sex life.

1 What are the two most important events in the menstrual cycle? Explain your answer.
2 Describe the functions of the two pairs of labia.
3 Where does fertilization take place? Explain the process.
4 Which organ can the clitoris be compared with in the male reproductive system? Explain the ways in which the two organs are similar and different. For language help, see **Activate your English**, 'Comparing and contrasting', on the inside cover.
5 **Say it in German:** You now know all about the menstruation cycle in English. But can you describe it in German? Prepare a short presentation in German. These words will help you: *Eileiter – Gebärmutter – Scheide – Gebärmutterschleimhaut – Eierstock – Eizelle – Eisprung – Befruchtung – Regelblutung.*

6.5 Puberty demands a high level of personal hygiene

Both male and female bodies go through many changes during puberty, and one of these is that they begin to produce more secretions[1] than during childhood. For example, **sweat glands** develop, especially in the armpits, which is why people start to have daily showers and use a deodorant or antiperspirant[2]. Also, changing your underwear regularly is important to avoid odour[3] and germs.

For the same reason, both male and female teenagers must also take care to keep their **genital area** clean. For boys this means paying special attention to the head of the penis. In order to clean this area you have to pull back the foreskin all the way and wash with water and a mild soap. You should do this every day, otherwise bacteria can grow, which cause a bad odour and, in the worst case, lead to infections. Sometimes boys are **circumcised** at a young age, meaning the foreskin is removed for religious or medical reasons. The practice of **circumcision** does not have any effect on sexual functioning or on urinating.

Girls have to keep their genital areas clean, too. Besides producing the menstrual flow at the beginning of each cycle, the vagina continually cleanses[4] itself and small amounts of secretions frequently flow out. Since it is self-cleaning, you should not try to wash the inside of the vagina, but you should take care to clean the vulva every day with plenty of warm water and a mild unscented[5] soap.

During menstruation girls need something to absorb the blood. The two most common methods are sanitary pads (also called sanitary

Key terms

Puberty and hygiene
- circumcise sb. *jdn. beschneiden*
- circumcision *Beschneidung*
- genital area *Intimbereich, Genitalbereich*
- sweat glands *Schweißdrüsen*
- toxic shock syndrome (TSS) *Toxic-Shock-Syndrom*

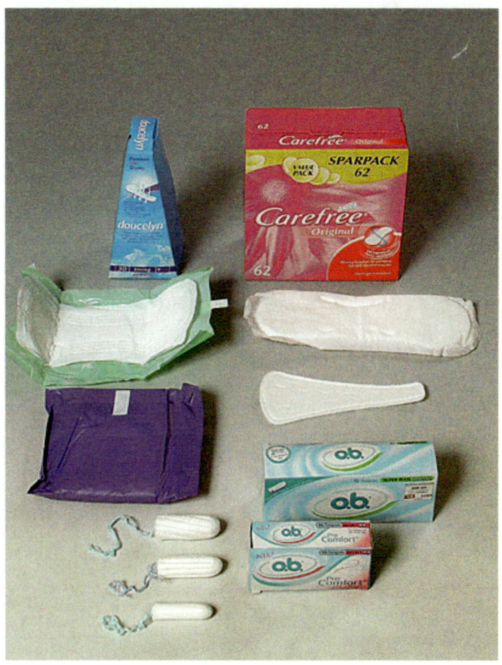

Figure 11 Pads and tampons

1 secretion *Absonderung*
2 antiperspirant *Antitranspirant, Deodorant*
3 odour *Geruch*
4 cleanse sth. *etwas säubern*
5 unscented *ohne Duft*

Figure 12 **Tampons make it possible to go swimming. However, during menstruation the uterus is less protected from infections.**

towels or sanitary napkins) and tampons. While pads are placed in the panties[1] and take up the blood outside of the vulva, tampons are placed inside the vagina. Some girls worry that they will not be able to get a tampon out again after inserting[2] it. However, it is impossible for tampons to enter the uterus because the cervix is much too small for them to enter. Also, tampons come with a string attached which you can pull at any time to remove them.

It is important to change pads and tampons regularly. Depending on how much menstrual blood you produce, you need to change pads every three to four hours, and tampons every four to six hours. Do not wait too long between changes, especially when you are using a tampon. When tampons are left in too long, it is possible for bacteria to grow and cause health problems. The most serious potential problem is **toxic shock syndrome (TSS)**, a rare but very severe[3] reaction to bacteria which is often connected with tampon use. If you experience a sudden high fever, vomiting, diarrhoea[4] or dizziness[5] while using a tampon, go to see a doctor immediately. You can buy pads or tampons at your local chemist's[6], and you will find instructions on how to use them on the packaging, along with information on TSS.

Allow some time to find out for yourself the best way to absorb your menstrual blood, and do not worry if you find tampons difficult to use at first: most women find them very convenient once they have learned to insert them. You will need to experiment to find the size of tampon that suits you best (try to use the smallest size that fits). If you have any questions, ask a parent, an older friend, the school nurse, or your biology teacher.

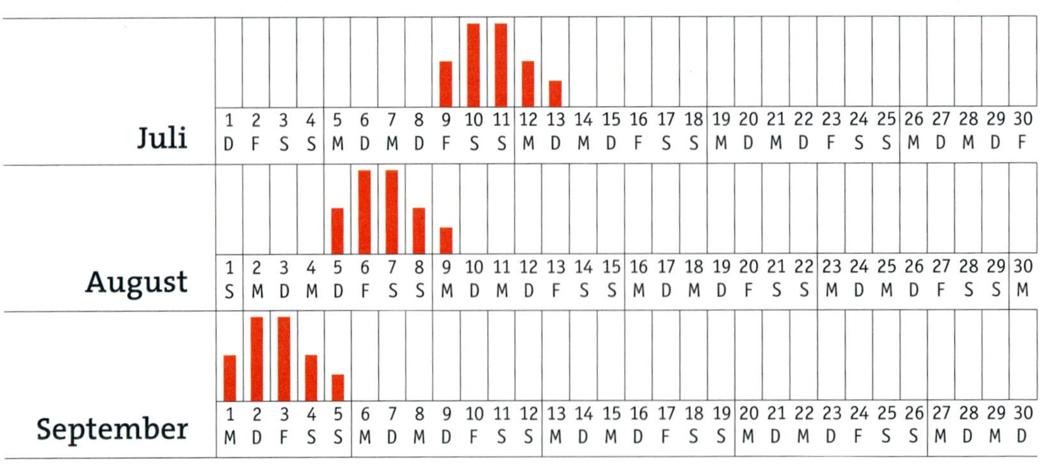

Figure 13 **Calendar showing a typical menstrual cycle**

1 panties *Schlüpfer*
2 insert sth. *etwas einführen, einlegen*
3 severe *heftig, schwerwiegend*
4 diarrhoea *Durchfall*
5 dizziness *Schwindel*
6 chemist's (BE) = (AE) drugstore *Apotheke, Drogerie*

1 Explain why it is a good idea for a girl to choose the smallest size tampon that suits her.

2 a For girls: discuss the advantages and disadvantages of the different methods of absorbing menstrual blood. For language help, see **Activate your English**, 'Describing advantages and disadvantages', on the inside cover.

2 **b** For boys: find out why people are circumcised and discuss possible advantages and disadvantages of circumcision. For language help, see **Activate your English**, 'Describing advantages and disadvantages', on the inside cover.

3 It is often recommended that girls keep a calendar in which they write down the dates of each menstruation. Why is this good advice? Take a look at Figure 13 and calculate the next date of menstruation.

6.6 If an egg cell is fertilized, a pregnancy can start

Key terms

Fertilization
- contraceptive *Verhütungsmittel*
- embryo *Embryo*
- fertile *fruchtbar*
- implantation *Einnistung*

In order for a new human life to develop, an egg cell needs to be fertilized. This happens when a sperm reaches an egg cell as the egg cell is travelling down the Fallopian tube after ovulation.
Sperms reach the egg through sexual intercourse. As you learned in Sections 6.3 and 6.4, when two partners become sexually aroused, the man's penis becomes erect and glands on the woman's labia and in her vagina secrete fluids which lubricate[1] the vagina and kill off bacteria. Only then can the penis be inserted[2] into the vagina. Moving it backwards and forwards then leads to an ejaculation (see Figure 14). The ejaculation releases millions of sperms and, if no **contraceptive** is used, they begin their journey through the vagina and the uterus into one of the Fallopian tubes, where they are attracted by the egg cell. If they reach the egg cell, a single sperm penetrates[3] the outer layer of the egg cell and fertilization occurs (see Figure 15).
It is important to know that sperms can be released from the penis even before ejaculation. This is one of the reasons why pulling the penis out of the vagina before orgasm is no protection against a possible pregnancy.

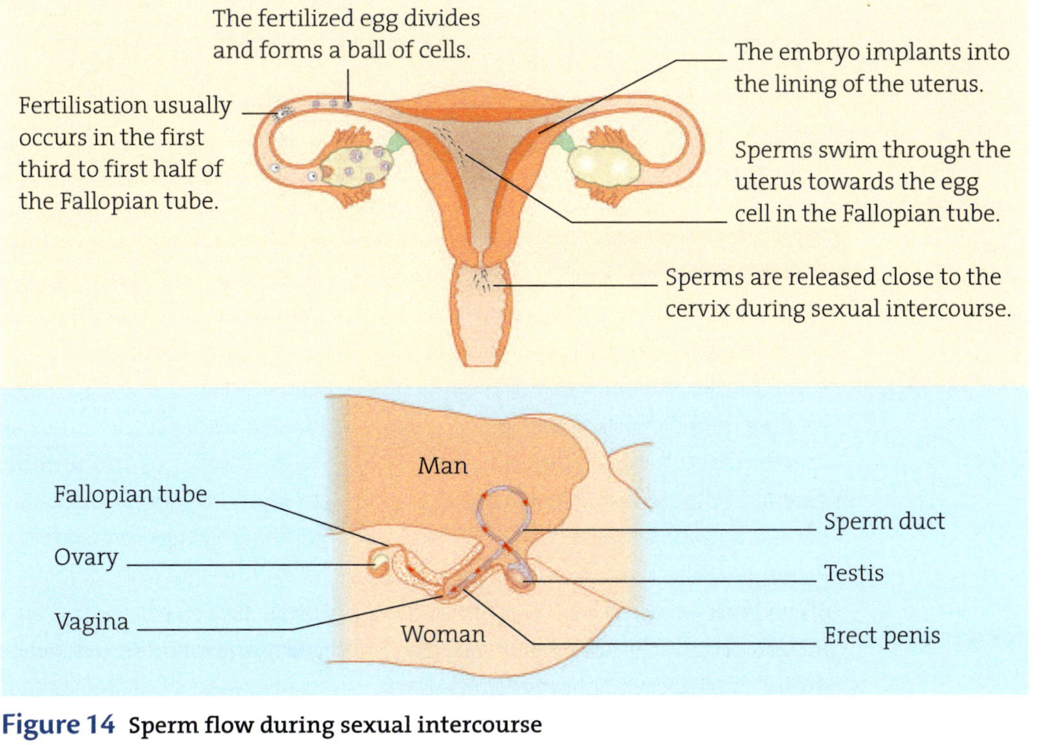

Figure 14 Sperm flow during sexual intercourse

1 lubricate sth. *etwas geschmeidig machen*
2 insert sth. *etwas einführen*
3 penetrate sth. *in etwas eindringen*

Did you know?

During orgasm a healthy male ejaculates 2–5 ml of semen containing 300–500 million sperms.

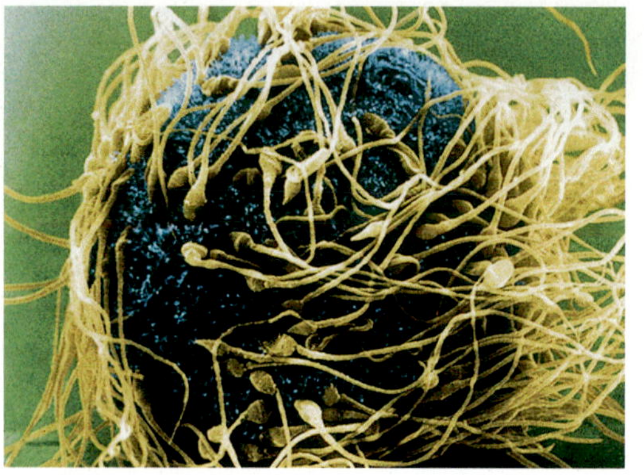

Figure 15 Sperms surrounding an egg cell

How long sperms are able to fertilize an egg depends on the environment. If semen has dried up on a blanket or warm skin, the sperms die quickly. But as long as there is liquid around them, they are still **fertile**, and even a single sperm can cause pregnancy. Once they have reached the Fallopian tube, the sperms can survive three days on average, with some living up to seven days. This means that even if ovulation occurs after sexual intercourse, the sperms in the uterus or the Fallopian tube can still cause a pregnancy.
After fertilization, dead sperms in the female body are broken down and the fertilized egg cell begins dividing[1], forming a ball of cells. About four days after fertilization, the developing **embryo** reaches the uterus. At this point it already consists of thousands of cells. It then sinks into the thick lining of the uterus in the process known as **implantation**, as explained on p. 98. A pregnancy has begun.

1 What functions do the secretions in the vagina have?
2 How long do sperms survive in a sperm stain on a blanket? How long do they survive in the Fallopian tube?
3 Draw a flow chart showing the route which sperms take, starting from the testis and ending with the fertilized egg cell. For help see 'Working with flow charts', p. 84

6.7 Pregnancy and birth are periods of rapid change

Once the embryo is **implanted** in the uterus, it develops rapidly (see Figure 16). In order to do this the **embryo** needs a lot of nutrients, which are provided by the mother's body.

Did you know?

For the first eight weeks of pregnancy, the new human life developing in the mother's womb is called an embryo. For the rest of the time until birth it is a foetus.
The heartbeat can often be detected by the end of the tenth week, when the foetus is about 5 cm long.
To find out whether a foetus is a boy or a girl, you often have to wait until week 16 or even longer. Of course, many parents prefer not to know until their baby is born.

At the spot where the embryo implants into the lining of the uterus, an organ called the **placenta** develops. This organ provides the developing **foetus** with all the food and oxygen it needs and also removes the foetus's waste products[2]. The placenta is connected to the foetus by the **umbilical cord**. Besides food and oxygen molecules, any other chemicals in the mother's blood, including viruses, can enter the unborn baby's bloodstream. In order to protect the foetus from shocks and bumps[3] when the mother moves about, it is surrounded by a liquid, the **amniotic fluid**. The membrane around the amniotic fluid is

1 divide *teilen*
2 waste product *Abfallprodukt*
3 bump *Erschütterung*

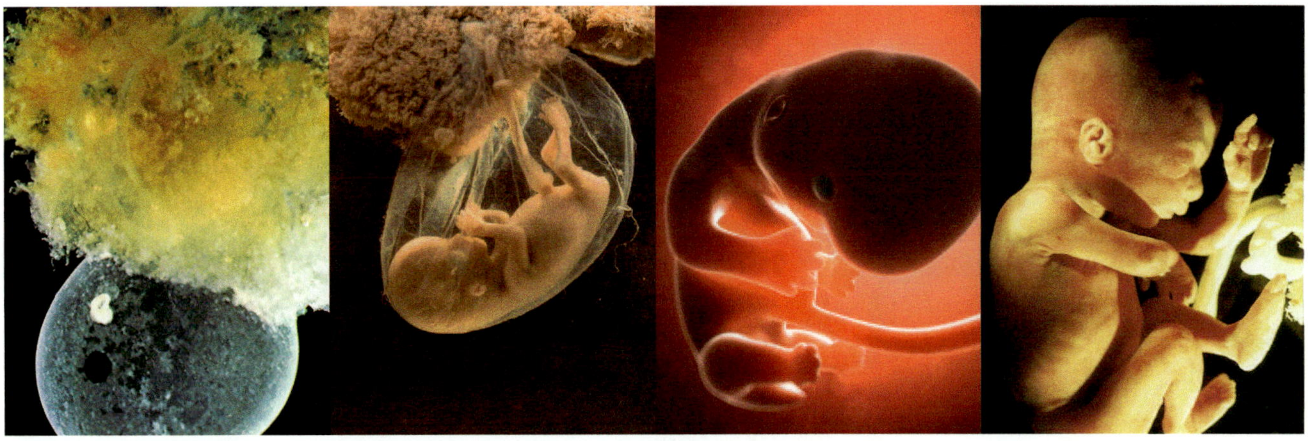

Figure 16 a The white body in the amnion is a two-week-old embryo **b** The embryo inside the amnion gets its food and oxygen via the umbilical cord **c** An eight-week-old embryo **d** A 20-week-old foetus

called the amnion[1] (see Figures 16 c and 17). At the end of the pregnancy, the amnion contains between 400 and 2000 ml of fluid.

During pregnancy, the mother's body has to provide for two organisms, which at times puts high demands on her **metabolism**. Regular appointments with a **gynaecologist** support the mother during pregnancy and ensure that both mother and baby are as healthy as possible. One of the standard ways of checking the development of the baby is with **ultrasound scans**. By placing a small scanner on the mother's abdomen, the doctor and mother are able to see on a screen an image of the baby in the uterus.

Thirty-eight weeks after the fertilization of the egg cell it is time for the baby to be born. When the mother experiences repeated **contractions** of the muscles in the uterus, this is the first sign that she is **going into labour**, another name for the process of giving birth. As the time periods between the contractions become shorter, the cervix opens up, or dilates[2], so that the child can pass through it. When the cervix is completely dilated, the amnion tears open and the amniotic fluid flows out. When this happens, the mother's 'water has broken'.

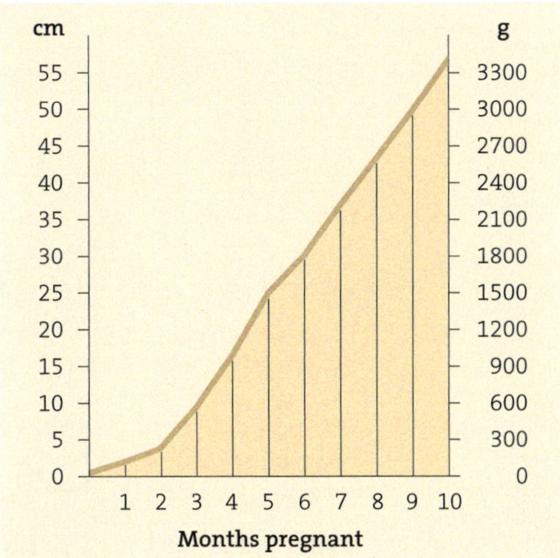

Figure 17 Changes in weight and height

Key terms

Pregnancy and birth

- afterbirth *Nachgeburt*
- amniotic fluid *Fruchtwasser*
- deliver a baby *ein Kind zur Welt bringen, ein Kind entbinden*
- Caesarean section *Kaiserschnitt*
- contractions (pl.) *Wehen*
- embryo *Embryo*
- foetus *Fötus*
- fraternal twins *zweieiige Zwillinge*
- go into labour *Wehen bekommen*
- gynaecologist *Gynäkologe, Frauenarzt*
- identical twins *eineiige Zwillinge*
- implant *sich einnisten*
- metabolism *Stoffwechsel*
- navel *Bauchnabel*
- placenta *Plazenta, Mutterkuchen*
- ultrasound scan *Ultraschalluntersuchung*
- umbilical cord *Nabelschnur*

1 amnion *Fruchtblase*
2 dilate *erweitern*

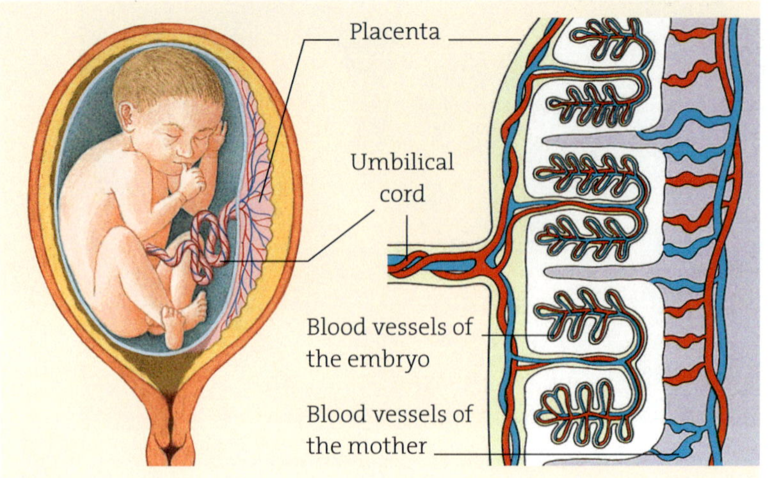

Figure 18 A foetus in the uterus (left) and the inner structure of the placenta (right)

Now it is time for the baby to be **delivered**. The contractions continue and the foetus is pushed out of the uterus and through the vagina, head first. When the baby is completely out of the vagina, the umbilical cord is clamped[1] close to the baby's body and cut. This does not hurt, because there are no nerves in the umbilical cord. The physical connection between the baby and the mother has come to an end. The stump of the cord later forms the **navel**. A short time later the placenta, which no longer has any function, is shed from the wall of the uterus and passes out of the vagina. This is called the **afterbirth**.

Now the baby's own organs have to provide for its metabolism, and this is why the first cry is so important. While it was in the uterus, the baby received its oxygen through the placenta. Now the lungs have to take over this task, and with the first cry the baby shows that its lungs are working properly.

Did you know?

Sometimes a baby cannot be delivered the normal way, for example when its shoulders are too broad or when it is lying in the wrong position. Doctors then surgically[2] open the abdomen of the woman and remove the child from the uterus, a procedure called a **Caesarean section**.

1 What is the function of the placenta?
2 Why is a baby's first cry so important?
3 Find out your weight and size at birth and whether you were born before or after your due date[3]. Compare your results with those of your class.
4 Research recommendations on what women should and should not do during pregnancy. Make a list.
5 In Figure 19 you can see identical twins and fraternal twins. Find out how twins develop during pregnancy and decide which of the two pictures shows identical twins and which shows fraternal twins.

Figure 19 Identical and fraternal twins

1 clamp sth. *etwas abklemmen*
2 surgically *chirurgisch*
3 due date *Geburtstermin*

6.8 Contraceptives can prevent pregnancy

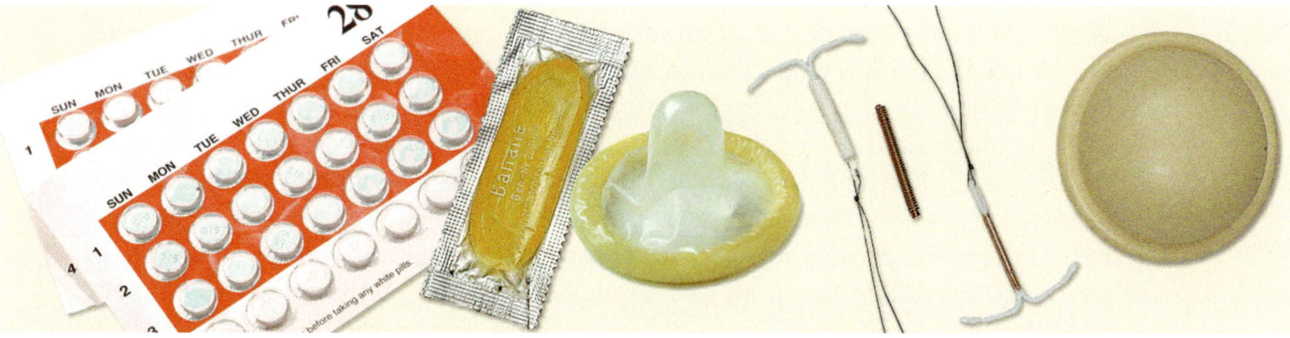

Figure 20 **Different contraceptives: birth control pill, condom, IUD and diaphragm**

A couple who love each other usually have a strong desire for physical intimacy, but they often also want to avoid pregnancy. One way for many couples to express their closeness without risk is by petting, which means kissing and touching one another without having full sexual intercourse. As long as no sperm enter the vagina, there is no danger of becoming pregnant.

Many couples, however, want to have sexual intercourse as their relationships develop. They can choose to prevent pregnancy by using contraceptives. It is important to note, however, that no contraceptive method is 100 per cent effective. In this book it is only possible to give a brief overview of some contraceptives, not a full description of the advantages and disadvantages of every method. If you want to know more about contraception, you can get information from **health-centre staff**, your doctor, books in the library, or from reliable websites.

One of the most common contraceptives is the **condom**, which is made of thin **latex rubber** and is pulled over the erect penis. When ejaculation occurs, the sperm is collected in the condom. It is very important to handle condoms carefully because even the tiniest[1] hole, caused by long fingernails, jewellery or any hard object, will reduce the protection. When having intercourse, a condom is the only way to protect yourself both from pregnancy and from STDs[2].

A contraceptive commonly used by women is the **birth-control pill**, often simply called 'the pill'. It contains hormones which hinder[3] an egg cell from maturing. Thus, no ovulation occurs and sperms in the Fallopian tube never reach the egg. Women who take the pill still have regular periods. The level of protection can be affected, however, by irregular intake, medication (for example against headaches) and diarrhoea[4]. At these times, a woman taking the pill should also use another form of protection, such as condoms, for the next seven days. The pill has to be

Key terms

Contraception
- birth-control pill ('the pill') *Antibabypille*
- condom *Kondom*
- health-centre staff *Angestellte einer Gesundheitsberatung (z. B. pro Familia)*
- intrauterine device (IUD) *intrauterines Pessar, ‚Spirale'*
- latex rubber *Latexgummi*

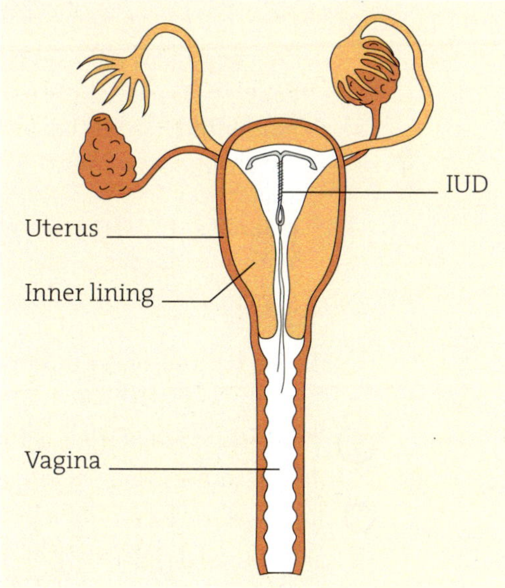

Figure 21 **Placement of an IUD in the uterus**

1 tiny *winzig*
2 STD *sexuell übertragbare Krankheit*
3 hinder sb./sth. from doing sth. *jdn./etwas davon abhalten, etwas zu tun*
4 diarrhoea *Durchfall*

prescribed[1] by a doctor, who will provide more information about how it works and how to take it.

Condoms and, even more so, the pill offer a high level of security against pregnancy if used correctly. (The pill, however, cannot provide protection against STDs: this is important to remember.) There are many other contraceptives, such as the diaphragm[2] and the **intrauterine device (IUD)**, which all have certain advantages and disadvantages.

Choosing whether and when to have sexual intercourse and choosing the right contraceptive are very important personal decisions. Often it is helpful to talk to an older person you can trust, for example a parent, teacher, doctor or health-centre staff member. The same is true if you have become pregnant (no matter whether you used contraception or not) or if you think you may have an STD and you are considering what to do. Remember, though, that people hold different opinions about sex and contraception, and you have the right to make the decision that you think is best for you.

Contraceptive	How it works	Reliability[3] (Pearl Index*)
Condom	Stops sperms from entering the vagina. Also protects from STDs.	2–12
Birth control pill	Prevents ovulation and makes it harder for sperms to travel up the Fallopian tube.	0.1–0.9
IUD (intrauterine device) – also known as 'the coil'	A small T-shaped plastic object, usually containing a copper spiral[4], which the gynaecologist inserts through the cervix and places in the uterus. It prevents pregnancy by changing the lining of the uterus. An IUD is not noticeable during intercourse and can stay in the uterus for 2–5 years.	0.4–1

*The Pearl Index tells you how many women out of 100 who use this method become pregnant in a year. Studies on this are difficult because some women have sexual intercourse more often than others, and some people make mistakes when using a method. For example, a woman can forget to take the pill one day, or a man might put on the condom too late. That is why the Pearl Index is only a measure of reliability rather than a hard fact. A Pearl Index of 2–12 means that with careful use two women in 100 become pregnant in a year. With less careful use about 12 women in 100 become pregnant.

(Adapted from: www.profamilia.de/article/show/9106.html)

Table 1 Three of the most common contraceptives

1 Work in groups of three. Each person chooses one of the contraceptive methods shown in Table 1 and explains its advantages and disadvantages to the other group members.

2 Describe the factors which can lower the effectiveness of the birth control pill.

3 Why does the Pearl Index always give you a range (for example 2–12) instead of a single number? Explain.

4 Extend Table 1 with information about other contraceptives. You can use the websites on p. 109. Compare your results with those of a classmate.

5 **Say it in German:** Prepare a quiz game by creating cards with a key term from this chapter on each card. Make cards for these categories: a) Male Reproductive System, b) Female Reproductive System and Menstruation, c) Puberty, d) Pregnancy and e) Contraceptives. For each category, give the cards a value of 10, 20, 30, 40 or 50 points, depending on their difficulty (10 = easy, 50 = difficult).
A player picks a category and the value of a card. The card is read out in English and the player must explain the term – but in German! If the player gives the right explanation but in English, he or she gets only half the points.

1 prescribe sth. *etwas verschreiben*
2 diaphragm *Diaphragma*
3 reliability *Zuverlässigkeit*
4 copper spiral *Kupferspirale*

6.9 Sexual intercourse can transmit diseases

Whenever two people have sexual intercourse they need to think not only about pregnancy (see Section 6.8) but also about **sexually transmitted diseases (STDs)**. When a couple has sex, sexual fluids are exchanged and these fluids can contain germs that cause illness.
You may have heard that **AIDS** (**acquired immunodeficiency syndrome**) is a disease that destroys the immune system, and that it is caused by a virus called the **human immunodeficiency virus (HIV)**. HIV is transmitted mostly by sexual fluids or contact with the blood of an infected person, but the virus also occurs in breast milk. If a person develops AIDS, their body becomes unable to fight off even small illnesses like colds. Many AIDS patients die of pneumonia, a disease which a healthy immune system would deal with easily. Although there are some drugs that help AIDS patients to live longer, there still is no cure for the disease.

Key terms

Sexually transmitted diseases (STDs)
- acquired immunodeficiency syndrome (AIDS) *erworbenes Immunschwäche Syndrom*
- antibiotic *Antibiotikum*
- gonorrhoea *Gonorrhö*
- human immunodeficiency virus (HIV) *humanes Immunschwäche Virus*
- human papillomavirus (HPV) *Human-Papilloma-Viren*
- rash *Ausschlag*
- sexually transmitted disease (STD) *Geschlechtskrankheit*
- syphilis *Syphilis*

Besides AIDS there are many other STDs, for example **syphilis**. Syphilis bacteria cause **rashes** on your genitals and your skin, and if it is not treated, syphilis can damage or even destroy your nervous system. Since syphilis is caused by bacteria, it can be treated with **antibiotics**. Another common STD is **gonorrhoea**, which causes an infection of the urethra and leads to pain while urinating. Gonorrhoea can also be cured with antibiotics. A third STD is the **human papillomavirus (HPV)**, which can cause cancer of the cervix. Although in recent years a vaccine has been found against the virus, it is still very difficult to fight cancer once it has developed.
Again, remember: the only contraceptive that offers protection from both pregnancy and STDs is a condom, although even condoms do not give an absolute guarantee.

1 What are the only two reliable ways to protect yourself against STDs? Explain.
2 Use the websites on p. 109 to answer the following questions:
  a What is so dangerous about HIV?
  b How does HIV affect the body, and how can it be treated?
  For language help, see **Activate your English**, 'Describing causes and effects', on the inside cover.
3 Further research: Find out about the vaccine against the human papillomavirus. What are its advantages and disadvantages?
4 Look at the advert in Figure 22. Together with a partner, think of a suitable slogan in English to replace the German one. Explain why your slogan is effective.

Figure 22 A German advert against AIDS

6.10 Sexual abuse takes many forms

Figure 23 **Nobody should have to tolerate sexual harassment or abuse.**

Key terms

Sexual abuse
- rape *Vergewaltigung*
- sexual abuse *sexueller Missbrauch*
- sexual harassment *sexuelle Nötigung*

Imagine these two scenarios: Paul, aged twelve, plays football in a club. After practice everybody is in the shower, including their coach, who continually makes jokes about Paul's naked body. Kathie, aged eight, does not like her teacher, Ms Mulbridge. When Kathie asks a question in class, the teacher always puts her hand on Kathie's shoulder and comes really close to her face. Last time, Kathie even thought that Ms Mulbridge had touched her bottom[1].

When adults (or other children) show verbal sexual behaviour towards boys or girls, it is called **sexual harassment**. If the behaviour becomes physical, like uninvited kissing, touching or, in the worst case, **rape** (unwanted sexual intercourse), then it is considered **sexual abuse**. Both forms of behaviour are illegal, but more importantly, they can deeply hurt the person who is abused.

Quite frequently the people who sexually abuse children are not strangers, but people the children know and trust, such as a family member or youth group leader. When you trust someone, it is hard to imagine that he or she could do something to hurt you, and so a lot of children do not realize at first that what is happening to them is actually abuse. Often they even blame themselves, imagining that they are responsible for the adult's behaviour.

The consequences of being abused can be devastating[2]: victims[3] often find it hard to trust someone again or to build new relationships. Many suffer from feelings of guilt and become very quiet and self-conscious around other people. These emotional scars[4] are often much worse than the physical ones that abuse can bring.

If you are in a situation where someone shows uninvited sexual behaviour towards you, it is important that you tell the person very clearly that you do not want this. You are allowed to be impolite, to run away, or even to scream. You are the only one who decides what you want to happen, and you are allowed to do everything that is necessary to get out of a situation you do not like.

If you or a friend have been sexually abused, tell a person you trust about it straight away. A lot of people have difficulties coping with[5] this topic, so if the first person you tell does not listen to you or does not believe you, do not give up. Search for someone who does believe you, and do not let anyone tell you that 'it can't be that bad'. There are many health centres and support services that can help you, and you can find their telephone numbers in the telephone book or on the Internet.

1 bottom *Hintern*
2 devastating *verheerend, vernichtend*
3 victim *Opfer*
4 scar *Narbe*
5 cope with sth. *etwas bewältigen*

1 Find out the phone numbers of hotlines in your city that you can call if you or a friend have been sexually harassed or abused.

2 In groups of two or three, discuss what Paul and Kathie (see the examples above) should do. Why might they feel afraid to do anything? What suggestions can you make to help them?

For language help, see **Activate your English**, 'Making suggestions', on the inside cover.

3 Why can a chat room be dangerous and how can you protect yourself? Discuss.

4 a Have a look at the two English-language websites in the box below. Read an article on the same topic on both sites and compare the opinions. For language help, see **Activate your English**, 'Comparing and contrasting', on the inside cover.

b Now look up the same topic on one of the German-language websites and see if the advice is the same or different. Together with a classmate, think of possible reasons for the differences.

Figure 24 **Using chat rooms is fun, but you need to be careful about the information you give out.**

Did you know?

Chat rooms are a great place to meet people. Unfortunately, sometimes chat rooms contain adults disguised as children who sexually harass kids. Here are a few short guidelines to help you to chat safely:

- Never tell anyone your real name, home address or e-mail address, and never give anyone your school, home or mobile phone number.
- Have two e-mail addresses: one that you give to your close friends, and one that you use in chat rooms. There are a few websites that give you a 'disposable e-mail address' which you can use to log into a chat room but then never have to use again.
- When you write your profile, you are allowed to make up a character – do not give out real information. This is not lying but protecting yourself.
- Never give your password to anyone. Think of it like the key to your house – you would not give that to just anyone, would you?
- You never know who is behind a nickname. It could be an adult, it could be a person from your school who you have never felt comfortable with. So do not share personal details that you would usually only tell to your closest friends.
- If someone you chat with makes sexual comments, you can leave the chat immediately. Sometimes it is a good idea to write something like 'My parents are coming in a few minutes and will want to use the computer' – even though this might not be true. Then your chat partner might be more careful.
- Do not post a photo with your profile.

Here are some English and German-language websites on love, sex and abuse:
www.kidshealth.org – an American website for kids and teenagers on all health issues
www.bzga.de – the German government's website for all health issues
www.loveline.de – a government website for teenagers
www.profamilia.de – an organization that informs you about everything to do with sex and pregnancy
www.zartbitter.de – a German organization for people who have been sexually abused
www.kinderschutzbund.de – an organization concerned with the welfare of children in general, but also with abuse

Getting the most out of a film sequence

Figure 25 **Working together helps you to get the most out of watching a film.**

Watching a film can be a challenge when the information is presented very quickly. For example, in a sequence[1] about the development of a human embryo, there may be a lot of numbers (e.g. how many cells the embryo has at each stage), new words and interesting images. Trying to keep all these things in mind is almost impossible, and if you try to write them all down while watching you will not be able to enjoy the film and may miss some of the content.

Viewing skills

Before watching

- **Divide into groups and share the tasks.** One group member's job is to write down the different numbers mentioned in the film. Another focuses on important biological terms and their meanings. A third person writes down headlines for the different sections of the film.
- **Write down some questions that you would like the film to answer**. By doing this, you prepare your mind to look out for important information.

While watching

- **Write down only the things your group wants you to focus on.** This way you do not need to try to understand and memorize everything at once. It is then easier to take in[2] the images, sounds and words and get a better overall impression of the film.

After watching

- **Get together in your group and share your results.** In the end, everybody should have notes on all aspects of the sequence.
- **Talk about the contents**. Which things surprised you most? What did you find most interesting? Which of your questions were not answered? Taking notes alone is not enough to make sure you have understood the film. Talking about it helps you to think and reflect about different parts of the film and to remember what you have learned from it.
- **Analyse how the film was made.** Did the images always fit what was said? Was the narration[3] easy enough to follow? Was the film successful in getting your attention and answering your questions? Why or why not? What effect did the music and sounds have on you? What special techniques were used in the film, if any? Apart from the things you noted down while watching, what else did you notice and why was this interesting?

Test yourself

As with a text, you have understood a film only when you are able to summarize it in your own words and answer questions about it.

A multiple-choice quiz is one way you can test your understanding: Everyone in your group writes down four questions with four answers each. You can make it harder by having more than one correct answer to each question. Test each other and find out how much you know.

1 sequence *Filmausschnitt*
2 take sth. in *etwas in sich aufnehmen*
3 narration *Erzählung*

Activity 1

Saying 'no' when somebody asks you out[1]

Sometimes a boy wants to go out with a girl, but she does not want to, or a girl wants to go out with a boy, and he does not want to. Make groups of two boys and two girls each, and think about the following scene:

> Lisa is talking to her friend when Tom and his friend come up to them. Tom has liked Lisa for a while and now he wants to ask her if she would like to go to the cinema with him. Lisa thinks that Tom is a nice person, but she does not want to do anything with him alone.

Prepare it

- In your group, think of ways for Lisa to tell Tom that she does not want to go out with him. Consider how all four people feel.
- How can the two friends help? Should they help at all?
- Would the situation be different if Lisa liked Tom and had asked *him* to go to the cinema with *her*? Why, or why not?

Act it out

- One group acts out the scene. Someone else watches the time: it should not last more than three or four minutes.
- The class should focus on the following points:
  - Who speaks the most?
  - How do the different people feel throughout the conversation?
  - What roles do the friends play?
- Then act out the scene again, only this time Lisa asks Tom.

Talk about it

- First the actors tell the class how they felt during the scene.
- Then the rest of the class discusses what they observed. What was different when Lisa asked Tom? Why?
- Which of the things the players did and said would you also do in a real-life situation and what would you do differently? Why?

Activity 2

Boys will be boys and girls will be girls?

The saying[2] 'boys will be boys' means that no matter what you do, boys will always act like 'real boys', for example by playing with guns or pretending to be cowboys. When people say 'All men are like this' or 'All women are like that' we call it a stereotype[3], because it is almost never the case that people in a large group feel or do the same. But we all have stereotypes in our minds (although not all of them are bad) and they influence our way of thinking.
As you grow older, it is important to explore your ideas of what being a man or a woman means, and to consider how these fit into the stereotypes you see in the world around you.

1 Make a poster to present your ideas: find pictures in magazines and/or lines from songs or movies (or people) and put them together to show what you think a real man and a real woman is. Display your poster in the classroom and take time to find out about your classmates' ideas by looking at their posters, too.

2 When you look through magazines, listen to songs or watch TV you will notice that men, women and relationships are frequently portrayed in a stereotypical way. Find an example where you disagree with the way the sexes are presented and explain to the class why you think so.

1 ask sb. out *jdn. einladen, mit jdm. auszugehen*
2 saying *Sprichwort*
3 stereotype *Klischee, Stereotyp*

Summary

Puberty is the time between the ages of eight and twelve when boys and girls reach sexual maturity and become able to reproduce, which means that girls begin to menstruate and boys start producing sperm. Other physical, mental and emotional changes also occur during this period, one of which is the development of sexual attraction to other people.

When thinking about physical intimacy, teenagers need to think clearly about their values and what they really want, as this can affect all areas of a relationship. It is especially important to be aware of the possibility of pregnancy and STDs.

The male sex organs produce sperm and hormones in the testes. The sperm are ejaculated when sexual arousal leads to an orgasm. The female organs produce egg cells in the ovaries. This is a natural part of the menstrual cycle, which prepares a woman for the possibility of pregnancy. As young people's bodies change during puberty, personal hygiene, especially in the genital area, becomes very important.

Pregnancy can occur when a sperm enters an egg cell in the Fallopian tube. The fertilized egg cell then implants into the uterus, where the cells of the embryo start dividing rapidly. The baby is born around the end of the 38th week after fertilization. The risk of pregnancy can be reduced by using contraceptive methods such as condoms or the birth-control pill. The only contraceptive that also protects against STDs is the condom.

Sexual abuse is a serious danger which often has devastating consequences for its victims. When sexual abuse occurs, it is very important to take action against it as quickly as possible.

Key terms

See the boxes:

Reaching puberty 1, 2 .... pp. 90, 91
Love and sexual relationships .... p. 93
The male reproductive system ..... p. 95
The female reproductive system 1, 2 97, 98
Puberty and hygiene .... p. 99
Fertilization .... p. 101
Pregnancy and birth .... p. 103
Contraception .... p. 105
Sexually transmitted diseases .... p. 107
Sexual abuse .... p. 108

A Multiple choice

Tick the correct answers. Sometimes more than one answer is correct.

1. Which of these developments occur during puberty?
   a) growth spurt
   b) formation of the sex organs
   c) widening of the pelvis in men
   d) growth of the larynx
2. Which of these things can you find in the menstrual flow?
   a) sperm
   b) cells from the lining of the uterus
   c) blood from the vagina
   d) three or more egg cells
3. Fertilization can occur ...
   a) in the uterus.
   b) in the Fallopian tube.
   c) often between days 12 and 16 of the menstrual cycle.
   d) when the woman lies on top of the man.
4. AIDS can be transmitted by ...
   a) contact with infected blood.
   b) drinking from the same cup as an infected person.
   c) sexual intercourse.
   d) shaking hands with a patient.

B Complete the table

Copy the table below into your notebook and fill in the gaps.

Organ	*Function*
ovary (pl. ovaries)	*the place where egg cells are produced in women*
(a)	*blood that enters it makes the penis hard*

Organ	*Function*
sperm duct	*(b)*
(c)	*a big gland under the urinary bladder which contributes liquids to make semen*
(d)	*the sack in which the testes lie*
epididymis	*(e)*
urethra	*the tube that sperm and urine go through*
(f)	*the entrance to the uterus*
uterus	*(g)*
Fallopian tubes	*(h)*

C Critical thinking

1 A lot of teenagers are worried about their bodies: Is my acne turning people off? Are my breasts or is my penis a 'normal' size? Why does everybody seem to be looking at me? Think about how you would like your body to look and ask yourself why you feel this way. Who tells you what is 'right'? Is it your own feelings, your peers, a member of the opposite sex, your parents, a singer, model, actor or other famous person? Now make a list of things you like about yourself (not just your body). Doing this can help you to accept the things you do not like so much. Another exercise is to write down things you like about a friend's character and then give the list to him or her.

2 Why should a pregnant woman take special care with what she eats and drinks?

3 In a foetus the lungs are among the last organs that are fully developed before birth. From what you know about the conditions in the womb before the foetus is born, explain why it is an advantage for the lungs to develop at a late stage. In which situation can this become a serious disadvantage?

D Mini-project

This chapter only briefly looks at sexually transmitted infections. As an extra research project, create fact sheets about some of the more common diseases and put your fact sheets up in your classroom.

a) Working in small groups, each group chooses one of the following diseases for research: gonorrhoea – syphillis – human papillomavirus – chlamydia – HIV – hepatitis – herpes – trichomoniasis

b) Your fact sheet should include information on the following points:
- full name of the disease
- prevalence[1]
- cause
- means of transmission[2] and risk factors
- symptoms
- treatment
- prevention

c) When doing Internet research, remember to use several search engines, for example: Google, MSN, Yahoo or Ixquick. For results that only give you information on the disease, just add the word 'disease' or 'STD' to your query[3]. If you want to find websites especially for young people, try adding the word 'students' to your search.

1 prevalence *Verbreitung*
2 transmission *Übertragung*
3 query *(hier) Suchanfrage*

Chapter 7 **The Sense Organs**

Chapter objectives

- To **describe** the structure and function of the eye and the ear
- To **explain** how the eye adjusts[1] to changes in light intensity, distance and colour
- To **investigate** how functional defects[2] of the eye and ear can be prevented and corrected
- To **understand** how perception[3] results from the combined work of the sense organs and brain
- To **compare** the performance[4] of the senses in humans and animals

Figure 1 Enjoying the scent of a flower

Everything we know about the world in which we live comes to us through our **sense organs**. The eyes, ears, skin, nose and tongue are our windows to the world around us and they keep us informed about the state of our body as well.

7.1 All our sense organs work in a similar way

The eyes, ears, skin, nose and tongue contain special **receptors** called **sensory cells**, which respond to **stimuli** and provide our brain with a steady stream of information. The sense organs help us interact[5] with the outside world. More than a dozen different types of sensory cells detect[6] changes outside and inside our body. The stimuli they receive are changed into **nerve impulses**, or electric signals, which travel along the **sensory neurons** and arrive at the **spinal cord** and the sensory areas in our brain. It is here in the **cerebrum** that **perception** takes place. The information from the sense organ is analysed and stored or leads to immediate or later responses. These responses travel to the muscles and glands by pathways[7] called **motor neurons**.

The receptors in our sense organs are highly specialized. The receptors for light in our eyes are not sensitive to sound, and the taste receptors on our tongue are blind to light and cannot be activated by vibrations. Sensory cells can only react properly to adequate[8] stimuli.

Visible[9] light is an adequate stimulus for the sensory cells in our eye, certain sounds are adequate for the sensory cells in our ears, and the receptors for taste and smell on our tongue and in our nose are **stimulated** by certain chemicals.

Key terms

Sensory cells and their functions

- neuron *Nervenzelle*
- nerve impulse *Nervenimpuls*
- pain receptor *Schmerzrezeptor*
- perceive sth. *etwas wahrnehmen*
- perception *Wahrnehmung*
- photoreceptor cell *Sehzelle, Lichtsinneszelle*
- pressure receptor *Druckrezeptor*
- reception *Aufnahme*
- receptor (cell) *Rezeptor(zelle); Sinneszelle, die einen Reiz aufnimmt*
- sense organ *Sinnesorgan*
- sensory cell *Sinneszelle*
- sensory neuron *sensorisches Neuron*
- stimulate sth. *etwas reizen*
- stimulus (pl. stimuli) *Reiz*
- temperature receptor *Temperaturrezeptor*
- touch receptor *Tastrezeptor*

1 adjust to sth. *sich an etwas anpassen*
2 functional defect *Funktionsstörung*
3 perception *Wahrnehmung*
4 performance *Leistungsfähigkeit, Leistungsvermögen*
5 interact with sb./sth. *aufeinander einwirken, mit jdm./etwas interagieren*
6 detect sth. *etwas aufdecken, aufspüren*
7 pathway *Übertragungsweg*
8 adequate *geeignet, passend*
9 visible *sichtbar*

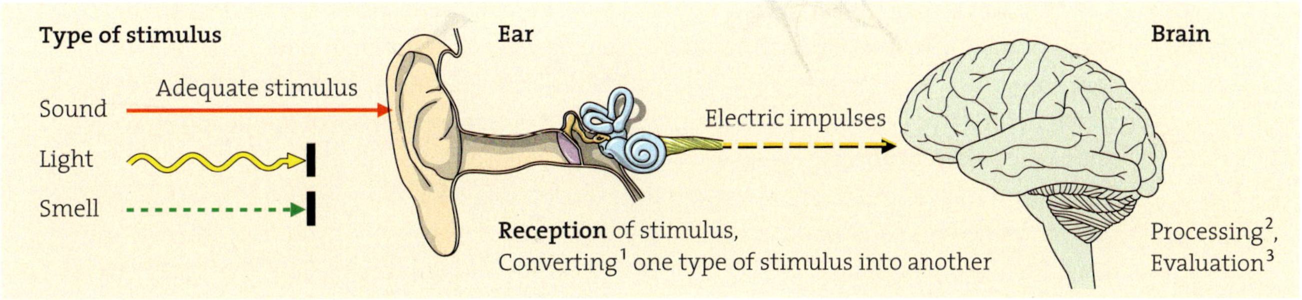

Figure 2 Adequate and inadequate stimuli for the human ear

1 convert sth. *etwas umwandeln*
2 processing *Verarbeitung*
3 evaluation *Auswertung*
4 intensity *Stärke*
5 dazzle sb. *jdn. blenden*
6 blurred *unscharf, verschwommen*
7 embed sth. in sth. *etwas in etwas einbetten*

Also the intensity[4] of the stimulus is important for the proper functioning of the sense organ, e.g. a sudden bright light in the eyes can dazzle[5] you and make your vision blurred[6]. If a sound is above or below a certain level, your ears cannot recognize it.

Various kinds of information about our environment reach us through our skin. Embedded[7] in our skin are a variety of different receptors, which are sensitive to touch, temperature, pressure and pain.

Did you know?

One square centimetre of your skin contains about 25 **touch** and **pressure receptors**, 200 **pain receptors**, 15 **temperature receptors** (13 for cold and 2 for heat), 4 metres of nerve fibres and 1 metre of blood capilliaries.

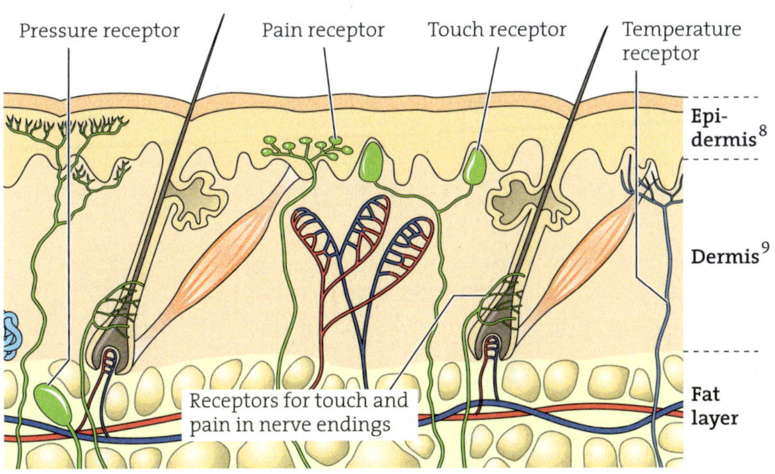

Figure 3 Sense organs in a small section of the skin

The receptors inside our body inform the **central nervous system** (i.e. the spinal cord and the brain) about the condition of the inner organs and the position of the body in space (you will read more about this later in this chapter). Receptors in our joints, our tendons and our muscles detect changes in the position of our limbs[10].

As humans we are only able to **perceive** a small range of changes in our environment, which has to do with the type and structure of our sense organs. Many species of animals can see or hear much better than we can. The **photoreceptor cells** in our eyes, for example, are not sensitive to ultraviolet[11] light, which bees can see. Dolphins produce clicking sounds of a very high frequency, not audible[12] to us, which they use to orientate themselves and to hunt for fish. Also dogs can hear sounds of a very high pitch[13], while elephants can hear very deep sounds. None of these sounds can be detected by the human ear.

We have no sensory cells that respond[14] to x-rays[15] or radioactivity[16], nor do we have organs for perceiving radio waves[17], electricity or magnetism. However, by using special technical instruments we can detect and measure stimuli which our sense organs cannot perceive.

Key terms

Pathways of signals
- central nervous system *zentrales Nervensystem*
- motor neuron *motorische Nervenzelle*
- spinal cord *Rückenmark*

Parts of the brain
- cerebral cortex *Großhirnrinde*
- cerebrum *Großhirn*

8 epidermis *Epidermis, Oberhaut*
9 dermis *Lederhaut, Haut*
10 limb *Gliedmaße*
11 ultraviolet *ultraviolett*
12 audible *hörbar*
13 pitch *Tonhöhe*
14 respond to sb./sth. *auf jdn./etwas ansprechen*
15 x-ray *Röntgenstrahl*
16 radioactivity *Radioaktivität*
17 radio wave *Funkwelle*

If all messages sent out by the sense organs take the form of nerve impulses, how does the brain know if an impulse it receives comes from a shining light, a sound or an odour? The answer is that there are special pathways for nerve impulses from your eyes, your ears and your nose, which transfer impulses to specific sensory areas in the **cerebral cortex** of the brain, the outer region of the **cerebrum** (see Figure 4).

But **neurons** may also send impulses from the eye to the brain for other reasons. When you press your finger tips against the corners of your eyes or when something hits your eye, you may 'see stars'. The reason for this is that your brain interprets any impulse coming from the eye as light.

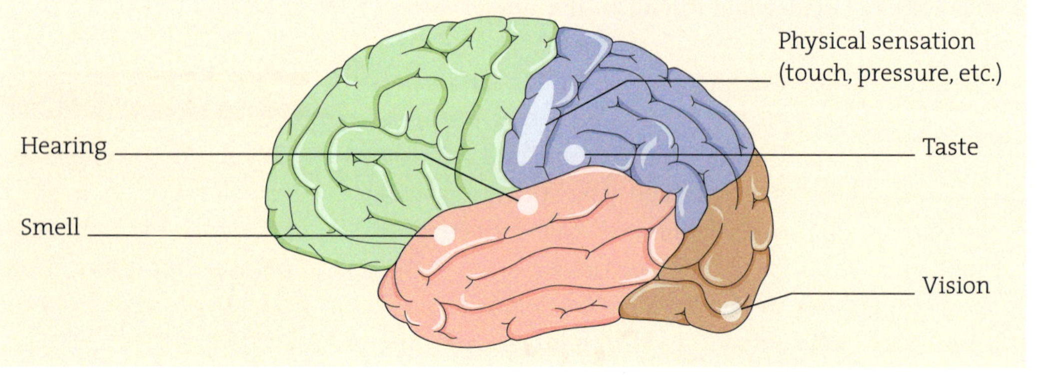

Figure 4 The localization of sensory functions in the left half of the cerebral cortex

1 Put the information in Figure 2 into words by writing a short report on the adequate stimuli for your ear.
2 In which respect do all our sense organs function in a similar way?
3 Explain why our skin is a very versatile[1] sense organ.

7.2 The eyes are complex and sensitive optical instruments

Since much of the information you gather about the environment reaches your brain through your eyes, it is important to know how the eyes function and how you can keep them healthy and working properly.

As the eyes are very sensitive[2] organs, they have to be protected against any damage and infections. The **eyebrows** and **eyelashes** play important roles in this: the eyebrows direct sweat[3] to the side of the head, while the eyelashes keep dust and dirt out. Your **eyelids** close automatically when anything touches the eyelashes or when a puff of wind[4] strikes the eye. The movement of

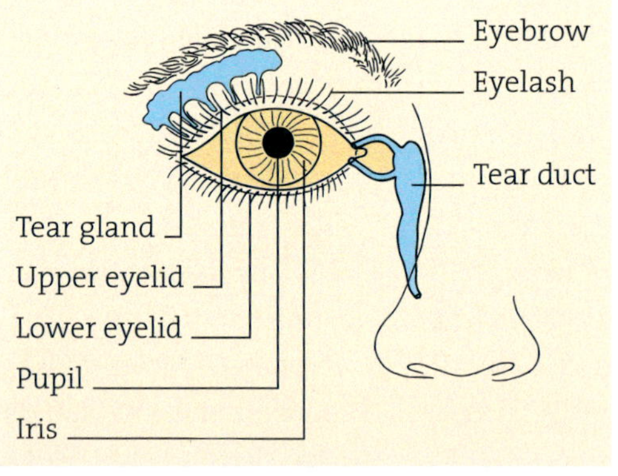

Figure 5 A tear gland and tear duct

Key terms

Outer-eye structures
- conjunctiva *Bindehaut*
- cornea *Hornhaut*
- eyeball *Augapfel*
- eyebrow *Augenbraue*
- eyelash *Augenwimper*
- eyelid *Augenlid*
- eye socket *Augenhöhle*
- sclera *weiße Augenhaut*

1 versatile *vielseitig*
2 sensitive *empfindlich*
3 sweat *Schweiß*
4 puff of wind *Windstoß*

the eyelids not only protects the eye from mechanical injury[1], but every time you blink[2], tear fluid[3] from the tear gland[4] under the upper eyelid is spread over the surface of the eye. In this way the surface of the eye is kept moist. Tears wash away dust and prevent bacterial infections[5], as they contain substances that kill bacteria. If you look into a mirror and pull down the lower eyelid, you will see the opening of the **tear duct**, which carries surplus[6] tear fluid into the nasal cavity[7]. It is no wonder that you have to blow your nose when you cry.

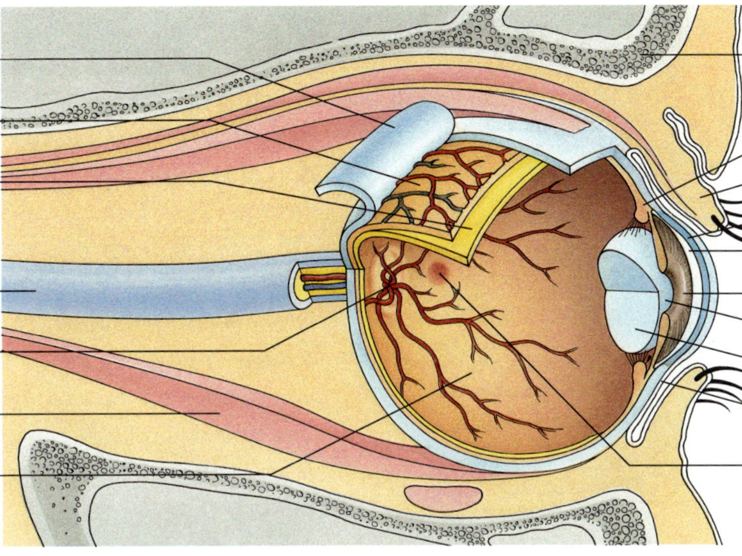

Figure 6 **The structure of the eye and its position in the skull**

The **eyeball**, the main part of the eye, rests in the bony **eye socket** in the skull, which provides strong protection on three sides. The wall of the hollow[8] eyeball consists of three layers: the tough protective **sclera**, the **choroid** and the **retina**.

The sclera, which is thick and white, covers the sides and the back of the eyeball. Six muscles are attached to it at different angles[9], which allow the eye to be moved within its socket. The muscles are coordinated by the brain, so that both eyes always move in the same direction. At the front of the eye the sclera is transparent and is called the **cornea**. It is covered by a thin membrane, the **conjunctiva**, which acts as a barrier against infections, secreting[10] oils and mucus.

The dark coloured choroid, the second layer of the eye, lies just inside the sclera and is filled with blood vessels which provide the eye with food and oxygen. The dark pigments in the choroid prevent reflection, which would disturb vision.

The **iris**, which is part of the choroid, lies behind the cornea. It is the iris which gives the eye its colour. It is shaped like a disc, with an opening in the middle, the **pupil**, through which light enters the eye. The iris is composed of muscle fibres which are arranged both radially[11] and in a circular[12] way. When the circular muscles contract, the pupil becomes smaller, allowing less light to enter the inside of the eyeball. When the radial muscles contract, the pupil becomes wider, letting more light in.

Key terms

Inner-eye structures

- aqueous humour *Kammerwasser*
- blind spot *blinder Fleck*
- choroid *Gefäßhaut*
- ciliary muscle *Ziliarmuskel*
- cone *Zäpfchen*
- iris *Regenbogenhaut, Iris*
- lens *Linse*
- optic nerve *Sehnerv*
- pupil *Pupille*
- retina *Netzhaut*
- rod *Stäbchen*
- suspensory ligament *Linsenband*
- tear duct *Tränennasengang*
- vitreous humour *Glaskörper*
- yellow spot *gelber Fleck*

1 mechanical injury *mechanische Verletzung*
2 blink *blinzeln, zwinkern*
3 tear fluid *Tränenflüssigkeit*
4 tear gland *Tränendrüse*
5 bacterial infection *bakterielle Infektion*
6 surplus *überschüssig*
7 nasal cavity *Nasenhöhle*
8 hollow *hohl*
9 angle *Winkel*
10 secrete sth. *etwas absondern, ausscheiden*
11 radially *strahlenförmig*
12 circular *kreisförmig*

Just behind the pupil lies the **lens**, which is curved[1] and transparent. It can change its shape like a bag filled with water. The **ciliary muscle**, a ring of muscle fibres, surrounds the lens, and fine threads, the **suspensory ligaments**, run from the edge of the lens to the ciliary muscle like the spokes[2] of a wheel. The contraction and relaxation of the muscle fibres loosen and tighten the suspensory ligaments and change the thickness and curvature[3] of the lens.
The space between the cornea and the lens is filled with a watery fluid, the **aqueous humour**, which nourishes the cornea and the lens and gives the front of the eye its form and shape.

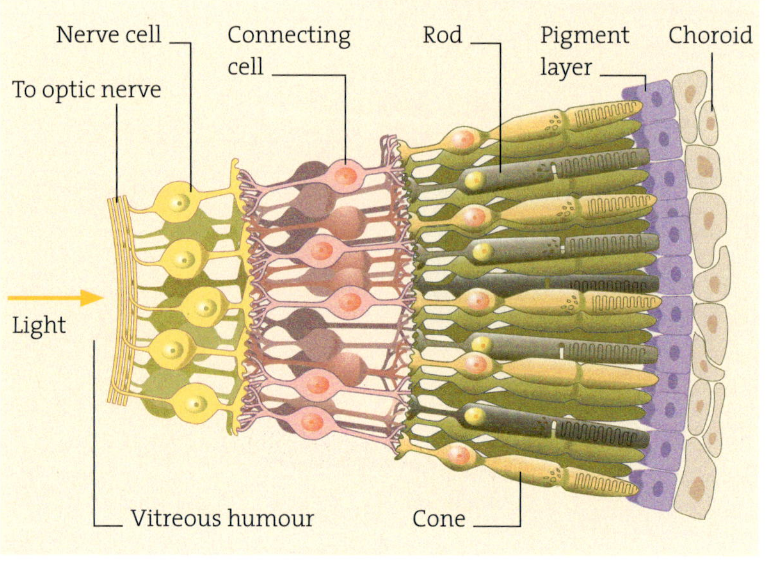

Figure 7 **The structure of the retina**

After passing through the lens, the light travels through a colourless jellylike substance, the **vitreous humour**, which fills the eyeball and helps to keep its shape. The light rays then fall onto the **retina**, which is the place where perception takes place.
The retina consists of several layers of cells. The light-sensitive[4] receptor cells form the layer next to the choroid. Further to the inside are the connecting cells and the nerve cells whose fibres are bundled in the **optic nerve**. The light rays that enter the eye have to pass first through these layers before reaching the receptor cells.
The optic nerve penetrates[5] the sclera and carries the visual information from the eye to the brain (see Figure 6) in the form of electric impulses.

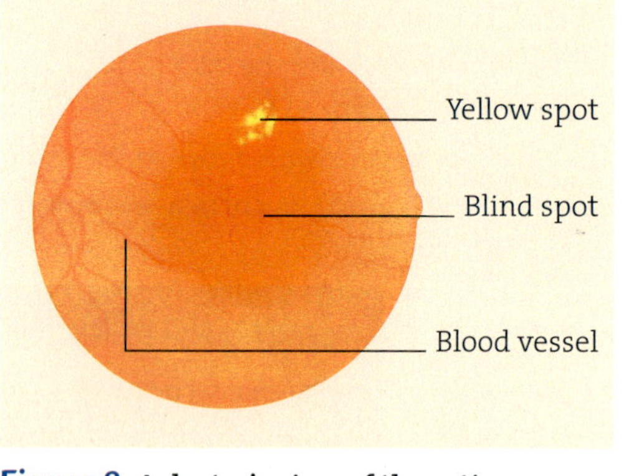

Figure 8 **A doctor's view of the retina**

The **yellow spot**, an area in the centre of the retina, is densely packed with **cones**, which are receptors responsible for colour vision. Because this area gives us the clearest vision, we always move our eyes in such a way that the image we want to see clearly is produced in just this area. Further away from the yellow spot there are fewer cones and more **rods**, receptors which have a higher sensitivity to light but can only distinguish between black and white. Where the optic nerve and blood vessels enter the eyeball, there are no light receptors at all. This area is called the **blind spot**.
When you look at an object, for example a tree or a person, the light rays are reflected from the object and enter your eye. They pass through the cornea and lens, where they are bent to the

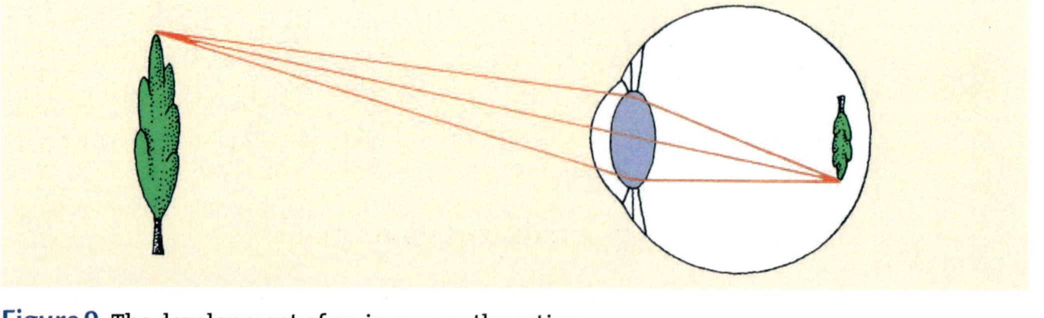

Figure 9 **The development of an image on the retina**

1 curved *gewölbt*
2 spoke *Speiche*
3 curvature *Wölbung*
4 light-sensitive *lichtempfindlich*
5 penetrate sth. *etwas durchdringen*

inside, so that they meet on the retina and produce an image of the object that is upside down and reduced in size.
However, we do not see our environment upside down, because in our brain the images we perceive are adjusted[1] and we see the objects in an upright position. This process of turning the images round by 180 degrees starts shortly after birth when a baby's eyes receive the first pictures from the outside world.

1 Make a table with the name, the description and the function of each part of the eye.
2 Which muscles in your eyes become active when you step out of the sunlight into the shade of a tree? Explain.
3 If you look at a faint[2] star in the sky at night, you can see it better if you look slightly to one side of it than straight at it. Explain why.

7.3 Your eyes can adjust to changing conditions

Adaptation of the pupils

When you open the front door of your home and step from the dark hall into the bright sunlight, it takes a short time for your eyes to adjust to the new situation. When you were in the dimly[3] lit hall, your pupils were wide open to let as much light in as possible, but now they become much smaller to protect the eye from damage from the high light intensity and to allow the retina to produce a clear image.
When you enter a dark room, on the other hand, the pupils dilate[4] automatically. This reflex action of adjustment by the pupil to varying light intensities is called **adaptation**. The size of the pupil is regulated by the circular and radial muscle fibres in the iris (see previous page). You can see these fibres when you look at the iris of your eyes in a magnifying[5] mirror.

Key terms

Functions of the eyes
- accommodation *Akkommodation*
- adaptation *Adaptation, Anpassung*
- field of vision *Gesichtsfeld*
- focus (on) sth. *etwas anvisieren, etwas scharf sehen/einstellen*
- stereoscopic vision *räumliches Sehen*

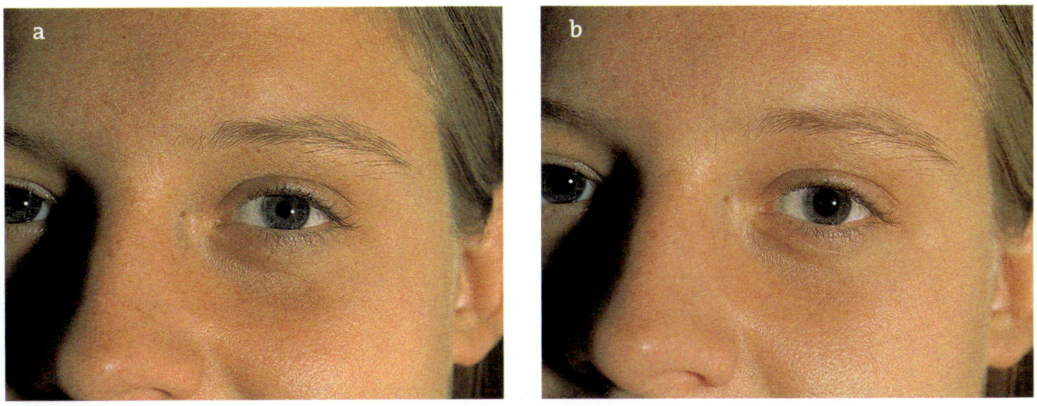

Figure 10 a and b **The eye of the same person when in the sun and when in a dark room**

Focusing on near and distant objects

When looking through a camera lens, near and distant objects cannot be seen clearly at the same time. It is the same with your eyes: you can only see an object clearly if the light rays

1 adjust sth. *etwas anpassen*
2 faint *schwach*
3 dim *dämmerig*
4 dilate *sich dehnen, sich erweitern*
5 magnify sth. *etwas vergrößern*

coming from a certain point of an object meet at the same point on the retina. The eye **focuses** on a near or distant object by changing the curvature of the lens. The lens in your eye is not rigid[1] but changes its shape so that the eye can adjust to near and distant objects. The ability of your eyes to do this is called **accommodation**.

When you focus on an object in the distance, the ciliary muscle relaxes, the suspensory ligaments tighten[2] and the lens is pulled into a flat shape. The light rays coming from the object meet in one place on the retina and you get a clear image. A nearby object cannot be seen clearly in this situation because the light rays coming from that object are only slightly bent by the lens and meet behind the retina. The retina can only produce a blurred image of the object.

When you focus on an nearby object, the ciliary muscle contracts, the suspensory ligaments loosen[3] and the lens takes on its natural shape, becoming rounder. With this shape the light rays are bent more and meet at the same point on the retina.

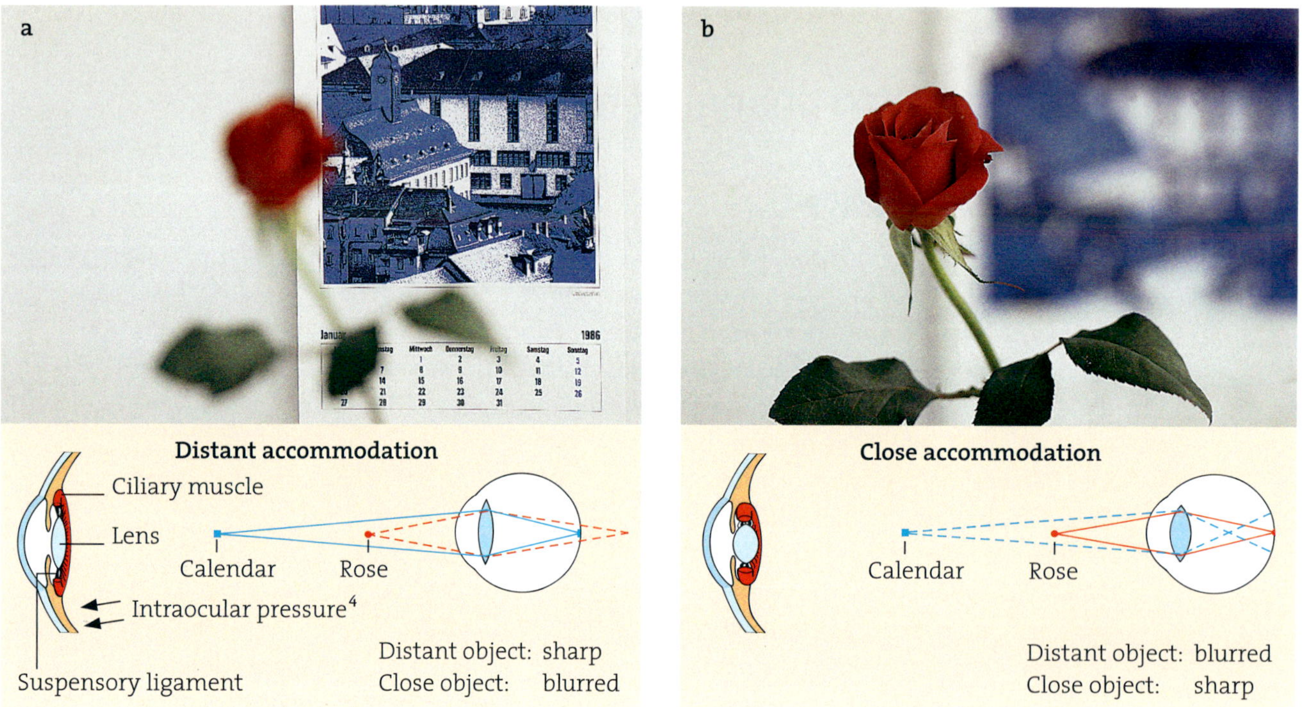

Figure 11 a and b How the lens changes shape when you focus on distant or near objects

Stereoscopic vision

As your eyes are situated side by side on your face, each of them gets a slightly different view of the object you are looking at. Two sets of impulses are sent to the brain, where the two images are combined to produce a three-dimensional impression[5] of the object. In certain situations, for instance under the influence of alcohol, the impressions from the two eyes cannot be correlated[6] properly and the person 'sees double'.

Stereoscopic vision helps you to reach for things and to safely judge[7] how far away objects are. For example, if you want to cross the road and a car is approaching in the distance, stereoscopic vision allows you to determine[8] roughly how far away the car is.

A disadvantage of the position of our eyes is our limited **field of vision** compared with animals, whose eyes are positioned on both sides of the head. The field of vision is the area that you can see when you hold your head still and only move your eyes from side to side.

1 rigid *rigide, unflexibel*
2 tighten *straffen*
3 loosen *sich lockern, sich lösen*
4 intraocular pressure *Augeninnendruck*
5 impression *Eindruck*
6 correlate *zueinander in Beziehung setzen*
7 judge sth. *etwas beurteilen*
8 determine sth. *etwas feststellen, bestimmen*

Figure 12 **Overlapping circles of colours are projected[1] onto a white wall**

Colour vision and the ability to see in dim light

As you have already learned, there are two types of photoreceptor cells in the retina: rods and cones (see Section 7.2). They both produce nerve impulses when they are stimulated by light. There are nearly 125 million rods compared with between six and seven million cones in a single retina. Rods are very sensitive to light and can detect shades of grey, even in dim light. But they cannot distinguish colours, for which the cones are responsible. For this function cones need more light than rods. This is the reason why in dim light we can see objects and movements but not colours. All objects appear in shades of grey.
There are three different types of cones, which react to different wavelengths of visible light: one type is more sensitive to red light, the second to green light and the third to blue light. Depending on the degree of stimulation, the three types of cones send out smaller or larger numbers of nerve impulses. The impression we have of the colour of an object is the result of the processing of the nerve impulses coming from the different cones.

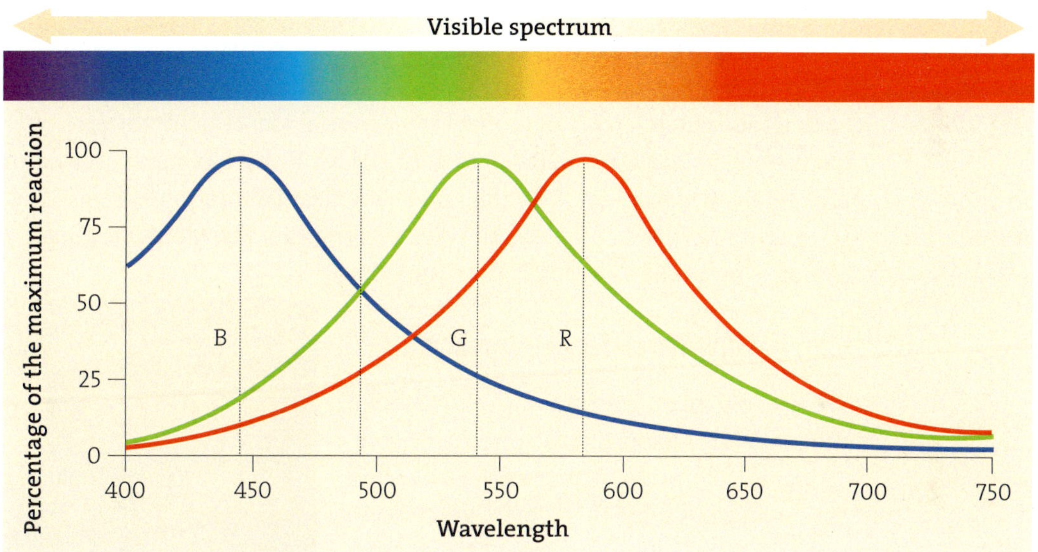

Figure 13 **The relative sensitivity of the three types of cones to the different wavelengths of sunlight**

This happens first in the different layers of nerve cells in the retina and then in the brain. As you can see in Figure 13, the sensitivity of the three types of cones for the wavelengths of light partly overlaps. So the colour we actually see depends on the degree of stimulation of the different types of cones. For example, if the cones for red and green are equally stimulated, you have the impression of seeing the colour yellow. An empty page appears to be white because it reflects the wavelengths of green as well as red and blue light. All three types of cones are activated equally, producing the impression of white (see Figure 12). If no light is reflected, we have the impression of black.

1 project sth. onto sth. *etwas an etwas projizieren*

1 Describe how the eye reacts to bright light and dim light.
2 Use Figure 13 to determine which cones in the retina are stimulated when we have the impression of light blue and orange.
3 Explain why it is relaxing to watch clouds drifting by in the sky, but after reading something for a while (a book, etc.) your eyes feel tired.
4 **Say it in German:** Work with a partner. You have just told a friend that you now understand how our eyes can see things clearly in both bright as well as dim light and how we can focus on objects both nearby and far away. He or she wants to know how. Explain.

7.4 You see with your eyes and your brain

What you actually see depends not only on your eyes, but also on your brain. When nerve impulses from your eyes reach your brain via the optic nerve, the brain turns them into an image. The brain influences the way we perceive things in many ways.

When you look at Figure 14, you will have no problem recognizing a cube[1], although we can actually only see a few black lines going in horizontal, vertical and diagonal directions. So why do we all agree that the drawing shows a cube? When the cornea and the lens produce an image of the black lines and the cream background on your retina, the receptor cells are activated to a different extent and electric impulses run along the sensory neurons in the optic nerve to the visual region of the brain (see Figure 4, p. 116). The image 'black lines on a cream background' is quickly compared with former visual experiences in our memory and is then recognized as a cube.

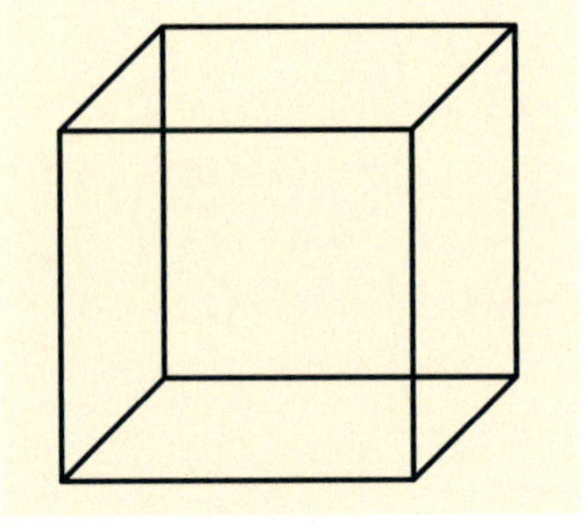

Figure 14 **A cube**

Figure 15 **Only black spots?**

Looking at Figure 15, you will probably recognize a rider on a motorbike. In reality, however, the picture only shows black spots in irregular shapes. If you had never seen someone on a motorbike before, you would not be able to see any clear picture. In a case like this your brain fills in the missing parts and completes the picture using your knowledge from past experience of how someone on a motorbike looks from the side.

Can we always trust our eyes?

We all know that there is no way you can hold the sun in your hand. But the photo in Figure 16 seems to show that it is possible. Your eyes have sent information to the brain which contradicts[2] all previous[3] experience. This is called an optical illusion[4]. Such illusions trick us into perceiving something other than what actually exists. What we see does not correspond[5] to physical reality.

Figure 16 **Taking the sun into your own hand**

Figure 17 **A similar sequence of pictures is produced on your retina.**

Perceiving movements

Your eye can only distinguish between a certain number of images that the lens projects onto the retina. If more than 18 pictures per second fall onto the retina, they fuse[6] into a stream of images and your brain receives the impression of a moving object. In Figure 17 this is demonstrated in the movement of a beach footballer, which is split into a sequence of single phases. Cartoons, flip books[7], computer games and movies make use of moving pictures in a similar way.

1 cube *Würfel*
2 contradict sb./sth. *jdn./etwas widersprechen*
3 previous *vorhergehend*
4 optical illusion *optische Täuschung*
5 correspond to sth. *übereinstimmen mit etwas*
6 fuse *miteinander verschmelzen*
7 flip book *Daumenkino*

1 Working in small groups, follow these steps:
  a Look closely at Figures 18–21 below and describe what you see in each picture.
  b Discuss your ideas about why your brain interprets the pictures the way it does. The **Activate your English** box and the following impulses will help you:
  - Figure 19: Cover one corner of the triangular object with your thumb. Now lift your thumb and then lower it again. Describe what happens. What is wrong with this triangle?
  - Figure 20: What happens when you look at the picture for a while after you have realized what it shows?
  - Figure 21: Estimate the height (on the page) of both dogs. Then take a ruler and compare the measured size with your estimate. Explain the phenomenon[1].

2 **Say it in German:** When Julian went home and showed his younger sister the optical illusions in Figures 18–21, she was fascinated. She wanted to know how our brain can be tricked into seeing things this way. Write down in German what he told her.

3 Activity: Make your own flipbook using an exercise book or note pad and a pen.
  a Start on the last page and draw a person, animal or other moving object.
  b Going from one page to the next, draw your object again, changing its place on the page as well as the position of any moving parts, e. g. the arms and legs, very slightly.
  c When you have filled at least 10 or 12 pages, run your thumb along the open side of the exercise book from bottom to top. Your drawings will appear to come to life.

Activate your English

Interpreting optical illusions

The … confuse(s)/irritate(s) your brain because …
Your brain tries to complete the … whenever part of it becomes visible
Your eyes jump from one (object) to the other because …
Because you are used to seeing an object against a certain background, your brain …

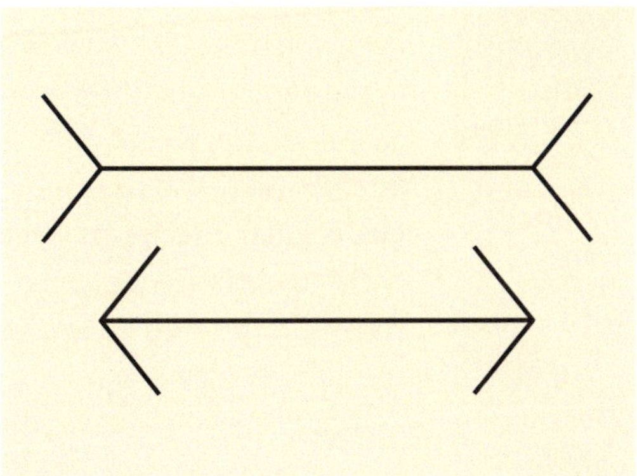

Figure 18 Compare the lengths

Figure 19 The impossible triangle

Figure 20 One object or two?

Figure 21 Bigger, smaller or the same size?

1 phenomenon. *Phänomen, Erscheinung*

7.5 Take care of your eyes

Figure 22 **Special ski goggles protect your eyes against UV radiation.**

There are many ways in which you can protect your eyes and keep them healthy.

Avoid eye strain[1]. You can do this by making sure that there is adequate light when you read, study or use the computer. Look up from your books or your computer from time to time and relax your eyes by looking at distant objects. Keep a safety distance of 50 cm from the monitor. Wear glasses or contact lenses if they have been prescribed[2] for you.

Avoid looking directly into bright light. This includes the sun or any other bright light, for instance the flash of a photocopier. Ultraviolet radiation[3], which is especially intense in mountainous regions, is dangerous for your eyes. Use good sunglasses or a hat with a visor[4] or large brim[5] to protect your eyes from the rays of the sun. For skiing, ski goggles[6] with UV filters to protect your eyes against the dangers of the harsh ultraviolet light are essential.

Figure 23 **Wearing safety goggles during a laboratory experiment**

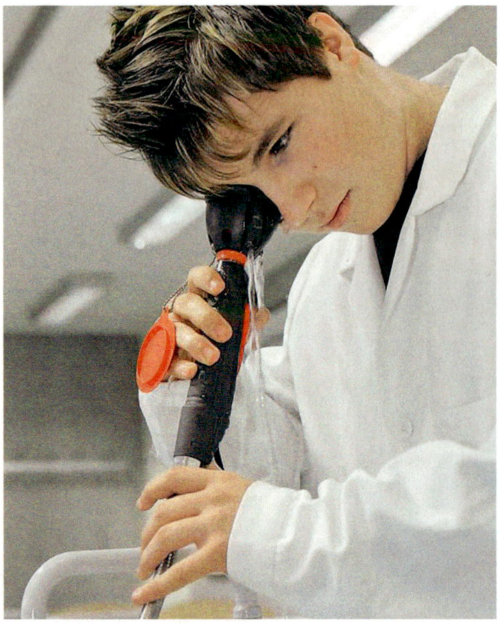

Figure 24 **Using a laboratory eye wash**

Protect your eyes against chemical substances and small objects. If you work with dangerous substances in school or at home, and if you use any tools or machinery which might throw particles of wood, metal, glass or stone into the air, make sure you use protective glasses or safety goggles[7].

If you get hit in the eye by a ball or any other fast moving object, the result is usually a 'black eye', which is caused when the blood vessels in the area around the eye are injured. In such a case, an eye doctor must decide if there are internal injuries to the eye.

Get regular check-ups. Regular eye examinations[8] are important and can prevent or detect many eye problems. An eye specialist looks through the pupils with a special instrument and can detect changes in the blood vessels of the retina, which are often warning signs of general health problems and illnesses.

What to do if an irritating chemical gets into your eye:

1. Hold the eye open and flush[9] it with water (away from the nose) for at least 15 minutes. If possible use an eye wash[10].
2. Make sure that the chemical does not flow into the other eye.
3. Cover the eye with a clean, dry tissue.
4. Visit an eye doctor and tell them the name of the substance that got into your eye.

1 eye strain *Augenbelastung*
2 prescribe sth. *etwas verschreiben*
3 ultraviolet radiation *ultraviolette Strahlung*
4 visor *Schutzschirm, Blendschutz*
5 brim *Krempe*
6 ski goggles *Skibrille*
7 safety goggles *Schutzbrille*
8 examination *Untersuchung*
9 flush sth. *etwas spülen*
10 eye wash *Augendusche*

What to do if a small object gets into your eye:

1. Do not rub the eye.
2. If the object is under the upper eyelid, gently pull the lid away from the eyeball and down over the lower lid. Hold it in this position. The freely flowing tears will wash the object out of the eye.
3. An object under the lower lid can be removed by pulling down the lower lid while standing in front of a mirror. Using the moist edge of a clean handkerchief[1], try to brush the particle out of the eye.
4. If the eye is injured and hurts, cover it with a loose, sterile[2] bandage[3] and see a doctor as soon as possible.

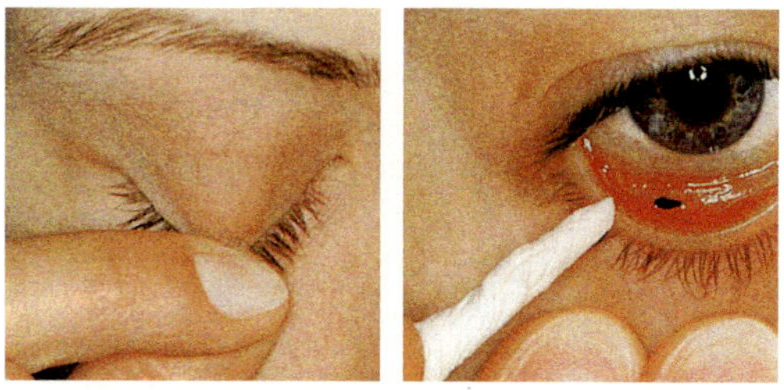

Figure 25 Removing an object from the eye

1 Name three ways you can keep your eyes healthy while working on the computer.
2 Why can cheap sunglasses be dangerous for your eyes?
3 Make a poster explaining safety regulations for the eyes, and display it in the school biology lab or in your classroom.

7.6 Disorders of the eyes can be treated[4]

Key terms

Eye disorders

- astigmatism *Astigmatismus, Stabsichtigkeit*
- cataract *grauer Star*
- concave lens *Zerstreuungslinse*
- convex lens *Sammellinse*
- disorder *Funktionsstörung*
- glaucoma *grüner Star*
- intra-ocular pressure (IOP) *Augeninnendruck, Intraokulärer Druck (IOD)*
- long-sighted *weitsichtig*
- long-sightedness, far-sightedness *Weitsichtigkeit*
- night blindness *Nachtblindheit*
- red-green colour blindness *Rot-Grün-Blindheit*
- short-sighted *kurzsichtig*
- short-sightedness, near-sightedness *Kurzsichtigkeit*

Many people have eye problems of one sort or another. People with **short-sightedness** can see things that are close to them clearly, but have problems focussing on objects that are further away. **Short-sighted** people have an eyeball that is too long. The light rays meet before they reach the retina. Short-sightedness can be corrected by contact lenses or glasses with **concave lenses** that bend[5] the light rays outwards before they reach the eye.

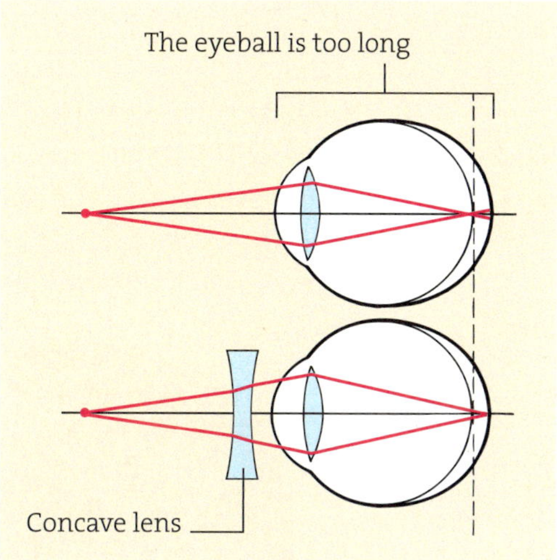

Figure 26 Short-sightedness and how it can be corrected

1 handkerchief *Taschentuch*
2 sterile *keimfrei, steril*
3 bandage *Verband*
4 treat sth. *etwas behandeln*
5 bend sth. *etwas beugen*

Long-sighted people can see things clearly that are far away, but have problems when trying to focus on objects that are nearby. The light rays meet behind the retina because either the eyeball is shorter than normal or the lens doesn't bend the light rays enough. **Long-sightedness** is corrected by glasses or contact lenses with **convex lenses**. These bend the light rays inwards so that they meet directly on the retina.

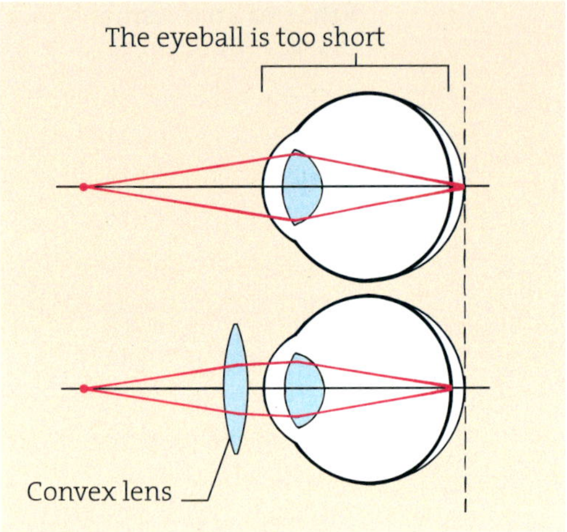

Figure 27 Long-sightedness and how it can be corrected

Have you ever noticed how older people often have to wear glasses for reading? This is because the lens has lost some of its elasticity and does not become as round as it should when the ciliary muscle contracts and the suspensory ligaments loosen their grip on the lens (see p. 120). Distant objects can still be seen clearly, but reading causes problems. Only by wearing glasses with convex lenses can many older people focus on nearby objects.

Figure 28 A test picture for detecting astigmatism

The most common eye **disorder** is **astigmatism**, which occurs when the curvature of the cornea or the lens is uneven. If, for instance, the curvature from top to bottom is greater than that from side to side, the eye cannot clearly see horizontal and vertical lines at the same distance from the eye. Astigmatism is often accompanied by short-sightedness or long-sightedness. It can be corrected by wearing cylindrical[1] glasses or contact lenses, or by surgery.

Uncorrected astigmatism can give you headaches or cause eye strain and distort[2] or blur your vision. A recent study carried out at a school in the US state of Ohio found that more than 28 per cent of the investigated students had astigmatism. As this defect can affect childrens' ability to see well in school and during sports, it is important that not only adults but also children have their eyes examined[3] regularly for astigmatism.

More than 8 per cent of people in Western Europe have problems identifying certain colours – in other words, they are colour blind. In rare cases colours cannot be seen at all, so everything looks black and white or shades of grey. A more common condition[4] is the inability to distinguish between red and green. **Red–green colour blindness** is inherited[5] and occurs in 4 per cent of men, but in only 0.4 per cent of women. Test pictures can be used to detect red-green blindness (see Figure 29).

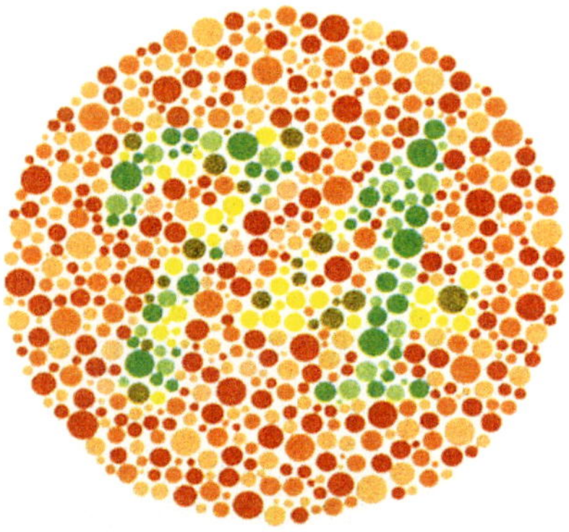

Figure 29 A test picture for detecting red-green blindness

Night blindness is the inability to see well in dim light. The reason can be that the rods in the retina are not functioning properly. Night blindness can be either inherited or caused by lack of vitamin A. In some cases it is an early sign of other diseases of the retina.

1 cylindrical *zylinderförmig*
2 distort sth. *etwas verzerren*
3 examine sb./sth. *jdn./etwas untersuchen*
4 condition *(hier) Leiden*
5 inherit sth. *etwas erben*

A **cataract** is a disease of the eye which is very common in older people, but in persons with diabetes[1] it can occur at any age. When a cataract develops, the lens gradually becomes clouded[2]. Surgery is the only effective treatment[3]. The cloudy lens must be removed and replaced with an artificial[4] lens. Exposure to intense sunlight over the years can contribute to the development of a cataract, so you should protect your eyes by wearing sunglasses.
A more serious disease is **glaucoma**. The sensory cells in the centre of the retina die off gradually. This disease leads to blindness if it is not treated. Increased pressure in the fluid between the cornea and lens, referred to as **intra-ocular pressure (IOP)** (see Figure 11a), is a main cause of glaucoma. It damages the fibres in the retina and the optic nerve, resulting in a gradual but permanent loss of vision. However, early detection by an eye doctor and immediate treatment can slow or even stop the progression of the disease.

1 Explain why a short-sighted person has blurred vision when looking at distant objects.
2 What kind of glasses must a long-sighted person wear to be able to read without difficulty? Explain.
3 Why do you think a person with red-green colour blindness is allowed to drive a car?
4 Explain why it is important to have your eyes checked regularly.

7.7 Hearing is only one function of your ears

In contrast to your eyes that have one vital function, your ears contain organs of hearing as well as organs of balance. This section looks at the part of your ear concerned with hearing.
When you strike a tuning fork[5], it starts to vibrate and sends out sound waves which travel through the air. When they reach your ear fractions of a second[6] later, the **pinna** directs the sound into the **ear canal**. The sound waves travel through this passage to the **eardrum**, which separates the **outer ear** from the **middle ear**.
When they hit the ear drum, the membrane of the ear drum starts to vibrate. This vibration moves the three **ear ossicles** in the middle ear chamber forwards and backwards.
The ear ossicles – the hammer, anvil[7] and stirrup[8] – are arranged in such a way that they amplify[9] the vibrations and transfer them onto the elastic membrane of the **oval window**. This is the entrance to the inner ear. The **inner ear** consists of the **cochlea**, the hearing part of the ear, and the **vestibular apparatus**, which helps us to keep our balance and informs us about the position and movements of the head. The inner ear is safely embedded in a bony cavity[10] of the skull.

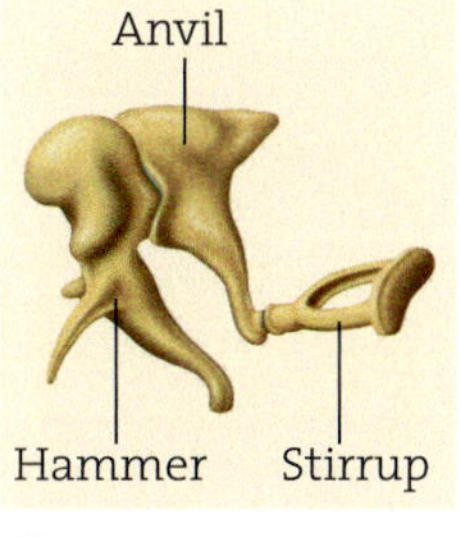

Figure 30 The ear ossicles

1 diabetes *Zuckerkrankheit*
2 clouded *trüb, unklar*
3 treatment *Behandlung*
4 artificial *künstlich*
5 tuning fork *Stimmgabel*
6 fractions of a second *Bruchteile einer Sekunde*
7 anvil *Amboss*
8 stirrup *Steigbügel*
9 amplify sth. *etwas verstärken*
10 cavity *Höhle, Raum*

Did you know?

The ear ossicles are the smallest bones in the human body. All three of them would fit inside a pea.

Key terms

Outer ear *äußeres Ohr*
- ear canal (auditory canal) *Gehörgang*
- eardrum *Trommelfell*
- earwax *Ohrenschmalz*
- pinna (auricle) *Ohrmuschel*

Middle ear *Mittelohr*
- ear ossicles *Hörknöchelchen*
- Eustachian tube *Ohrtrompete*
- oval window *ovales Fenster*

Inner ear *Innenohr*
- auditory nerve *Gehörnerv*
- basilar membrane *Grundmembran*
- cochlea *Schnecke*
- cochlear fluid *Flüssigkeit in den Schneckengängen*
- median canal *Schneckengang*
- round window *rundes Fenster*
- tectorial membrane *Deckmembran*
- vestibular apparatus *Vestibularapparat*

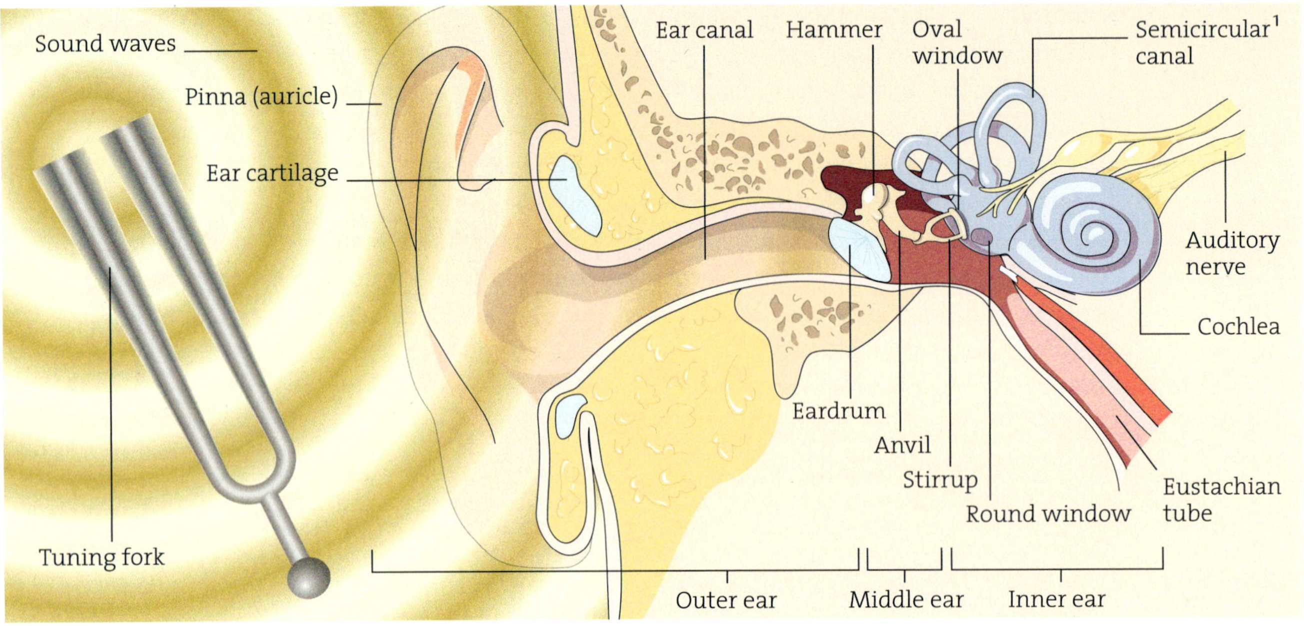

Figure 31 The structure of the ear

The vibrations of the oval window send waves through the fluid-filled canals of the cochlea, which are coiled[2] like the shell of a snail[3]. As you can see in Figure 32, the inside of the cochlea is divided lengthwise by two membranes into an upper and a lower canal with the **median canal** in between. On their way through the upper and lower canals, the waves cause the **basilar membrane** in the median canal to vibrate. Special receptor cells, so called 'hair cells', attached to this membrane are connected to the **tectorial membrane** by thin hairs (see Figure 33). Any vibration of the basilar membrane stimulates the receptor cells, which send off impulses that run along the nerve fibres of the **auditory nerve** to the auditory region in the cerebral cortex of the brain (see Figure 4, p. 116). It is here that the sound of the tuning fork is recognized.

The pressure which develops in the **cochlear fluid** when the waves run along the canals is taken up by the **round window**, which bulges[4] slightly in and out. The **Eustachian tube** connects the middle ear with the cavity of the pharynx. Most of the time the tube is closed, but when you

Did you know?

Although the cochlea is about the size of a pea, it contains around 24,000 sensitive hair cells.

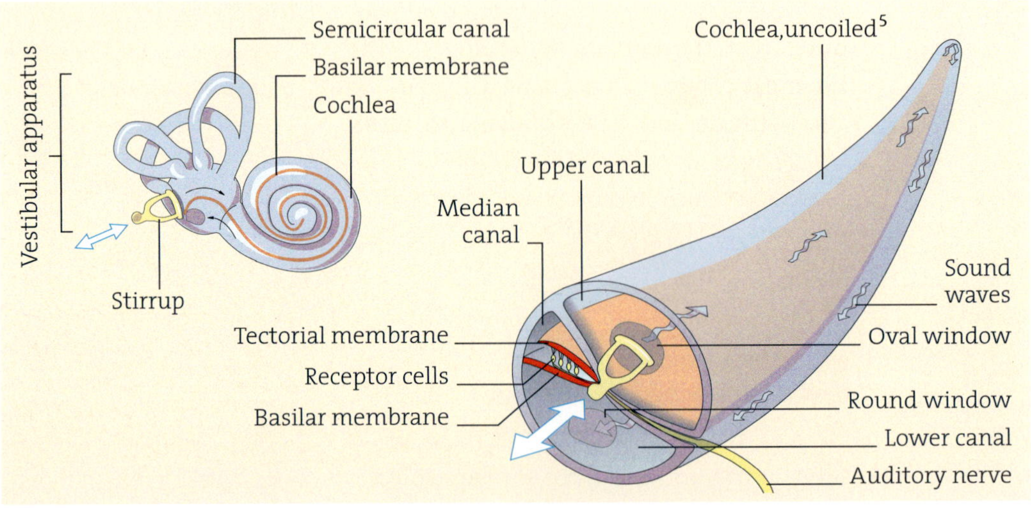

Figure 32 The structure and function of the cochlea

1 semicircular *halbkreisförmig*
2 coiled *aufgerollt*
3 snail *Schnecke*
4 bulge *ausbeulen*
5 uncoiled *abgerollt*

yawn, swallow or cough, it opens and pressure differences between the middle ear and the outside are equalized[1].

The louder a sound is, the stronger are the vibrations of the **basilar membrane** and the more impulses are sent along the auditory nerve to the brain.

Earwax is produced by glands in the wall of the ear canal. The wax protects the ears by catching dust particles before they reach the ear drum. It is transported towards the opening of the ear automatically when you move your jaws. If you are healthy, there is no need to clean out the auditory canal, for instance by using cotton sticks[2]. With cotton sticks you only push the wax towards the ear drum. If a large piece of wax is blocking the auditory canal and affects your hearing, see your doctor. He or she will remove the wax by rinsing out the ear.

Figure 33 Groups of receptor cells in the cochlea (coloured electron micrograph[3])

Perceiving the direction of sound

When sound waves reach your left ear just fractions of a second before your right ear, the nerve impulses from your left ear arrive in the auditory region of your brain slightly earlier than those from your right ear. From this small difference in time your brain can determine that the sound came from the left. The ability to recognize which direction a sound comes from is important in everyday life, for instance in a discussion with several people involved, or when walking or cycling along a busy road.

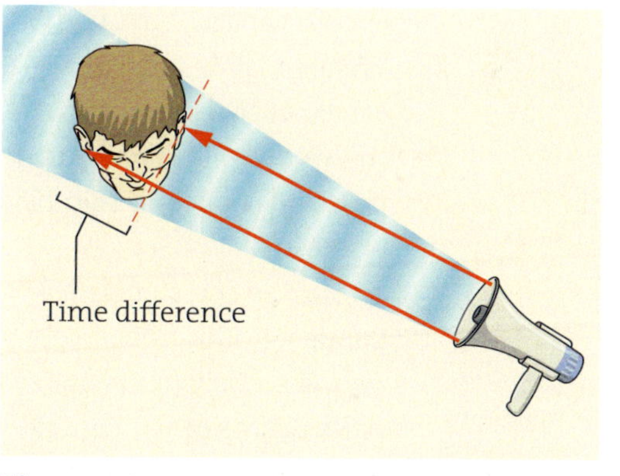

Figure 34 Recognizing the direction of a sound

1 equalize sth. *etwas ausgleichen*
2 cotton stick *Wattestäbchen*
3 electron micrograph *elektronenmikroskopisches Foto*
4 inflammation *Entzündung*

1 Make a table with two columns: one for the different parts of your outer ear, middle ear and inner ear, and one for their functions. Complete the table using the information in this section.

2 A bacterial infection of the nose-throat region can lead to inflammation[4] of the middle ear. This kind of inflammation can be very painful and can spread to the ear ossicles if it is not treated.

a Explain how bacteria can get into the middle-ear cavity.

b Speculate about the consequences of the eardrum or the ear ossicles getting infected.

3 Think of other everyday situations in which it is important to know the direction from which a sound comes. Discuss them in class.

7.8 Your sense of balance lies in the inner ear

The vestibular apparatus is located[1] close to the cochlea in the inner ear. It consists of three **semicircular canals** that are filled with a fluid, **endolymph**, and two saclike[2] structures, the **utricle** and **saccule**.

Each semicircular canal lies at a right angle to the other two and ends with a bulblike[3] extension, the **ampulla**, which is connected to the utricle. Each ampulla contains a group of receptor cells with tiny hairs held together by a jellylike[4] **cupula**, which sticks out into the endolymph (see Figure 35). When you turn your head or your whole body, the fluid in at least one of the semicircular canals starts to flow, bending the cupula and its hairs and stimulating the receptor cells, which send electric impulses to the brain along the **vestibular nerve**. By comparing the messages from the different semicircular canals, your brain can tell exactly how your head is moving.

We are mainly used to horizontal movements in daily life, which involves certain semicircular canals, but we are not used to vertical movements along the body axis, which sets the fluid in the vertical semicircular canal in motion. When such movements occur, for instance in a fast-moving lift or on a ship pitching[5] in rough seas, we may experience motion sickness[6] or sea sickness[7].

The utricle and saccule help us to know the position of our head and to keep our balance. They contain groups of hair cells positioned horizontally and vertically whose tips are enclosed in a cupula containing crystallized[8] minerals (see Figure 36). These are pulled down by gravity[9] and press against other hair cells which, when stimulated, send impulses along sensory nerve fibres to the brain.

Key terms

Balance and the inner ear
- ampulla *Ampulle*
- cupula *Cupula*
- endolymph *Endolymphe*
- saccule *kleines Vorhofsäckchen, Sacculus*
- semicircular canals *Bogengänge*
- utricle *großes Vorhofsäckchen, Utriculus*
- vestibular nerve *Vestibularnerv*

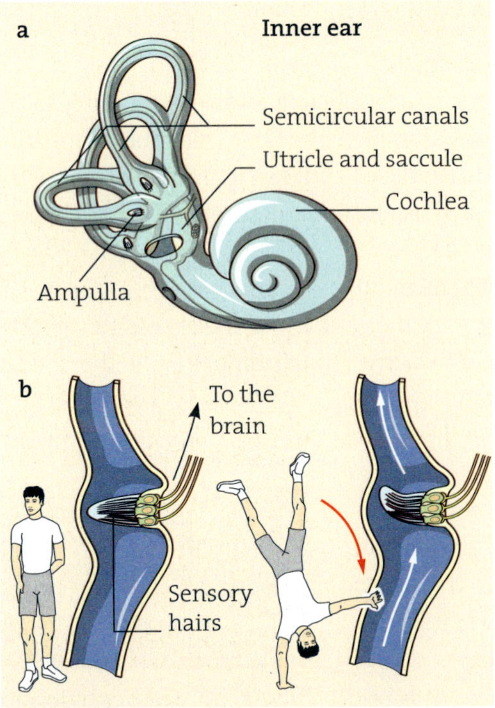

Figure 35 How our sense of balance functions

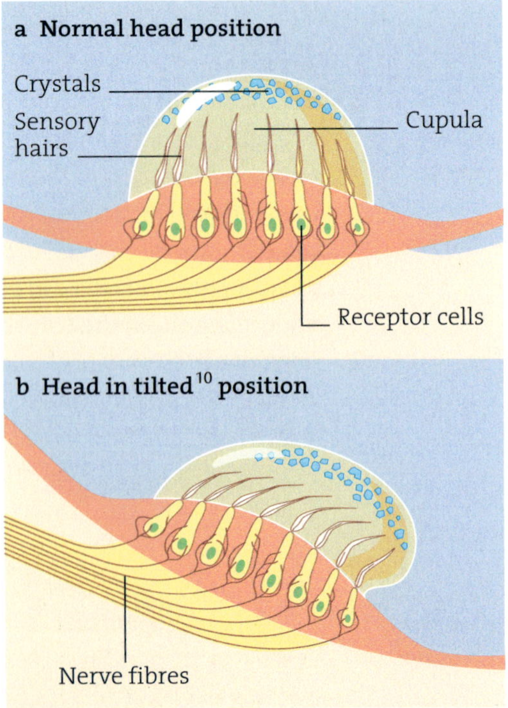

Figure 36 The function of the organs in the utricle and saccule

1 be located *sich befinden*
2 saclike *sackartig*
3 bulblike *knollenartig*
4 jellylike *geleeartig*
5 pitch *(bei einem Schiff) sich auf und ab bewegen*
6 motion sickness *Reisekrankheit*
7 sea sickness *Seekrankheit*
8 crystallize *kristallisieren*
9 gravity *Schwerkraft*
10 tilted *geneigt*

When you change the position of your head, different hair cells are stimulated. In this way you are able to perceive the direction of gravity, which is important for your sense of orientation and balance, for example when walking or turning around.

1 Describe the types of information that we receive from the different parts of our ear.
2 Describe how your brain is informed when you move your head.
3 Why do you think people who experience seasickness on a moving boat are often less affected when they lie down?

7.9 Noise can make you ill

The boy on the train is listening to his favourite music and has turned the volume[1] on his MP3 player all the way up. He does not realize that what he is doing is as dangerous for his ears as if he were standing beside a road worker using a jackhammer[2] without any ear protection. For your ears it makes no difference whether it is music or any other kind of sound. Loud or continuous noise can damage your sense of hearing permanently.

Figure 37 **If other people can hear the music you are listening to, your MP3 player is turned up too loud.**

Key terms

Sound and noise
- decibel (dB) *Dezibel (dB)*
- expose sb./sth. to sth. *jdn./etwas einer Sache aussetzen*
- exposure *Belastung, Einwirkung*
- noise/sound level *Geräuschpegel*
- sound energy *Schallenergie*
- sound intensity, loudness *Lautstärke*
- threshold of hearing *Hörgrenze*
- threshold of noise pain *Schmerzgrenze*

High **sound energy**, which affects the sensitive receptor cells in the cochlea, can destroy the cells forever. Figure 39 on the next page demonstrates the different **sound levels**, from the rustling of leaves to a starting rocket. The **intensity** or **loudness** of a sound is measured in **decibels (dB)**. The quietest sound the human ear can detect is called the **threshold of hearing** and is given the value[3] of zero decibels, whereas a sound level of about 120 decibels is the **threshold of noise pain** for the human body. Besides the power of the sound, the length[4] of **exposure** is important for the effect which sound has on your ears. If your ears are **exposed** to any noise at or above about 85 to 90 decibels for periods longer than 15 minutes, you risk gradual[5] loss of

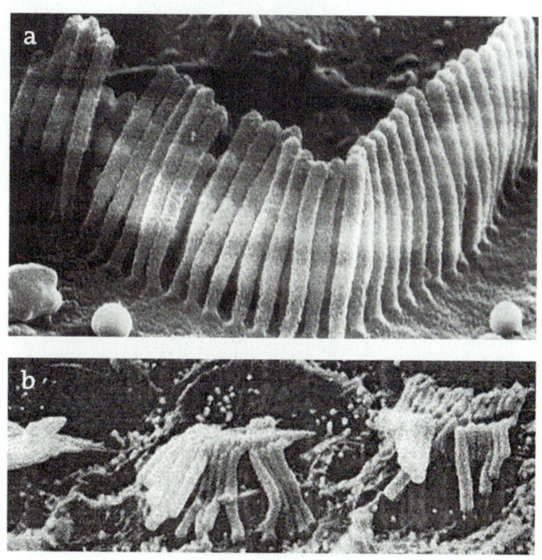

Figure 38 **Sensory hairs of receptor cells in the cochlea: healthy (a) and partly destroyed (b)**

1 volume *Lautstärke*
2 jackhammer *Pressluft-hammer*
3 value *Wert*
4 length *(hier) Dauer*
5 gradual *allmählich*

hearing. People who use or work near loud machines like lawn mowers, electric saws or motorcycles, have to wear ear plugs[1] or earmuffs[2] for protection. Continuous noise can not only damage your sense of hearing, but also affects different functions of your body (see Figure 40).

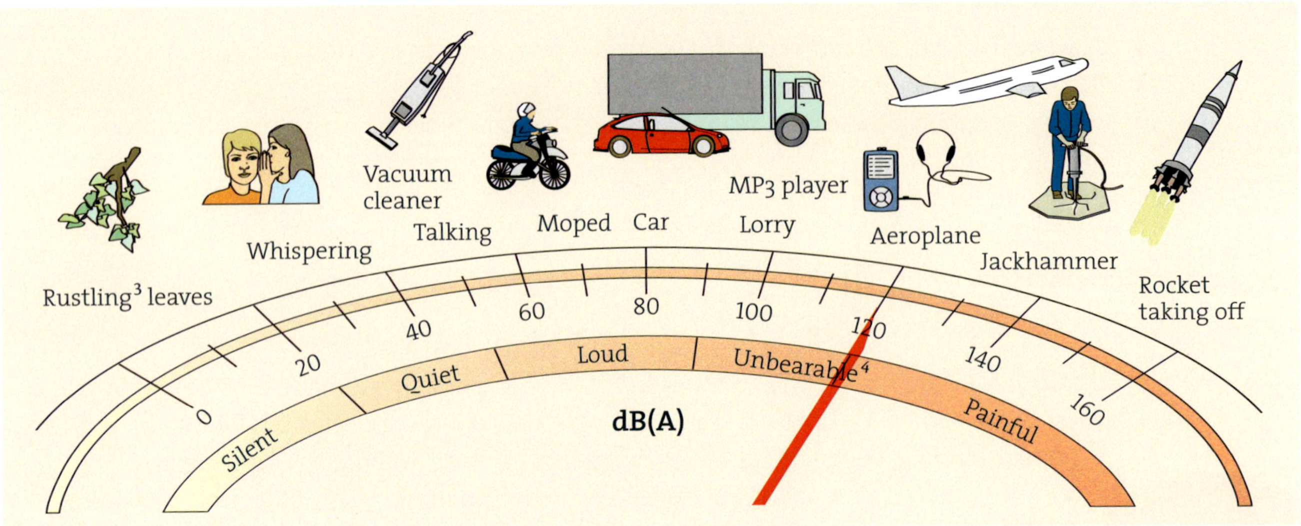

Figure 39 Different sound levels on the decibel (A) scale, written as dB(A)

Figure 40 The effects of noise on your whole body

1 ear plug *Ohrstöpsel*
2 earmuffs *Ohrenschützer*
3 rustle *rascheln*
4 unbearable *unerträglich*
5 short-term *kurzfristig*
6 nausea *Übelkeit*

1 Name some everyday situations in which your body is exposed to different noise levels. Figure 39 will help you.
2 Look at Figure 40 and write a short text describing the different effects which noise can have on a person's body. Write in full sentences and use as many verbs as possible.
3 What possible consequences can a gradual loss of hearing have for the affected individual? Discuss in class.
4 In small groups, discuss why people in today's society are perhaps more exposed to dangerous noise levels than ever before. Make a list of ways in which people can be protected and can protect themselves from high noise levels. Present your results in class.

Making an educational poster

In biology you often have to present information to other people. One way of doing this is to make a poster – real scientists often do this at conferences where there are many people. Here are some points to consider when presenting the results of group work or of a report in the form of a poster:

1. **Materials:** a large piece of white or coloured cardboard, glue, a pair of scissors, coloured pencils or felt-tip pens of different thickness, and drawing pins.
2. **Collecting information:** Look through books, magazines and the Internet for suitable information and pictures for your poster. Avoid copying texts. Instead write down the most important information in your own words.
3. **Using visual information:** Photos, drawings, graphs and other picture material, which you can also make yourself, are important for presenting your topic in an attractive and effective way; but be careful not to use too many pictures, as this can confuse the viewer. Only use pictures that help people understand what you are trying to say. If you use pictures from books or the Internet, do not forget to name the sources.
4. **Writing texts:** Use short texts and do not give too many details. The writing should be easy to read.
5. **Arranging texts and pictures:** Make sure that your information is presented clearly on the cardboard. It should be easy to read the texts in their proper order and find the pictures which go with them.
   The topic of the poster should be written in bold letters. Subtitles are useful for structuring the texts. Each text should have a heading.
6. **Deciding on the final form of the poster:** Before you glue the texts and pictures to the cardboard, first position them loosely and check to see if any changes are necessary.

1 Get together in small groups and follow the steps above to make an educational poster on one of the biology topics you have learned about in this chapter. Here are some suggestions:
- How can we keep our eyes and ears healthy and safe from dangerous influences like dirt, chemicals or noise?
- What possible defects and diseases threaten our eyes and ears, and how can they be treated?

Figure 41 An example of a poster

Activity 1

Making a model eye

Build a model as shown in Figure 42 below.

1. Compare the different parts of your model with those of the human eye.
2. Adjust the image distance[1] in such a way that you get a clear picture of the candle on the screen[2]. What is the position of the candle in the image?
3. Place the candle further away from the screen. How does the image change?
4. Use a lens with a greater or smaller curvature. When does the image improve?

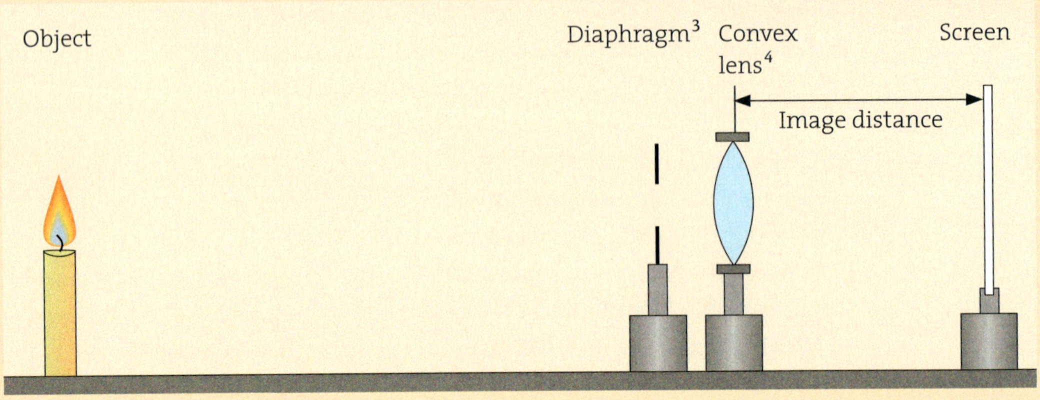

Figure 42 A model eye

Activity 2

Testing the capacities and limitations of the eye and ear

1 image distance *Bildweite*
2 screen *Schirm*
3 diaphragm *Blende*
4 convex lens *Sammellinse*
5 be in / out of focus *scharf / nicht scharf eingestellt sein*
6 blurry/fuzzy *unklar, undeutlich, verschwommen*

A Detect your blind spot

Figure 43 Where is the mouse?

1. Hold your book about 60 cm away from you.
2. Close the right eye and concentrate on the figure of the cat (above) with your left eye.
3. Slowly bring the book closer to your face.

1 Describe what happens in the experiment. For language help see **Activate your English** on the right.
2 Explain your observation.

4. Now repeat the experiment with both eyes open.

1 What happens this time?
2 How would you explain the difference?

Activate your English

Talking about what you see

At first I have no problems seeing/ recognizing ...

When I move the ... closer to / nearer to / further away from ..., the ... is/goes suddenly out of focus / comes into focus[5]

The ... is/gets/becomes bigger/smaller/ sharper/blurrier/fuzzier[6]

The reason for this (effect) is (that) ...

The difference between the two / the first and second time is that ...

B Determine[1] the near point of your eyes

Follow these steps together with a partner:

1. Close your left eye and hold a pencil upright in your right hand while keeping your arm stretched.
2. Move the pencil slowly towards you and stop when the tip of the pencil is in sharp focus.
3. Your classmate now uses a ruler to measure the distance between your right eye and the pencil.
4. Follow steps 1–3 to determine the near point of your left eye.

Figure 44 **Determining the near point**

Change places with your partner and have him or her carry out the same procedure.

1 Note down the results and compare the measurements (in cm) of everyone in the class. Determine the average.

2 Do you expect longer or shorter distances in short-sighted pupils when they take off their glasses? Explain.

C Demonstrate stereoscopic vision

1. Place a book on the table in an upright position about 30 cm away from you. Look at the spine[2] of the book first with your left eye, then with your right, then with both.
2. Hold two pencils in each hand at arm's length so that they point towards each other. They should be about 50 cm apart. Now close one eye and move the pencils closer together to make their points touch. Repeat the experiment with both eyes open. Explain the difference.

D Demonstrate the coordinated functioning of the eye muscles

1. Close one eye and keep the eyelid closed with your fingertip.
2. Now look around in all directions with your open eye.

1 What do you feel?

2 Why do you think both eyes should be bandaged if one eye is injured?

E Test the function of the auricles[3]

1. Place a loud ticking alarm clock or timer on a table a few metres away from you. Listen to the noise it makes.
2. Now observe and discuss how your perception of the sound changes when you carry out the following experiments:
   a) Use your thumbs to press the auricles against the sides of your head without covering the ear canals.
   b) Bend your auricles forward and hold your hands behind your ears, thus increasing the size of the auricles.
   c) Cover both auricles with your hands from the front.

Figure 45 **Listening with bent auricles**

1 How important are your auricles for hearing? Discuss.

2 Some animals can turn their auricles in different directions independently of each other. What advantage can this have for these animals?

1 determine sth. *etwas bestimmen*
2 spine (of a book) *Buchrücken*
3 auricle *Ohrmuschel*

F Judge[1] the direction a noise is coming from

Carry out this experiment with a partner.

Materials:
- rubber tube (4 metres long)
- wooden stick or pencil

1. One person sits on a chair with their back towards the table on which the rubber tube is lying.
2. The other person hands them the two ends of the tube, which they hold near their ear openings. The centre of the tube is left lying on the table.
3. The other person takes the stick or pencil and taps on the rubber tube directly in the middle.
4. Then they tap on the tube further away from the middle, first to the right, then to the left.

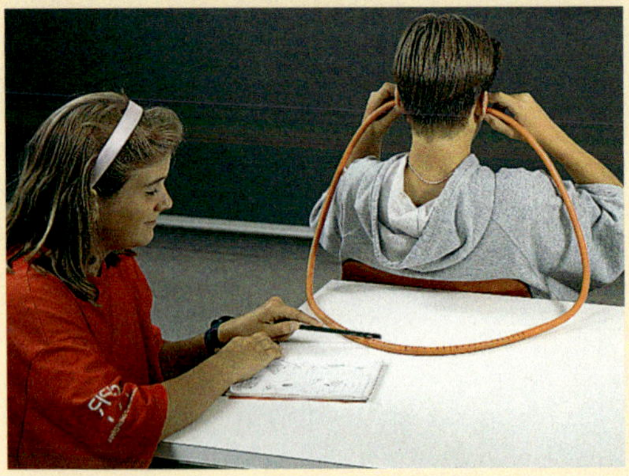

Figure 46 Experiment with a rubber tube

After each tap, the person listening in the tube must indicate where the tube was tapped (in the middle or on the right or left). Note down your conclusions and explain.

G Find out how your skull influences your sound perception

Carry out this experiment with a partner.

1. Close the openings to both of your ears with your fingers or with ear plugs.
2. When both ears are closed and you cannot hear anything, your partner strikes a tuning fork and places it on your head in the following position:
   a) next to your ear
   b) right on top of your head
   c) at the side or at the back of your head.
3. Give a sign as soon as you hear the sound.

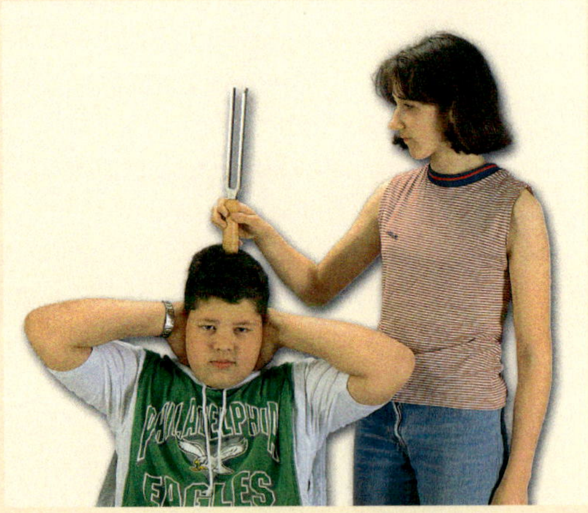

Figure 47 Hearing with closed ears

1 judge sth. *etwas beurteilen*

1 Describe how your perception changes when the tuning fork is placed in the different positions on your head.
2 Why do you think you can still hear the sound although your ears are completely closed?
3 You may have experienced that your voice sounds different to you when you hear it on a recording. What could be the reason for this?

H Keep a sound diary

Noise can be experienced quite differently depending on the specific situation in which it occurs and on your general attitude or mood. People have different degrees of sensibility towards noise.

1 Think of situations in which certain noises or sounds were unpleasant or bothered you. In each case:
   a Estimate the approximate sound level of the noise by using the diagram in Figure 39.
   b Note down what effect the noise had on your body (headache, aggressive feelings, nervousness, etc.).
   c Compare your experiences with those of your classmates.
2 Discuss how people can protect their ears from damage by noise.

Activity 3 Testing the skin's sensitivity to touch and temperature

A. Examine the sensitivity to touch in different areas of the skin

Carry out this experiment with a partner.

1. Blindfold[1] your partner and use a thin (water soluble) felt pen to draw a square of $1\,cm^2$ on the back of their hand and on the inside and outside of their forearm.
2. Hold a pig bristle[2] or a short piece of horsehair (e.g. from a clothes brush) by a pair of forceps or fasten it to a bottle cork. This can be done by using a needle or a pin to make a hole into the cork and by gluing the bristle into the hole.

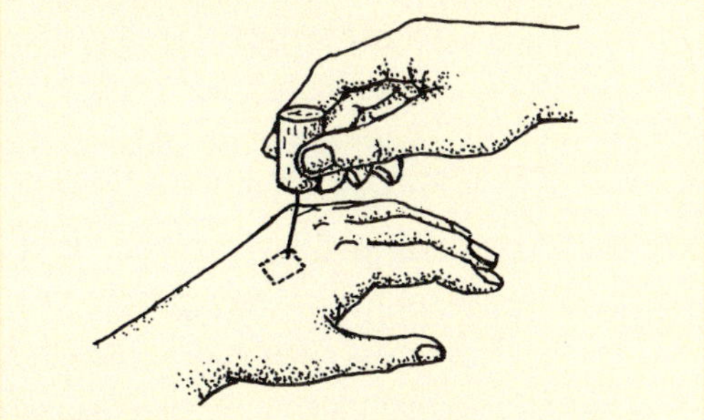

Figure 48 Testing the sensitivity to touch

3. Ask your partner to tell you if they can feel the stimulus when you press the bristle or the horsehair onto their skin within the three marked squares.
4. Then use the bristle to scan each of the three squares at intervals of[3] about 1 mm. When your partner feels the stimulus, use the felt pen to mark the spot with a dot.
5. Change places with your partner and repeat the procedure.

1 Write your results into a simple table and compare them with those of other teams in your class.

2 What do the results tell you about the relative concentration of touch receptors in the investigated areas of the skin? Note: The sensitive spots do not correspond to single touch receptors (see Figure 3, p. 115).

B. Investigate your perception of temperature

1. Fill three flat bowls or troughs[4] half-full: one with warm water (35 °C), one with cold water (10 °C) and one with lukewarm[5] water. Place the bowl with lukewarm water between the other two.
2. Now hold your left hand in the warm water for 30 seconds and at the same time your right hand in the cold water (see Figure 49).
3. Immediately afterwards hold both your hands in the lukewarm water for 30 seconds.

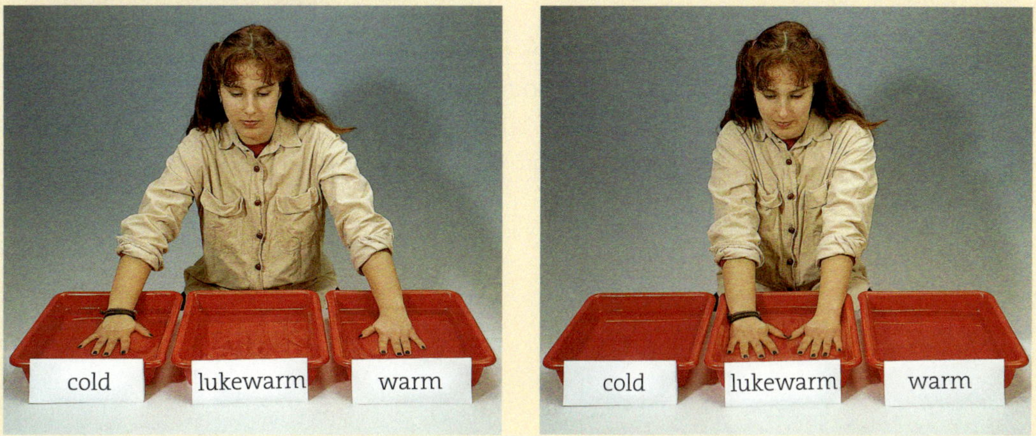

Figure 49 Testing the sensitivity to temperature

1 Describe and compare how you perceived[6] the different temperatures of the water in the first and second experiment. Note down your observations.

2 Speculate about what the reasons for your results might be and what the temperature receptors in the skin actually respond[7] to.

1 blindfold sb. *jdm. die Augen verbinden*
2 bristle *Borste*
3 at intervals of *im Abstand von*
4 trough *(hier) Wanne*
5 lukewarm *lauwarm*
6 perceive sth. *etwas empfinden, wahrnehmen*
7 respond to sth. *auf etwas reagieren*

7.10 Animals see and hear differently

There are specialists in the animal world whose senses of seeing and hearing are much more acute[1] than those of humans. Some animals have senses that we do not possess at all. Scientists have found that migratory birds[2] can use the magnetic field of the earth as a source of **navigation** on their long flights across seas and continents. Research shows that magnetite[3], an iron oxide[4] crystal which is found in certain brain cells of migratory birds, aligns[5] with the magnetic north like a compass needle[6]. Information sent out from these cells may give the brain a sense of direction.

Key terms

Animal senses and capabilities
- compound eye *Komplexauge*
- field of view (vision) *Gesichtsfeld*
- heat-sensitive *hitzeempfindlich*
- navigation *Orientierung, Navigation*
- pit organ *Grubenorgan*
- ultrasonic *mit Ultraschallqualität*
- visual resolution *Auflösungsvermögen, Sehschärfe*

Humans rely[7] very much on their eyesight, but birds of prey[8] can see much better. The density of photoreceptors in the centre of an eagle's[9] retina is one million per mm^2 compared to 160,000 per mm^2 in humans. The high **visual resolution** (the ability to see tiny objects that are very close to each other as separate units) enables birds of prey to recognize an animal as small as a mouse from high up in the sky.
An owl can see an object as small as a one-euro coin from a distance of 400 metres, whereas humans can only recognize a football at that distance. With its large pupils, hardly any cones and a high density of rods all over the retina, an owl can see sharply even in dim light.

Figure 50 The huge eyes of an owl

Most animals have eyes that look sideways and not straight ahead like yours. This means each eye has a different view. Some can even see what is happening behind them. This is especially useful for animals that are hunted by others for food.

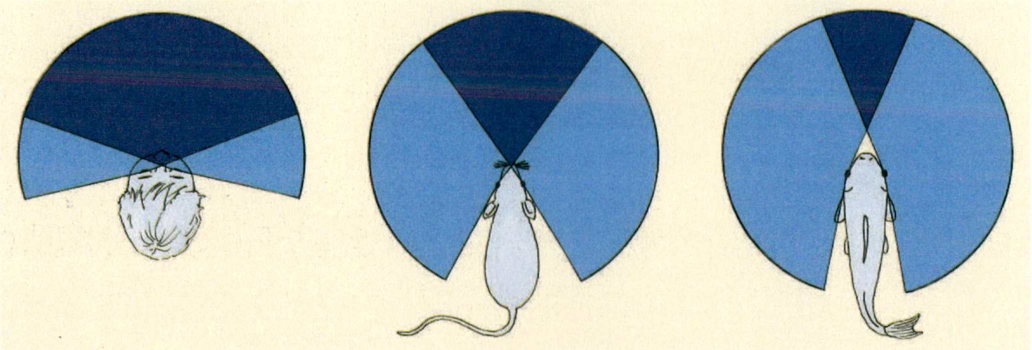

Figure 51 The fields of view for a human, mouse and fish

1 acute *scharf, fein*
2 migratory bird *Zugvogel*
3 magnetite *Magneteisenstein*
4 iron oxide *Eisenoxid*
5 align with sth. *sich ausrichten auf etwas*
6 compass needle *Kompassnadel*
7 rely on sb./sth. *sich auf jdn./etwas verlassen*
8 bird of prey *Raubvogel*
9 eagle *Adler*

Many insects do not have the same type of lens eyes as vertebrates[1]. Some species of insect, however, have extremely well-developed senses of vision. Their **compound eyes** consist of hundreds or even thousands of single eyes. These eyes produce pictures that are like mosaics[2], as

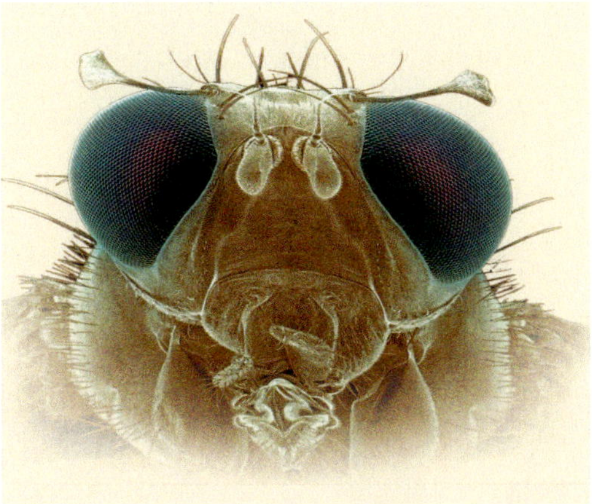

Figure 52 The head of a fly

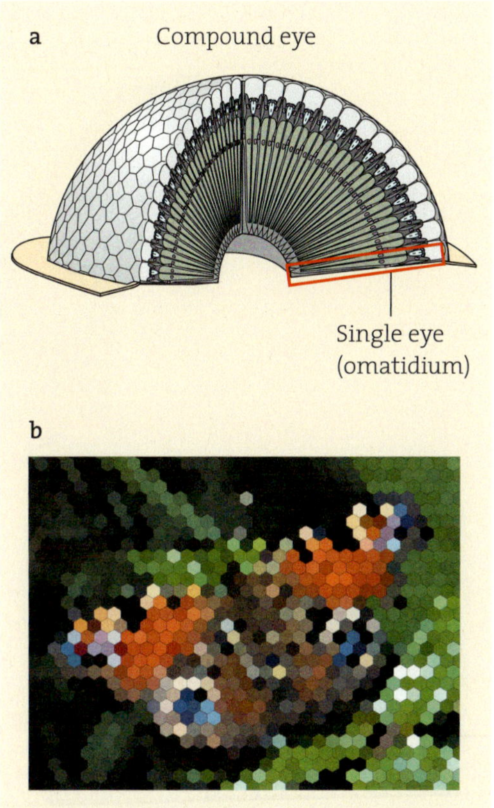

Figure 53 The structure of the compound eye of an insect (a) and the type of image it produces (b)

each single eye reproduces only a section of the object which is observed. The visual resolution is not as good as in vertebrate eyes, but insects can see fast movements much better and can therefore react more quickly than vertebrates. Bees can see ultraviolet light as colours, which the human eye cannot perceive[3]. The flowers which they visit for nectar therefore appear to them in different colours than we see them.

As their compound eyes are dome-shaped[4] and cover large portions of their head, insects also have a much greater **field of view** than humans and can see in all directions at the same time. This is very useful for a carnivorous[5] insect like a dragonfly when it is looking for prey[6]. For insects which may be possible prey, these types of eyes are equally an advantage, as they have a better chance to recognize the predator[7] before it is too late to hide or escape.

1 vertebrate *Wirbeltier*
2 mosaic *Mosaik*
3 perceive sth. *etwas wahrnehmen, erkennen*
4 dome-shaped *kuppelartig*
5 carnivorous *fleischfressend*
6 prey *Beute*
7 predator *Räuber*
8 marsh marigold *Sumpfdotterblume*
9 bat *Fledermaus*
10 utter a cry *einen Schrei ausstoßen*

Figure 54 a and b The flower of a marsh marigold[8] as we see it (a) and what it may look like to a bee (b)

Bats[9] communicate by sounds which we can hear quite well, but to orientate themselves in the dark and to hunt insects, they utter cries[10] at **ultrasonic** frequencies. These sounds are reflected by objects in the area around them. The further away the object is, the longer it takes until the acoustic echos[1] reach the very sensitive ears of the bat. By using special detectors you can listen to the bat's signals, which cannot otherwise be heard by the human ear.

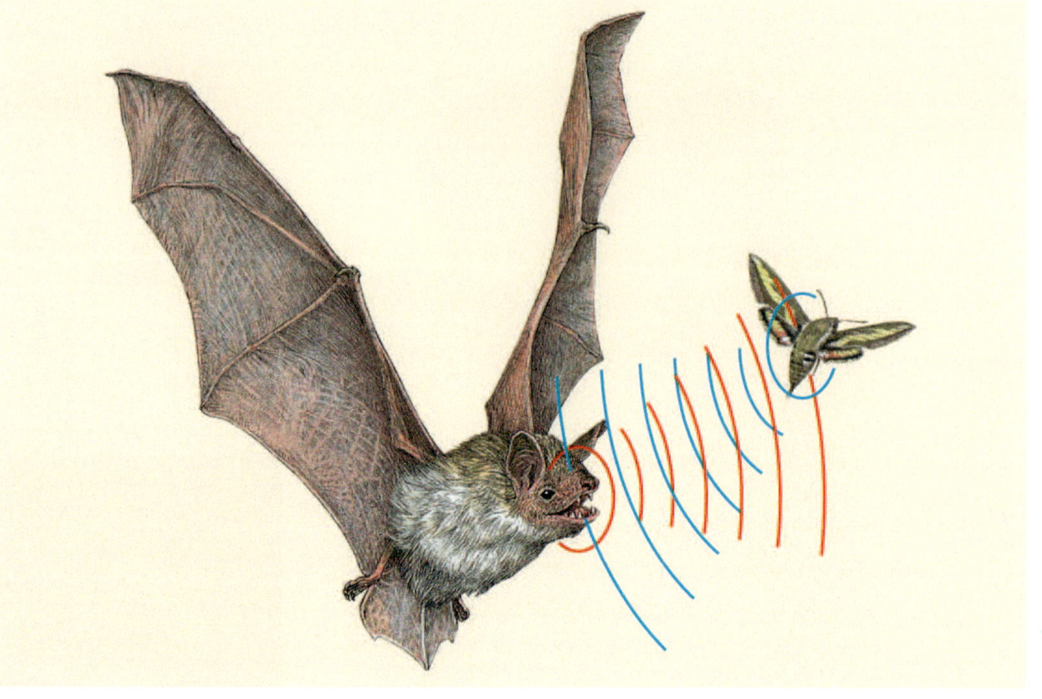

Figure 55 An echolocating[2] bat

Some snakes, for example pit vipers[3], boas[4] and pythons[5], have special **heat-sensitive** organs on their heads. The **pit organ**, as it is called, looks like a small hollow[6] or pit[7] between the nostril and eye. It contains extremely sensitive heat receptors which direct the snake to its prey. Even in complete darkness the pit organ gives the snake a three-dimensional heat picture of any warm-blooded animal nearby. Some snakes also use these organs to find cool places when it gets too hot in the sun.

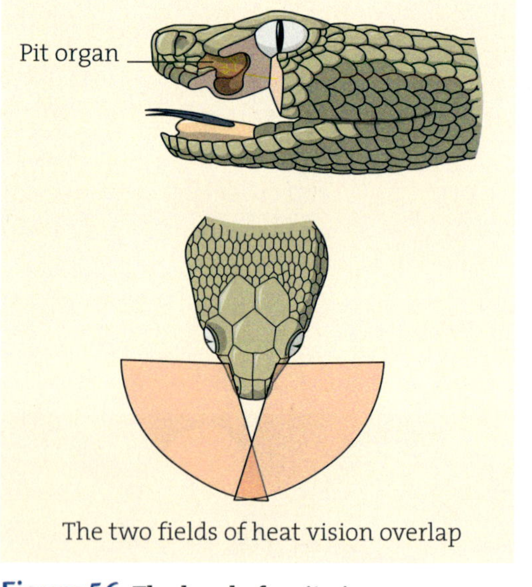

Figure 56 The head of a pit viper

Locusts[8] produce high-pitched[9] sounds to attract their mates. These chirping[10] sounds are made by rubbing the inner side of the back leg, which has fine spines on it, against the edge of the forewing[11]. This action can be compared to the playing of a violin. The locust's ears are located on both sides of the abdomen. There is no auditory canal, but the ear starts on the outside with the flat membrane of the eardrum. The larger green grasshopper[12] has its ears situated on the upper part of the front legs.

Figure 57 A locust

1 acoustic echo *hörbares Echo*
2 echolocate *sich durch Echoortung orientieren*
3 pit viper *Grubenotter*
4 boa *Boa*
5 python *Python*
6 hollow *Einbuchtung, Vertiefung, Mulde*
7 pit *Grube*
8 locust *Heuschrecke*
9 high-pitched *hoch*
10 chirp *zwitchern, zirpen*
11 forewing *Vorderflügel*
12 grasshopper *Heuschrecke, Heupferd*

1 Match the following descriptions with the animals shown in Figure 58.
 a They have a sense of magnetism which enables them to find their way home from very far away.
 b They are deaf but can feel even slight vibrations of the ground and have a special organ that is highly sensitive to warmth and thus helps them find their prey.
 c They make high clicking sounds for orientation and for hunting their prey.
 d They compensate for their poor eyesight by having an extremely good sense of smell.
 e They can see well in dim light, which makes them good nocturnal hunters.
 f Their very sharp eyes enable them to recognize their prey from high above the ground.

Figure 58 **Clockwise from top left: cat, eagle, pigeon, dog, pit viper, dolphin**

2 Which of the animal senses described in this section are extraordinary[1] compared to the capabilities[2] of human sense organs? Explain how they are useful for the animals in their natural environment.
3 Explain what is meant by the statement 'Bats see with their ears'.
4 The body of some types of moths[3] is covered with fur. Find out why these moths are rarely eaten by bats.
5 When a cat's eyes are caught in a beam of light, they show the characteristic green night-time glow[4]. This happens because a layer at the back of the eye acts like a mirror and reflects the light back to the retina. What is the advantage of this reflecting layer in the cat's eyes?
6 Use the Internet and other sources to find out about other animals with 'super senses'.

1 extraordinary *außergewöhlich*
2 capability *Fähigkeit*
3 moth *Nachtschmetterling*
4 glow *Leuchten*

Summary

Sense organs take up stimuli from the environment, change them into electric impulses and transfer them along the nerves to the brain. The eye is our most important sense organ. The main part is the complex eyeball consisting of several layers of cells, of which the most important layer is the retina with its sensory cells. The very sensitive rods are for black and white vision, whereas the less sensitive cones perceive colours. The light is bundled by the lens and produces a smaller image on the retina, which is upside down but is corrected by the brain.
The eye can adapt to changing conditions in the environment by adjusting the opening of the pupil and the curvature of the lens. As both eyes are arranged at the front of the head, we can see objects three-dimensionally. What we actually see is always the result of a combined action of our eyes and our brain.
Our ears take up sound waves through the outer ear. The vibrations of the eardrum are magnified by the ossicles and cause waves in the cochlear fluid. These waves stimulate receptor cells, whose signals are interpreted in special areas of the cerebrum.
As the eyes and the ears are very sensitive organs, we must take good care of them. Some disorders of the eyes and ears can be corrected, especially if they are detected at an early stage.
Many animals have sensory organs which are more efficient than those of humans; some animals can even perceive stimuli that our sense organs cannot detect.

A Study tasks

1. Develop a flow chart which describes in short steps how the visible impression of an orange on a table in front of you is produced in the sensory region of your brain.
2. Describe the path of a musical note from the instrument to the point of perception in the brain.
3. Make a poster with copies of photos, drawings and texts about the structure and functioning of the human eye or ear.

B Quiz

Are these statements about the eyes and ears true or false? Rewrite the false ones so that they are true.

	True	*False*
1. *Sense organs only react properly to adequate stimuli.*		
2. *Humans can see and hear better than most species of animals.*		
3. *The only function of the iris is to give the eye its colour.*		
4. *The light-sensitive receptor cells lie in the retina.*		
5. *The yellow spot is the area where the optic nerve enters the eyeball.*		
6. *Our eyes have three different types of cones for detecting colours.*		
7. *Adaptation is the ability of the eye to focus on near and distant objects.*		
8. *It is most relaxing for the eyes when you read a book.*		
9. *People who are short-sighted have to wear glasses with convex lenses.*		
10. *The receptor cells for hearing are on the basilar membrane of the cochlea.*		
11. *The vestibular apparatus is responsible for your sense of balance.*		
12. *Bats can hunt insects at night because of their excellent eyesight and sense of smell.*		

C Label the picture

In this chapter you have learned about the structure and functioning of your eyes.

1. Name the numbered parts in Figure 59.
2. Briefly explain the function(s) of each.

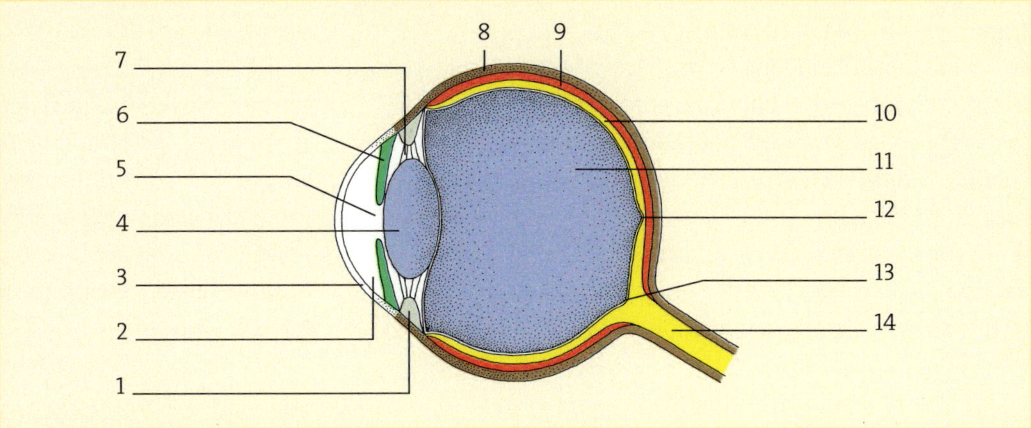

Figure 59 **The structure of the eye**

D Fill in the gaps

Complete the sentences using key terms introduced in this chapter (see the list on the right).

1. The eyeball is covered by a tough skin called the … .
2. When you read something, your … muscles contract and the … becomes rounder.
3. The ability of the eye to adjust to changing amounts of light is called … .
4. The receptor cells for colour vision are the … .
5. When sound waves reach the …, it starts to vibrate.
6. In the middle ear the vibrations are magnified by the … .
7. … can use the magnetic field of the earth for navigation.
8. With their huge … eyes insects see objects like mosaics.

Key terms

See the boxes:

Sensory cells and their functions … p. 114
Pathways of signals … p. 115
Parts of the brain … p. 115
Outer-eye structures … p. 116
Inner-eye structures … p. 117
Functions of the eyes … p. 119
Eye disorders … p. 125
Outer/middle/inner ear … p. 127
Balance and the inner ear … p. 130
Sound and noise … p. 131
Animal senses and capabilities … p. 138

E Critical thinking

1 What is the biological explanation for the English proverb[1] 'All cats are grey by night'?

2 Look at Figure 60.

a Do not read out the words, but quickly name the colours in which the words are printed, one after the other. Can you do it without making any mistakes?

b Preschool children who cannot read yet have no problems doing this task. Why?

3 When you have a cold, you can get earache if you travel by plane or use a fast-moving lift. Explain why.

blau	gelb
grün	grün
rot	rot
gelb	blau
blau	grau

Figure 60

1 proverb *Sprichwort*

Chapter 8 Drugs, Addiction and Prevention

Chapter objectives

- To **explain** what a drug is
- To **list** the short-term and long-term effects of alcohol
- To **describe** what blood alcohol concentration means
- To **look at** the long-term effects of heavy drinking
- To **explain** the difference between a substance addiction and a behavioural addiction
- To **identify** the warning signs of addiction
- To **describe** the effects of different drugs
- To **learn** strategies to prevent addiction

'You must try this – it's fun!' Has a friend ever offered you an alcoholic drink or a cigarette saying something like this? Discovering and trying out new things is an important part of teenage life: new and different clothes, hairstyles, films, music, and new ideas about how to live. But sometimes teenagers also begin more dangerous experiments by getting involved with alcohol, tobacco or other unhealthy substances and behaviours. This chapter will help you to understand how these substances work, so that you can choose a safe and healthy lifestyle.

8.1 Drugs come in many forms

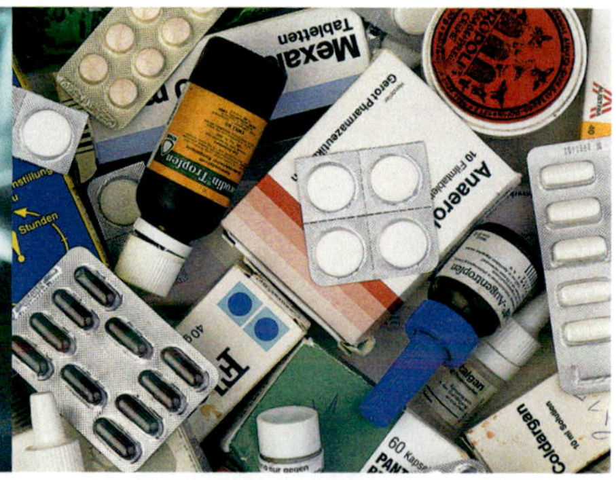

A **drug** is any chemical a person takes that causes changes in their body or behaviour. That is why substances as different as **painkillers** (e.g. aspirin), alcohol, **marijuana** and **ecstasy** are all called *drugs*.

Many people take drugs when they have a headache, want to stop coughing or cannot sleep. These substances are part of everyday life in many countries, and, if taken correctly, are usually not harmful to the user. A doctor might **prescribe** a certain drug to you to heal your bronchitis[1], for example (these are called **prescription drugs**). Other drugs, like certain pills against headache, you can simply buy at the chemist's[2] without seeing a doctor first (these are called **over-the-counter drugs**). All of these substances, which are also called **medicines**, are legal.

Key terms

Types of drugs

- drug *Droge, Medikament*
- ecstasy *Ecstasy*
- illegal drug *illegale Droge*
- marijuana *Marihuana*
- medicine *Medikament*
- over-the-counter drug *nicht verschreibungspflichtiges Medikament*
- painkiller *Schmerzmittel*
- perception *Wahrnehmung*
- prescribe sth. *etwas (ein Medikament) verschreiben*
- prescription drug *verschreibungspflichtiges Medikament*

Other drugs have such dangerous effects that there are laws to stop them being sold or used. These **illegal drugs** can damage or destroy important organs in your body. They can also alter[3]

1 bronchitis *Bronchitis*
2 chemist's (BE) = (AE) drugstore *Apotheke*
3 alter sth. *etwas verändern*

your **perception** – sometimes even making you see things that are not there. Although some people take drugs because they *want* to experience these effects, drug-taking can easily have serious consequences, for example when you drive a car under the influence of marijuana.

In the next section we will look at a legal drug which is easily available and widely used by teenagers and adults alike: alcohol.

Did you know?

In English, when someone is ill the doctor might ask if he or she has already taken any drugs. In this context the word 'drugs' simply means 'medicine'. In German, however, the word *Drogen* only refers to *illegal* drugs.

1 Name the main types of drugs.
2 Why do you think you can buy certain drugs over the counter but have to get a prescription from a doctor for others?
3 Together with a partner, make a list of any over-the-counter drugs you can name. Are they all completely safe to take? Who can advise you about how to take them?

8.2 The short-term effects of alcohol depend on how much alcohol is in the blood

You may not think of alcohol as a drug because people everywhere use it. However, alcohol has tremendous[1] effects on your body and your behaviour, so it is important to know what it does and how you can react when you are offered a drink.

Unlike food, alcohol is not digested, i.e. broken down in your stomach or intestine. It is absorbed through the mucous membranes of these organs directly into the bloodstream, which is why you can feel the effects of alcohol sometimes only minutes after you have taken a drink. Still, even 30 to 90 minutes after your last drink, alcohol can continue to pass into the bloodstream from your intestine. This is why people can get even more drunk after they have had their last drink, and sobering up can sometimes take longer than expected. Once in the blood, alcohol travels into all the cells of your body. When it reaches the liver, it is broken down so that it loses its **intoxicating** effect.

The amount of alcohol in your blood is called **blood alcohol concentration**, or **BAC**. Table 1 and Figure 1 show you some of the effects of an increasing BAC on a person.

1 tremendous *enorm*

Key terms

Short-term effects of alcohol
- blood alcohol concentration (BAC) *Blutalkoholkonzentration*
- driving under the influence of alcohol (DUI) *Alkohol am Steuer*
- gender *Geschlecht*
- hangover *Kater*
- intoxicating *berauschend*
- intoxication *Rausch, Berauschtsein*
- rate of consumption *Geschwindigkeit, mit der etwas konsumiert wird*
- sober *nüchtern*

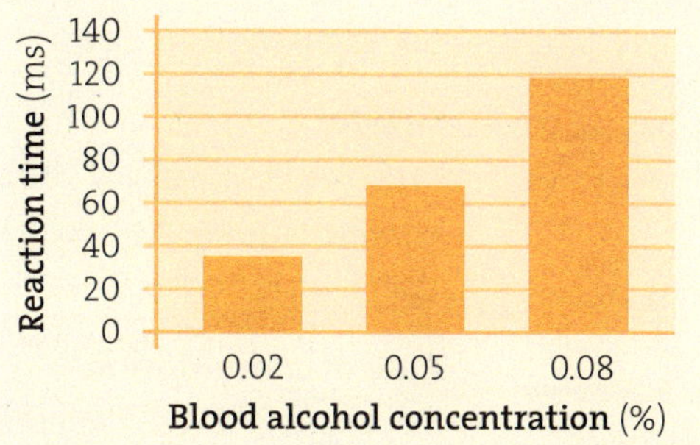

Figure 1 Changes in reaction time under the influence of alcohol

0.02–0.04 % BAC
• Muscles relax, reflexes become slower and resistance[1] to further drinking declines.
0.05–0.09 %
• Muscle coordination decreases, reactions deteriorate[2].
0.12–0.15 %
• Some drinkers become aggressive and reckless[3], others tired and depressed. • Vomiting[4] often occurs. • Reactions and vision deteriorate; readiness to take risks decreases.
0.15–0.24 %
• Dizziness[5] and blackouts[6] are common. • Speech and vision are impaired[7]. • Movements become uncoordinated.
0.25–0.34 %
• Orientation and memory become seriously impaired. • Loss of consciousness[8] may occur.
0.35–0.39 %
• Loss of consciousness often occurs. • Serious alcohol poisoning[9] occurs.
0.4 %
• Alcohol coma usually occurs.
0.5 %
• Death usually occurs.

Table 1 **The short-term effects of alcohol (according to blood alcohol concentration)**

Blood alcohol concentration is affected by several factors:

Body size and weight

Alcohol disperses[10] throughout the body. Bigger people have more cells to take up the alcohol than smaller people, so their BAC drops more quickly. However, fatty tissue does not take up much alcohol, so having more fat on your body does not automatically help to keep the BAC low.

Amount of food in the stomach

Since alcohol is taken up through the mucous membranes of the stomach and the intestine, a full stomach slows down absorption[11] into the bloodstream.

Gender

In males, BAC usually drops more quickly than in females, because they have more of the enzyme that breaks down alcohol. Also, male bodies normally have a higher muscle/fat ratio[12] than female bodies.

Rate of consumption[13]

Of course, BAC also depends on how much you drink and how quickly you drink it. The liver has a certain rate of breaking down alcohol (in males it is usually 0.1 g of alcohol per hour, in females 0.09 g) and there is nothing you can do to change it. Drinking coffee, taking a shower or going for a walk in the fresh air might keep you awake, but it will not decrease your BAC.

1 resistance *Widerstand*
2 deteriorate *verschlechtern*
3 reckless *leichtsinnig*
4 vomit *sich erbrechen*
5 dizziness *Schwindel*
6 blackout *Ohnmachtsanfall*
7 impaired *beeinträchtigt*
8 consciousness *Bewusstsein*
9 alcohol poisoning *Alkoholvergiftung*
10 disperse *sich verteilen*
11 absorption *Aufnahme*
12 muscle/fat ratio *Verhältnis von Muskel- zu Fettgewebe*
13 consumption *(hier) Verzehr*

Figure 2 **How much alcohol does a drink contain?**

Even mild **intoxication** can have life-threatening[1] effects. **Driving under the influence** of alcohol often results in car crashes, because reaction time becomes longer even at a legal BAC. The only way to drive safely is to stay **sober**, i.e. to not drink at all.

Figure 2 shows you how much alcohol is contained in different drinks. Many people think that they will notice when they have 'had too much', or that they will pass out[2] before they reach a really dangerous state. But remember that alcohol can continue to be taken up into your bloodstream up to 90 minutes after your last drink. Even after a person passes out, their BAC can still increase.

Alcohol also interacts with a variety of medicines. Depending on the drug and the amount of alcohol in the body, the effects of the drug can become much more extreme. The effects of tranquilizers, for example, may lead to household or driving accidents and can also depress[3] your heart and breathing rate. In contrast, alcohol can inhibit the effect of certain antibiotics, causing a longer illness. It is important to always check the medication package insert[4] for information on such interactions.

The short-term effects of alcohol slowly fade as your liver breaks down the alcohol in your blood. People who drink heavily often experience a **hangover** the next morning. Hangover symptoms include an upset stomach[5], headaches and an increased sensitivity to noise.

1 How does alcohol get into your bloodstream?
2 Which factors influence a person's BAC?
3 Explain how the lower muscle/fat ratio in females might affect their tolerance of alcohol compared to males.
4 Compare the amounts of alcohol in different drinks (see Figure 2). What do you notice? What are the consequences of this for someone who drinks?
5 If an alcoholic drink like Irish cream liqueur (e.g. Baileys) has an alcohol content of 17 %, how much alcohol is in a 40 ml serving?

Did you know?

If you want to calculate how many millilitres of alcohol are contained in a drink, simply multiply the volume[6] by the alcohol percentage shown on the label and divide it by 100 (it is per cent, after all). For example, a mix of beer and lemonade (*Radler* in German or 'shandy' in English) contains 2.8 per cent alcohol in a 330 ml bottle. So your calculation would look like this: $330\,\text{ml} \cdot 2.8/100 = 9.24\,\text{ml}$.

1 life-threatening *lebensbedrohlich*
2 pass out *ohnmächtig werden*
3 depress sth. *(hier) etwas herabsetzen*
4 package insert *Beipackzettel*
5 upset stomach *verdorbener Magen*
6 volume *Volumen*

8.3 The long-term effects of heavy drinking can be devastating[1]

Adults who use alcohol responsibly are usually not at risk of any long-term health problems related to alcohol consumption. Those who drink heavily, however, damage their bodies in several ways, and it takes a long time for this damage to heal. Unfortunately, some of the damage caused by alcohol never heals, even if the person stops drinking alcohol completely.

Brain damage

Excessive alcohol consumption kills nerve cells in the brain. These cells usually cannot grow again, and eventually the memory is affected, the arms and legs feel **numb**, and the personality can change as well. This is especially dangerous for teenagers, because their brains are still growing and developing (see Figure 3 below).

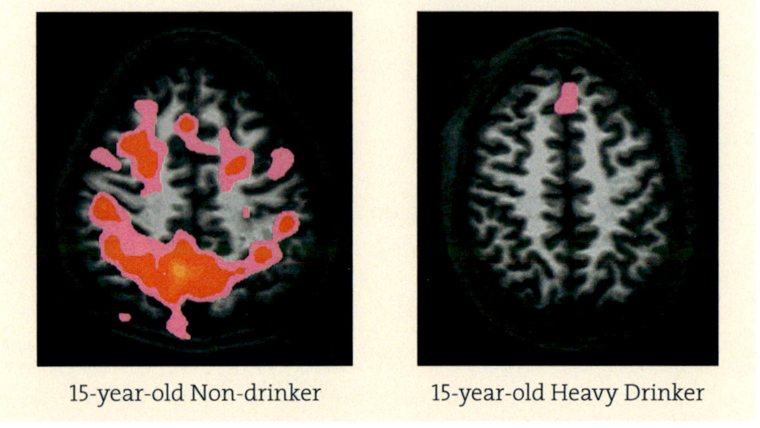

Figure 3 **Alcohol and brain activity. The red colour shows active regions in the brain.**

Key terms

Long-term effects of alcohol
- alcoholism *Alkoholabhängigkeit, Alkoholismus*
- cancer *Krebs*
- liver cirrhosis *Leberzirrhose*
- mental retardation *geistige Behinderung*
- numb *taub*
- relapse *Rückfall*
- scar tissue *vernarbtes Gewebe*
- withdrawal symptoms *Entzugserscheinungen*

Liver damage

The liver is the organ that breaks down alcohol in the body, and as a result it suffers most from the long-term effects of alcohol. When alcohol damages liver cells, the cells ultimately die, leaving behind permanent **scar tissue** which does not function any more. This disease, known as **liver cirrhosis,** can be lethal[2] (see Figure 4).

Heart and blood vessels

Alcohol leads to a constant widening of blood vessels, which causes high blood pressure and raises the danger of a heart attack or a stroke. The heart itself can also be affected, for example by inflammation.

Cancer

Since alcohol inhibits[3] the body's repair mechanisms, people who drink excessively are in greater danger of developing cancer. The organs that are most often affected are the mouth cavity, the gullet, the stomach, the pancreas, the colon and, in women, the breasts.

Unborn children

Drinking alcohol during pregnancy can have devastating consequences for the baby. An unborn child is not able to break down the alcohol, which means that almost all of its organs can be damaged, including the brain. This damage can lead to **mental retardation**.

There are many other negative consequences that heavy drinkers suffer from, such as falling school grades[4], and problems with relationships and money. The stress caused by these problems often leads people to drink even more. People who cannot control the amount of alcohol they drink, and who suffer from **withdrawal symptoms** when they stop drinking, suffer from a

1 devastating *verheerend*
2 lethal *tödlich*
3 inhibit sth. *etwas hindern*
4 grade *Schulnote*

disease called **alcoholism**. It is possible to treat this disease, but recovery from alcoholism is a long and painful process, and a high percentage of people who become dependent on alcohol experience a **relapse** when the treatment ends.

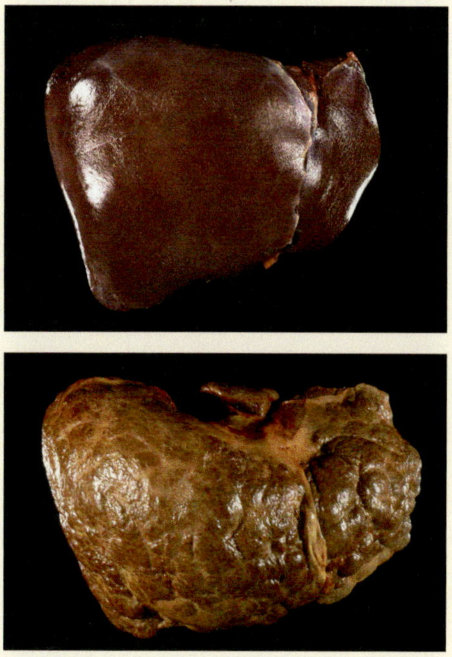

Figure 4 A healthy liver (above) and a liver damaged by cirrhosis (below)

1 The text names some of the social problems that affect heavy drinkers. Brainstorm more possible problems on a blank piece of paper under headings like *Work/School*, *Money*, *Family* and *Friends*.
2 What is the difference between 'heavy' drinking and 'responsible' drinking? Working with a partner, make a list of warning signs which would show that a person is drinking more than they should.
3 What is cirrhosis of the liver? Write an explanation, defining the disease and describing how it is caused.
4 Why is excessive drinking especially dangerous for teenagers? Think about the effects on the body as well as other possible dangers when someone has drunk too much (see also the box below). Discuss your ideas in small groups.

Further facts

Binge drinking[1]

Many teenagers like to get together and drink until they are intoxicated. So-called binge drinking is often defined as consuming five or more alcoholic drinks on one occasion. In Germany the percentage of 12–17 year-olds who engage in binge drinking rose from 20 per cent in 2005 to 26 per cent in 2007. In 2006, UK adolescents had an average of 28 binge-drinking nights a year (in Germany, it was 21).

You have already learned about some of the consequences of heavy drinking. Binge drinkers also face other problems: they are more likely to damage property, to be violent towards other people, to have problems with the police, to miss classes, to engage in unplanned and unsafe sexual activity – which can result in STDs or unwanted pregnancy (see Chapter 6, p. 107) – and they suffer more injuries than people who do not binge drink. Surprisingly, they often do not see themselves as problem drinkers.

If you or any of your friends show questionable drinking behaviour, it is important to have the courage to talk about it. Try explaining the problem to a close friend, a member of your family, a teacher or somebody else you trust. There are also some good websites you can consult[2], such as *www.drugcom.de*, *www.bist-du-staerker-als-alkohol.de* (from the German Bundeszentrale für gesundheitliche Aufklärung) or *www.thecoolspot.gov* (a US website run by the National Institutes of Health).

1 binge drinking *Besäufnis, Komasaufen*
2 consult sth. *etwas heranziehen, zu Rate ziehen*

8.4 People can become addicted to substances or behaviour

Maybe you know someone who spends hours a day playing video games and sometimes even neglects[1] their homework because of the time spent in front of the computer. Is this an addiction? What about people who smoke cigarettes when they hang out with their friends at the weekend, but not during the week? Recognizing[2] when an addiction starts is not always easy, but there are a few things which are typical for the early stages of an addiction:

- You cannot stop yourself from starting the activity and you have a hard time stopping it;
- You have to do the activity more and more often in order to feel good;
- You lose interest in other activities you previously enjoyed, e.g. meeting friends;
- You continue doing the activity even though you know that it is not good for you.

Key terms

Addiction
- behavioural addiction *stoffungebundene Sucht*
- compulsive *zwanghaft*
- dose *Dosis*
- gambling *Glücksspiel*
- physical *körperlich*
- psychological *psychisch*
- substance addiction *stoffgebundene Sucht*
- tolerance *Toleranz*

Figure 5 A British public-service advert

When people show these tendencies in connection with a drug like alcohol or heroin, doctors say that they have a **substance addiction**.

In contrast, there are addictions which do not have anything to do with taking in[3] substances: for example, some people lose lots of money **gambling** but continue doing it, while others get into debt[4] because they cannot stop shopping. This kind of **compulsive** behaviour is called **behavioural addiction**.

Behavioural addictions, as well as some substance addictions, affect the person on a **psychological** level (for example, when someone is used to smoking a cigarette after every dinner). Taking a drug or behaving in certain ways creates a feeling of satisfaction[5], which leads to a very strong desire to repeat the action again and again.

Other substance addictions are **physical** (for example, the nicotine in cigarettes makes the body crave[6] more). As time goes by, people who suffer from these addictions have to increase the amount of the drug they take. If they do not, or if they try to decrease the **dose**, they suffer from withdrawal symptoms. This effect is called **tolerance**, which means that the body gradually gets used to a substance, so more of the drug is needed to produce the feeling of satisfaction that the person desires.

1 neglect sth. *etwas vernachlässigen*
2 recognize sth. *etwas erkennen*
3 take in sth. *etwas aufnehmen*
4 debt *Schuld*
5 satisfaction *Befriedigung*
6 crave sth. *sich nach etwas sehnen*
7 fix *(hier) Dosis*

Figure 6 An example of behavioural addiction

Figure 7 An example of substance addiction

1 Describe the difference between substance addiction and behavioural addiction.
2 Explain how a drug can affect a person on a psychological level.
3 Look at the pictures in Figures 6 and 7. Use the four guidelines above to determine[1] if either of these behaviours can lead to addiction. Then do the same with the following: smoking marijuana, drinking beer, taking painkillers, trying to stay thin.
4 Figure 5 shows a British poster which tries to raise awareness for the addictive potential of everyday drugs. How effective do you think it is? Explain.

8.5 No one plans to become addicted

When trying out a dangerous substance or doing a certain activity for the first time, no one usually expects or wants to become addicted. For example, if you ask addicted smokers why they started smoking, they might give any of the following answers:
- I wanted to feel good – I enjoyed the nice feeling after smoking.
- I wanted to relax – I was very stressed and anxious[2] about certain things and smoking helped me to forget these feelings.
- I wanted to do better – smoking helped me to calm down before an exam, so I got better marks.
- I was curious – all my friends did it, so I wanted to know what it was like.

These are all very normal desires – after all, doesn't everyone want to feel good? One of the problems with using drugs to satisfy these needs is that their positive effect does not last very long. As pointed out in the previous section, drugs lose their effectiveness once you get used to them and your body develops a tolerance for them. Drugs can quickly take over a person's life, influencing most of their decisions as they look for more time to take the substance and try to get and use larger amounts of it.

Figure 8 Relying on pills to make you feel better can lead to addiction.

1 determine sth. *etwas bestimmen*
2 anxious *ängstlich*

The positive feeling that drugs produce is created in the brain, which is why an addiction is a brain disease. That makes taking drugs especially dangerous for teenagers, because between the ages of about 10 and 20 the brain undergoes[1] major changes and develops very rapidly. Taking drugs during this time engraves[2] the need for the substance deeply into one's memory and can hinder some of the brain's development.
Addiction often happens without a person realizing it. The guidelines from Section 8.4 can help you to notice when a dangerous habit starts to develop. Paying attention to your behaviour and talking about these things with your friends will help you to spot the warning signs of addiction and to stop yourself before it is too late.

1 Why is it very risky to satisfy our desire to feel good by taking drugs?
2 Make a list of more reasons why people might start taking drugs, then find an argument against each reason. Practise arguing the reasons for and against drug-taking with a partner. In these role plays, each of you should argue as if you really believe what you are saying, even if you do not actually agree with it. For language help see **Activate your English**, 'Making arguments', on the inside cover.
3 Do you think teenagers are more likely to try illegal drugs than adults? Give reasons for your answer.
4 Look at the 'Further facts' box below. Is it always wrong to take sleeping pills? Think of things that Mark could do instead of taking pills to fall asleep, and then share your ideas with the other students in a class discussion.

Further Facts

Young people and sleeping pills
Between 2000 and 2004, the number of prescriptions for sleeping pills in the USA for children and adolescents went up by 85 per cent. Here is one boy's story:
When Mark, a 14-year-old eighth-grader, knew that he had a test in school the next day, he was always very nervous and anxious. On those nights he always had trouble falling asleep, which made him even more nervous, because he knew that he would not do well on the test if he was not well rested. He finally told his mum, who sometimes also had a hard time falling asleep, and she gave him some of her sleeping pills. That night Mark slept perfectly and did fairly well on the test.
From then on he started using sleeping pills every time he had a test the next day, but also when he knew there was a party and he wanted to be well rested for that. After a while he noticed that falling asleep was hard for him even when there was no test, so he started taking the pills more often. When his mum realized what he was doing, she told him that he could not have any more pills, which made him so angry that he tried to hit her, something he would never have done before. The next day she took him to a doctor.

1 undergo sth. *etwas durchmachen, etwas erleben*
2 engrave sth. *etwas einprägen*

8.6 Different substances have different effects

There are many different drugs that people abuse. The following is a brief overview of four of the most important drugs.

Key terms

Effects of different drugs
- mind-altering *bewusstseinsverändernd*
- nausea *Übelkeit*
- psychosis *Psychose*
- seizure *Anfall*

Marijuana

Description: Marijuana is a substance found on different parts of the hemp plant[1]. It is usually smoked but can also be eaten. The main active substance is called THC, but marijuana also contains some 400 other chemicals.

Effects (selection)**:** Marijuana changes processes in the brain. It has the effect of decreasing the functions of memory and learning, and making it hard to think or solve problems. Perception can be distorted[2], movements are harder to coordinate and the heart rate increases. It can also cause laziness and lack of motivation. Long-term marijuana use can lead to addiction and to mental problems.

Figure 9 The flower of a hemp plant

Ecstasy (MDMA)

Description: Normally taken in pill form, this drug has stimulating effects, which is why it is sometimes taken by people at parties. It is also **mind-altering**: users' perception[3] of time and space can be distorted.

Effects (selection)**:** When ecstasy is taken in large amounts the body can on rare but unpredictable[4] occasions lose its ability to control temperature. When this happens, the body can get too hot, leading to organ failure and sometimes death. Ecstasy increases the heart rate and blood pressure, which can lead to heart disease. Muscle tension, **nausea**, faintness, and chills or sweating can also occur. Users of ecstasy sometimes suffer from depression, sleep problems and severe anxiety[5].

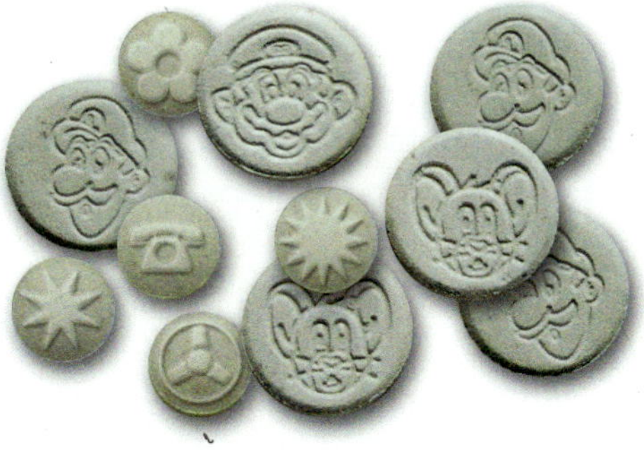

Figure 10 A variety of ecstasy pills

Cocaine

Description: A very addictive drug that is snorted[6], injected or smoked.

Effects (selection)**:** Cocaine makes the user feel euphoric[7] and energetic. This 'high' may last from 5 to 30 minutes, depending on the way the drug is taken. Heart attack, lung failure, stroke and **seizure** are all possible consequences of repeated use. **Psychosis** can also occur. Treating users of this powerfully addictive drug is difficult and takes a long time.

Figure 11 Cocaine

1 hemp plant *Hanfpflanze*
2 distorted *verzerrt*
3 perception *Wahrnehmung*
4 unpredictable *unberechenbar*
5 anxiety *Angst*
6 snort sth. *etwas (eine Droge wie Kokain) schnupfen*
7 euphoric *euphorisch*

Tranquilizers

Description: Tranquilizers are used to treat anxiety or sleeping problems. They need to be prescribed and are taken in the form of pills.

Effects (selection)**:** They have a calming effect, but are highly addictive. Improper use can result in difficulty in concentrating, slower breathing, memory loss and sleepiness. When taken with alcohol their effect is magnified, which can lead to very dangerous states, including coma.

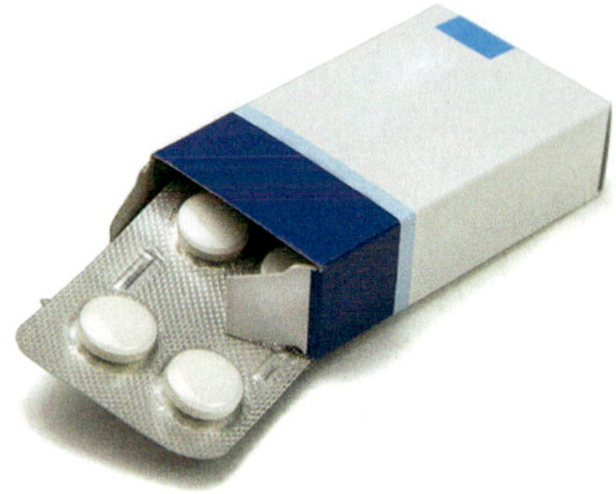

Figure 12 Tranquilizers

8.7 What is your antidrug?

Now that you have learned about different drugs and how they work, the question is how you can avoid them, even if you are under stress or if you know people who are taking drugs.

For many teens the answer is: 'By knowing my **antidrug**'. When you feel stressed out or down, your antidrug helps you to calm down or to feel good again without the short cut[1] of drug abuse. So if skateboarding is your antidrug, that is what you do when you have the urge[2] to smoke. If it is playing the guitar, you start playing as soon as you feel the desire to take a pill. Your antidrug may change as you get older, but you should always know what you can do to make yourself feel better without hurting your body.

It is also important to make a strong **commitment** against drugs. If you do not think about whether you want to try drugs until you are offered some, you may not have the strength to say no. The first time it is a choice, but the second time it might not be: once can be enough to become 'hooked'[3]. So think about it before you get into the situation and make a strong decision to have a lifestyle that is good for you. Doing this together with a friend or even a group of

Key terms

Strategies against drugs
- antidrug *Gegendroge*
- commitment *bewusste Entscheidung, Zusage*
- enabling behaviour *Verhalten, das dem anderen etwas ermöglicht*
- peer pressure *Gruppenzwang*

1 short cut *Abkürzung*
2 urge *ein starker Drang*
3 hooked *abhängig*

Figure 13 Some favourite antidrugs

friends can be a great encouragement. If you go to a party together you can look out for each other and each of you can help the other to stick to their decision. This is called positive **peer pressure**.

If you have a friend who takes drugs, one thing you should not do is ignore the problem. By ignoring your friend's problem, you are enabling[1] and even encouraging him or her to continue abusing drugs. **Enabling behaviour** can take many forms: laughing at your friend's problem, saying it is not unusual to take drugs, lending your friend money, cleaning up when he or she makes a mess, blaming yourself for your friend's addiction, or lying to his or her parents about it. You may think you are supporting your friend by behaving like this, but you are actually making the problem worse. Have the courage to stop enabling behaviour.

It can be hard to resist taking drugs or to keep a close watch on behaviour that might become addictive, especially when your friends do these things or when the media suggest they are 'normal'. However, a healthy lifestyle without drugs brings a lot more quality to your life.

Did you know?

Teens from all over the USA have written about their experiences with drugs and avoiding them on the website: *www.abovetheinfluence.com*.

1 What is your antidrug? Write down one experience where a certain activity helped you to overcome[2] an urge to do something unhealthy or dangerous. Then talk with a partner about how you felt before and after using your antidrug.

2 In small groups, brainstorm three situations where people show enabling behaviour, then write down ideas on how they could behave better. Appoint one member of your group as a speaker to present your ideas to the rest of the class.

3 Describe three situations where resisting taking drugs might be hard. Suggest ways that a person could handle these situations successfully. For language help, see **Activate your English**, 'Making suggestions', on the inside cover.

4 A report says that 68 per cent of teenagers would talk to a friend or brother or sister if they had a drug problem. Imagine that a friend is showing addictive behaviour and you want to talk to him or her. Plan what you would say and act it out in a role-play. The phrases in the **Activate your English** box will help you.

Activate your English

Talking about drugs

I'm concerned.
I feel like you're a different person when you ...
I think you should see a counselor.
I suggest/recommend ...
I care for you, that's why ...
I'd like to support you.
I'm there for you.
How do you feel when you ...?
Quitting is possible.

Further Facts

How to refuse a drug

Sometimes when you say that you don't want a drink or a cigarette (or any other drug), people will try to pressure you into taking it anyway. These guidelines can help you to stand your ground.

Do:
- keep eye contact
- stand up straight
- sound sure of yourself
- give reasons for saying no
- provide an alternative.

Don't:
- show that you are nervous or unsure
- try 'only one glass'. Drinking responsibly is possible, but never because other people pressure you to drink.
- think you're the only one refusing a drug.

1 enable sb. to do sth. *jdn. ermöglichen etwas zu tun*
2 overcome sth. *etwas überwinden*

Asking an expert

There are many different ways you can get information on a topic. The Internet, books, magazines and newspapers are some of the most obvious sources of information. Another way is to contact people who have experience in the area you are interested in and to interview them about their experiences and views. An Internet search or a telephone book will provide you with addresses of people who work in organizations that deal with your topic. Maybe somebody's parents know an expert or are experts themselves, so it is also a good idea to ask around in school.

The following suggestions will help you get the most out of an interview with such an expert:

Before the interview

Collect questions that you want to ask the expert. Begin if possible at least a few days before the interview by using one section in your notebook to write down questions as they come to mind. You can also review the material you have worked on so far and look for things you would like to know more about.

An expert can tell you a lot of things a book cannot tell you, so do not forget to ask about their experiences with certain aspects of the topic and about their opinions on controversial[1] issues[2].

In class, write your questions on cards and spread them out on the floor or on a big table where everybody can see them. Then arrange them into groups, take out any doubles, and write them down in a clear format[3].

The expert may like to receive the list of questions before the interview so that they can prepare their answers. With some questions, however, you might want to get their spontaneous[4] reactions.

Figure 14 **Students leading an interview**

During the interview

Make sure that only one person asks a question at a time, and give the expert time to answer. Ask him or her if it is alright to shoot a video of the interview or to record it, and respect their wishes if the answer is no. Choose which students will write down the answers and alternate from time to time. As with taking notes from a video (see 'Getting the most out of a film sequence', p. 110), it is good to always have more than one student taking notes, so that if one person misses an important fact, the information will not be lost.

After the interview

Discuss the information you have received. Which answers did you expect and which ones surprised you? Are there still unanswered questions? Where can you get more information? What impressed you most about the expert's knowledge?

Conclude your work by putting some of the questions and answers on a poster and hanging it up for other students to see. It is also nice to send the expert a thank-you card to show your appreciation[5] for the time they spent helping you.

1 controversial *umstritten, Kontrovers*
2 issue *Streitpunkt*
3 format *Format, Größe, Form*
4 spontaneous *spontan*
5 appreciation *Wertschätzung*

Activity 1

Organize a 'Drug Information Street'

Make one hallway in your school into a 'Drug Information Street' by decorating it with posters offering information on drugs, their effects and how to stay drug-free.

Step 1: Decide which substances and behaviours to cover[1]
Brainstorm in class to find out what information about addictions students need. Then divide into teams with each team working on one kind of addiction.

Step 2: Make a poster
In your team, think of questions you want your poster to answer. To get the information you need, consult books, experts (e.g. give a counselling centre[2] a call) and the Internet – but remember only to use trustworthy[3] websites (see the 'Did you know?' box below). Each poster should give information on: what the addiction involves, what the symptoms are, how you can tell if you might be addicted, and tips on how you can prevent this addiction from becoming part of your life. See also the skills page 'Making an educational poster', Chapter 1.7, p. 133.

Step 3: Decorate the street
Ask your head teacher for permission[4] to decorate a hallway with your posters. Then arrange the posters in an attractive way. Have one poster which summarizes all the advice[5] on staying drug-free.

Activity 2

Do a role-play

Divide into groups. Think of a situation in which it might be difficult to resist[6] taking drugs (e.g. when you feel down). Then do a role-play (see also Chapter 1.6, Activity 1, p. 111).

Step 1: Prepare the role-play
Make the setting[7] as real as possible and think about how the different people who are involved might feel. What can they do to resist? How might the scene end if the person does not resist?

Step 2: Act it out
Before you show your role-play to the class, practise it a few times. You can do one role-play where the person resists the drug and another one where he or she gives in[8].

Step 3: Talk about it
Discuss how the players and the audience felt during each performance. What enabled the person to resist in the first role-play, and what made them give in during the second? How could the person in the second role-play resist in the future?

1 cover sth. *etwas behandeln, sich mit etwas befassen*
2 counselling centre *Beratungszentrum*
3 trustworthy *vertrauenswürdig*
4 permission to do sth. *Erlaubnis, um etwas zu machen*
5 advice *Ratschläge*
6 resist sth. *etwas widerstehen*
7 setting *Szenario, Situation*
8 give in (to sth.) *(etwas) nachgeben*

Activity 3

Give a talk

Prepare a talk which compares the laws concerning alcohol in different countries. Some points to cover: the legal drinking age; the maximum acceptable level of BAC when you drive; which alcoholic drinks are popular; how much the average person drinks. Research not only German law but also the laws in English-speaking countries like the UK, the USA, Australia, New Zealand, South Africa, and so on. For language help, see **Activate your English**, "Comparing and contrasting', on the inside cover.

Did you know?

One sure indicator that a website is trustworthy is if it is government-funded. This means that the government pays for the website. This information will usually appear somewhere on the site. Check out these websites:

American	*www.justthinktwice.com*
	www.teens.drugabuse.gov
British	*www.talktofrank.com*
German	*www.drugcom.de*
	www.bzga.de

Summary

Legal drugs are also called medicines and can improve your health when taken properly, while illegal drugs are never good for your body. Alcohol, however, is an example of a legal drug which can damage your health, depending on the amount you drink. The effects of alcohol consumption range from muscle relaxation to uncontrolled behaviour and even death, depending on the blood alcohol concentration (BAC). While the short-term effects usually disappear when the liver breaks down the alcohol, there are many dangerous long-term effects such as liver cirrhosis. When people become addicted to alcohol or drugs, we speak of substance addiction, but addiction can also occur with certain types of behaviour. There are indicators for when an addiction begins and it is important to know when an activity starts to become unhealthy. One way to control the urge to take drugs is to have an antidrug.

Key terms

See the boxes:

Types of drugs .... p. 144
Short-term effects of alcohol .... p. 145
Long-term effects of alcohol .... p. 148
Addiction .... p. 150
Effects of different drugs .... p. 153
Strategies against drugs .... p. 154

A Fill in the gaps

1. At a pharmacy you can get two kinds of drugs: ... drugs, which you can only get after seeing a doctor, and ... drugs, which you can buy right away[1].
2. Alcohol is absorbed into the blood through the ... of your stomach and your intestine.
3. A liver damaged by alcohol suffers from ... .
4. Four of the problems that binge drinkers might face are ... .
5. Marijuana is found on parts of the ... .
6. When a person has the desire to take more and more of a drug, this is because their body has developed a ... to the drug.
7. When you show enabling behaviour, it means that you ... .

B Complete the sentences

Put a tick next to all the correct answers (there may be more than one).

1. Consequences of using marijuana can be
   - *... enhanced memory and learning capabilities.*
   - *... changes in perception.*
   - *... serious car accidents.*
   - *... addiction.*
2. When you have drunk a lot of alcohol, the only way to lower your BAC is
   - *... to take a cold shower.*
   - *... to drink lots of coffee.*
   - *... to get some fresh air.*
   - *... to sleep and wait until you are sober.*
3. You show signs of a behavioural addiction when
   - *... you continue with the activity although you know it is bad for you.*
   - *... you neglect your friends in order to do it.*
   - *... you really enjoy doing it.*
   - *... all your friends are doing it.*

C Critical thinking

1 Until studies proved differently, a lot of people thought that using marijuana did not lead to addiction. Do you think there can be such a thing as 'safe' drugs? Discuss.

2 **Say it in German:** In Section 8.5 you thought about situations in which teenagers might take drugs and then practised arguing against drug-taking in those situations. Choose one of the situations and do a role-play in German. How is this easier or more difficult than in English?

1 right away *sofort*

Unit II **Flowering Plants**

Chapter 1

The Organs of a Flowering Plant and Their Functions

Chapter objectives

- To **name** the characteristics of flowering plants
- To **describe** the structure and function of the main organs of a flowering plant
- To **recognize** the importance of leaves for photosynthesis
- To **compare** forms of sexual and asexual reproduction
- To **observe** and **describe** the development of a flowering plant

The 250,000 species of **flowering plants** that are known today have an incredible[1] variety. Most of them are **herbaceous** plants, like daisy[2], dandelion[3] and bluebell[4], which remain relatively small, because their stems are weak and contain no wood. **Woody plants** with strong **stems**, like hawthorn[5] or horse chestnut[6], grow into bushes and trees. But although flowering plants can look very different and live in all possible places on land and in water, they have some characteristic features in common: they all have roots and a stem with leaves and flowers. These main organs of a flowering plant have different functions. Starting with a look at the rape[7] plant, we will find out about these functions and then go on to study the great variety of reproductive strategies used by different plants.

Figure 1 A meadow in bloom

Figure 2 The main parts of the rape plant

Key terms

Plants and their parts

- broad-leafed tree *Laubbaum*
- coniferous tree *Nadelbaum*
- flowering plant *Blütenpflanze*
- herbaceous *krautig*
- stem *Stamm*
- trunk *Baumstamm*
- woody plant *Holzpflanze*

1 incredible *unglaublich*
2 daisy *Gänseblümchen*
3 dandelion *Löwenzahn*
4 bluebell *Glockenblume*
5 hawthorn *Weißdorn*
6 horse chestnut *Rosskastanie*
7 rape *Raps*
8 bud *Knospe*
9 node *Knoten (Ansatz der Blätter)*
10 internode *Stängel zwischen zwei Knoten*

1.1 Roots provide the plant with water and minerals

If you pull a rape plant out of the soil, you can recognize the long **main root**, a **taproot**, and many small **lateral roots**. The roots form the underground part of the plant. Their two main functions are to anchor[1] the plant in the soil and to provide the plant with water and minerals.

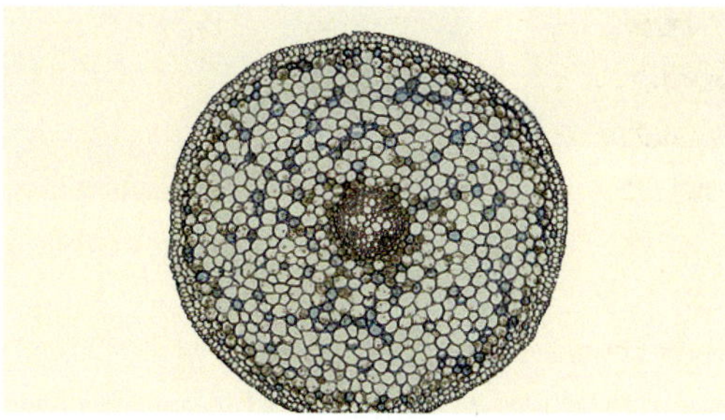

Figure 3 **Cross section through the root of a buttercup under the light microscope**

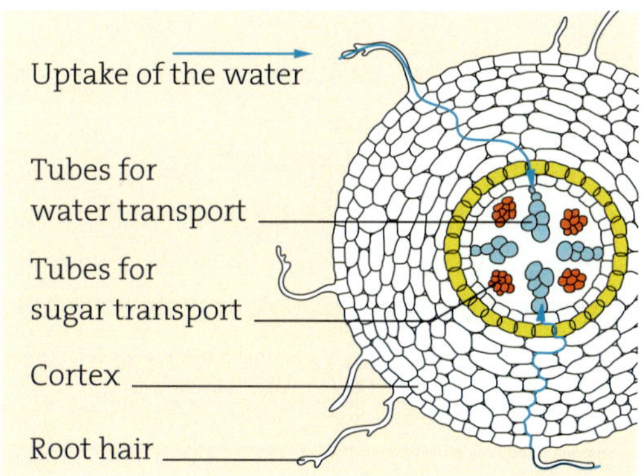

Figure 4 **Drawing of a cross section of a root in the area of the root hairs**

When a root grows, it squeezes[2] itself between the soil particles. The **growth centre**, where new cells develop by cell division[3], is at the tip of the root. It is protected by the **root cap**, a layer of dead cells. A few millimetres behind the tip of a growing root thousands of tiny **root hairs** form a kind of fur. The root hairs take up water containing dissolved mineral salts from cavities in the soil, which fill with water when it rains.

In a cross section[4] of a root you can see the tubes which transport water upwards into the stem and also the tubes which transport sugar into the root tissue (see Figures 3 and 4). In Section 1.2 we will look at these two important lines of transport in more detail.

Rape, like all plants that belong to the cabbage family[5] (see Chapter 2, pp. 194 ff.), has a taproot, a long, thick main root that grows vertically downwards. Trees like oak[6], pine[7] and fir[8] have main roots which can grow to a depth of 30 metres or more. In contrast, plants such as maize[9] and wheat[10] that belong to the grass[11] family, and also many others like the raspberry[12] plant, have a system of **fibrous roots** consisting of many main roots developing from the end of the stem, with lateral roots branching off. Such a widely distributed[13] network[14] of thin roots close to the surface of the soil allows these plants to take up a lot of water during short but heavy rain showers.

Trees with long main roots can resist strong winds, but their **trunks** (as the stem of a tree is called) may break in stormy weather, whereas trees with flat roots, like spruce[15] or poplar[16], are easily uprooted[17] and toppled over[18].

Trees in mixed **broad-leafed** and **coniferous** forests[19] show a better resistance to storms than trees growing in monoculture[20], because the various species of trees support each other, as they differ not only in their root systems, but also in the arrangement[21] and stability[22] of their branches.

Key terms

Roots
- cortex *Wurzelrinde*
- fibrous root *Faserwurzel*
- growth centre *Wachstumszentrum*
- lateral root / side root *Seitenwurzel*
- main root *Hauptwurzel*
- root *Wurzel*
- root cap *Wurzelhaube*
- root hair *Wurzelhaar*
- taproot *Pfahlwurzel*

Did you know?

The combined length of all the fibrous roots of a single rye plant can reach up to 80 km.

1 anchor sth. *etwas verankern*
2 squeeze sth. *(sich oder etwas) durchzwängen*
3 cell division *Zellteilung*
4 cross section *Querschnitt*
5 cabbage family *Familie der Kreuzblütler*
6 oak *Eiche*
7 pine *Kiefer*
8 fir *Tanne*
9 maize *Mais*
10 wheat *Weizen*
11 grass *Gras*
12 raspberry *Himbeere*
13 distribute sth. *etwas verteilen*
14 network *Netzwerk*
15 spruce *Fichte*
16 poplar *Pappel*
17 uproot sth. *etwas entwurzeln*
18 topple sth. over *etwas umkippen*
19 forest, wood *Wald*
20 monoculture *Monokultur, Anbau nur einer Art*
21 arrangement *Anordnung*
22 stability *Festigkeit*

Figure 5 Root tips with a furry coat of root hairs

Figure 6 Fibrous roots of a grass plant

1 a Use a botanical guidebook[1] to look up the names and pictures of the plant species that have been mentioned in this chapter so far. Tell your partner which ones you have seen before and where you found them.
 b Make a table in your exercise book with two columns and the headings 'Herbaceous plants' and 'Woody plants'. Fit the different species into the appropriate columns.
 c As you read the rest of this unit, add more plant names to your table.
2 How is a root able to push its way through the soil without damaging the sensitive cells in the root tip?
3 During a heavy storm some trees are uprooted, while others are bent or broken. Give an explanation for this and include examples of tree species likely to be affected in each way.
4 The fur of root hairs increases the surface area[2] through which water and minerals are taken up. Where in animals and humans do we find the same principle for increasing surface area, and what is the increased surface area used for? (You can discuss the question with your classmates or look back through previous chapters to help you remember.)

1.2 Stems provide transport and support

The stems of herbaceous plants have a different inner and outer structure from the stems of bushes and trees. However, both in herbaceous plants and in young woody plants you can see that the stem is divided by **nodes**, where the leaves are attached. The stem parts between the nodes are called **internodes**.
The water and minerals that the root hairs take up from the soil are transported upwards through tubes that branch out into leaves and flowers. This system of tubes or **vessels** is called **xylem**. Xylem vessels originate from living cells whose end walls have disappeared, forming long tubes. The strong cellulose walls of these vessels can withstand[3] the pressure of the water inside them: this is very important in high trees, where the tubes have to transport water from the soil to the tree's highest branches. The xylem also supports the plant and enables it to grow tall and straight.

Key terms

Stems
- bark *Borke*
- cambium *Kambium*
- cork *Kork*
- cotyledon *Keimblatt*
- dicot *zweikeimblättrige Pflanze*
- growth ring *Jahresring*
- internode *Stängelabschnitt, Stängelglied*
- monocot *einkeimblättrige Pflanze*
- node *Knoten*
- phloem *Siebteil*
- sieve tubes *Siebröhren*
- sugar solution *Zuckerlösung*
- vascular bundle *Gefäßbündel*
- vessel *Gefäß, Ader*
- xylem *Holzteil, Gefäßteil*

1 botanical guidebook *Pflanzenführer, Pflanzenbestimmungsbuch*
2 surface area *Oberfläche, Grundfläche*
3 withstand sth. *etwas aushalten, widerstehen*

Other much thinner tubes, or rather lines of cells, whose connecting walls are perforated[1] like a sieve[2], carry **sugar solution** from the leaves to all other parts of the plant where sugar is needed for energy production. This transport system of **sieve tubes** is called **phloem**. Unlike the xylem vessels, phloem consists of living cells. Xylem and phloem tubes are usually found close together, forming **vascular bundles** which lie in the centre of a root (see Figure 7).

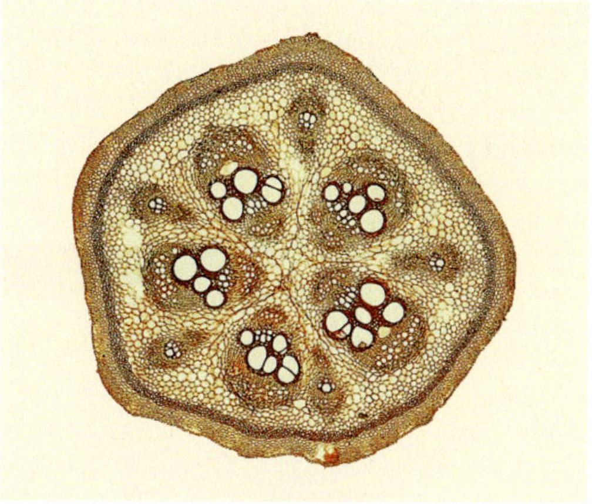

Figure 7 Vascular bundles in the stem of a herbaceous plant (cross section)

Figure 8 Bark and growth rings in the stem of a broadleafed tree (cross section)

Plants can be divided into two groups: the **dicots**, which have two **cotyledons** when they develop from a seed, and the **monocots**, which start their lives with only one cotyledon. A cotyledon is part of the embryo in the seed of a plant, which usually develops into the first leaves of a new plant. In dicots, such as rape, the vascular tubes are arranged in a ring, with the phloem near the outer edge. In monocots, by contrast, the vascular bundles are distributed irregularly across the stem. Monocots include plants that grow from a bulb, like daffodils[3], but also grasses, irises[4], orchids[5] and palms[6].

In woody plants, which are all dicots, a ring of embryonic cells called the **cambium** runs through the centre of the vascular bundles. Every year the cambium adds another ring of xylem surrounded by a ring of phloem tissue. These **growth rings** gradually increase the diameter of the stem. Together with the phloem, a layer of water resistant[7] **cork** is produced which forms **bark** around the stem. The bark protects the tree against injuries and damage by insects.

By producing a new ring of xylem and phloem every year, trees form woody stems. When you look at a cross section through the stem of a lime tree[8] or a climbing plant like the clematis[9], you can recognize the xylem and phloem.

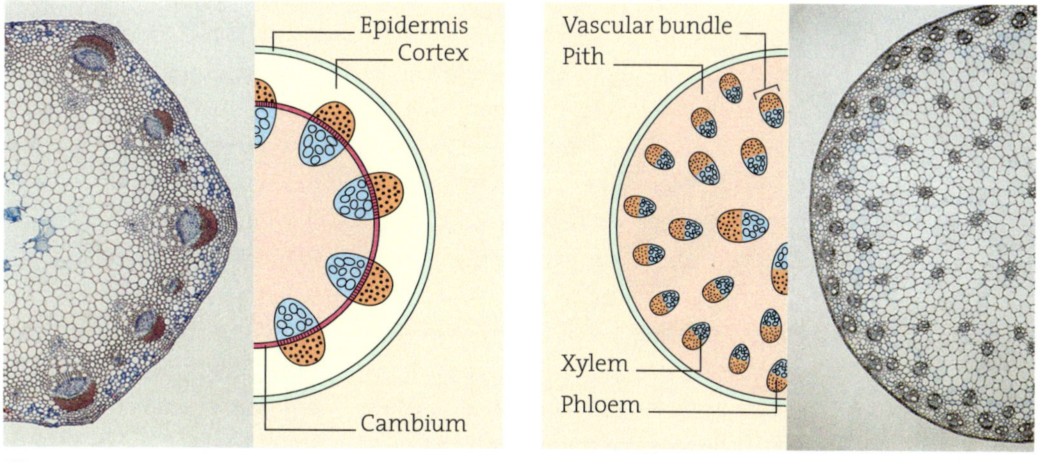

Figure 9 Cross sections of a dicot (left) and a monocot (right)

1 perforate sth. *etwas durchlöchern*
2 sieve *Sieb*
3 daffodil *Osterglocke, Narzisse*
4 iris *Schwertlilie, Iris*
5 orchid *Orchidee*
6 palm *Palme*
7 water resistant *wasserfest, wasserundurchlässig*
8 lime tree *Linde*
9 clematis *Clematis, Waldrebe*

1 Using the information in the text, think about and describe how the water-carrying vessels are arranged in the stem of: a cherry tree, a stalk[1] of meadow grass[2], a gooseberry[3] bush and a tulip[4]. Explain your answers.
2 How is the function of xylem vessels supported by their structure?
3 Explain how sieve tubes form in contrast to xylem vessels.
4 Look in botanical guidebooks and on the Internet for information about the growth rings of trees. Why are the rings wider in some years than in others?

1.3 Leaves have many different shapes

Most flowering plants have clearly visible green leaves. These develop out of buds at the tip or the sides of the **shoot** (the young stem of a developing plant). The leaves of most herbaceous plants and broadleafed bushes start to wilt[5] and drop off after a short time. The leaves are shed[6] in late autumn and new ones are produced in spring. Plants like laurel[7] and rhododendron are exceptions, as their thick and leathery leaves survive through the winter in Mediterranean climates. Coniferous trees also keep their hard and often needle-shaped leaves for several years.

Key terms

Parts of stems and leaves
- blade *Blattspreite*
- compound leaf *zusammengesetztes Blatt*
- leaf base *Blattgrund*
- leaf margin *Blattrand*
- leaf stalk (petiole) *Blattstiel*
- shoot *Spross*
- tendril *Blattranke*
- vein (of a leaf) *Blattader*

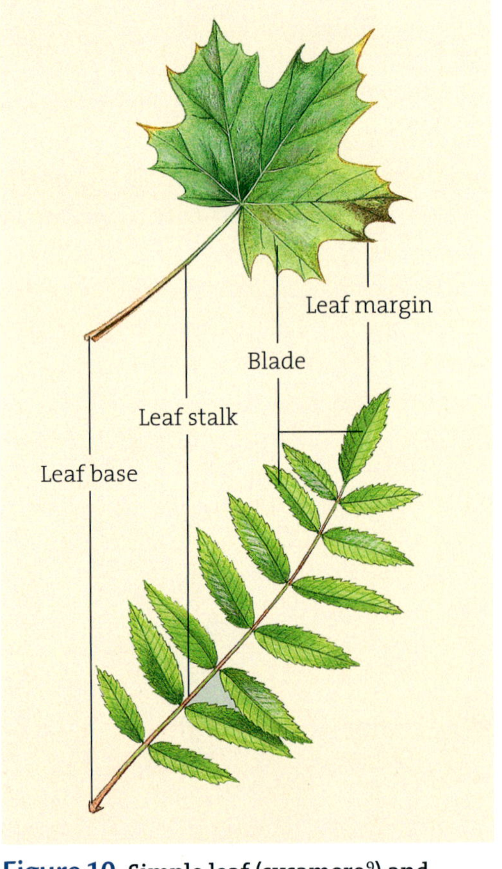

Figure 10 **Simple leaf (sycamore[9]) and compound leaf (mountain ash[10])**

Leaves vary[8] greatly in shape and colour, but most of them have a broad part called the **blade**, a **leaf stalk** and the **leaf base** where the leaf stalk is attached to the stem. The tubes in the leaf continue in the stem and are responsible for transporting substances to and from the leaves, which contain xylem and phloem tubes. In grasses and other monocots these **veins** run parallel, while in most flowering plants, which are dicots, they are branched and form a network.

Some plants have simple, undivided leaves, while others have leaves that are divided into many parts, called **compound leaves**. The shape and the **margin** of a leaf can also vary widely. Closely related plants can be recognized by the similarity of their leaves. You often only need a single leaf to know the name of a tree, which is why leaves are very useful for identifying plants. This is especially useful when plants have no flowers. Figures 10–13 show the main differences in the shape and outer structure of leaves. There are also different ways in which leaves are arranged on a stem, but the type of arrangement is constant in each species.

1 stalk *Halm, Stiel, Stängel*
2 meadow grass *Wiesenrispengras*
3 gooseberry *Stachelbeere*
4 tulip *Tulpe*
5 wilt *welken*
6 shed sth. *etwas abwerfen*
7 laurel *Lorbeer*
8 vary *variieren*
9 sycamore *Berg-Ahorn*
10 mountain ash *Eberesche, Vogelbeere*

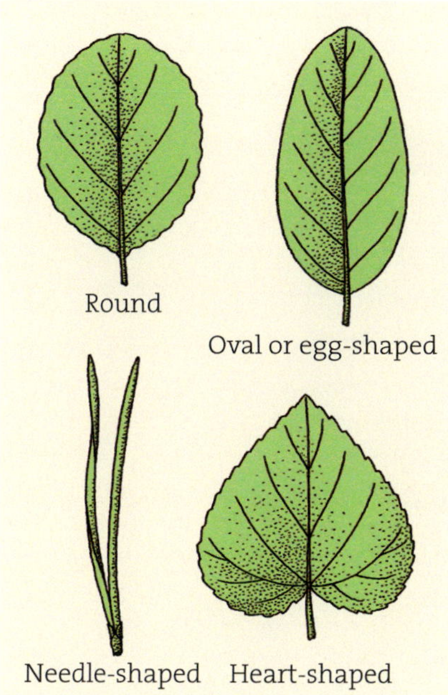

Figure 11 Different leaf shapes

Figure 12 Different leaf margins

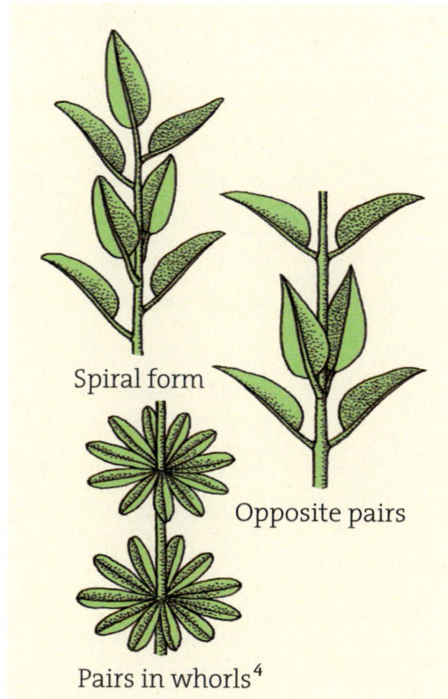

Figure 13 Arrangement of leaves on the stem

1 What kind of leaves can stay on a tree in winter? Give examples of trees that do not shed their leaves.

2 Collect two different kinds of leaves and describe them to a partner using words from Figures 10–13. Then use a botanical guidebook to find out which species of trees the leaves belong to. The phrases in the **Activate your English** box will help you.

Activate your English

Describing leaves

... is shaped / has a shape like ...
... the margin of this leaf is ... (shaped)
... the leaves are arranged ...
This is / looks like / appears to be a(n) ... leaf / a leaf from a(n) ... tree.

Did you know?

The hard spines[5] of cacti[6] and the **tendrils** of sweet peas[7] are modified leaves. Leaves are adapted to the conditions under which plants have to live, so plants that grow in dry climates have developed small, leathery leaves or leaves with an extremely reduced surface, like those of cacti. By having some of their leaves change into tendrils, sweet peas can climb upwards and provide their leaves with enough sunlight for photosynthesis (see Section 1.5) to take place.

Spines on the skin of a cactus (left); the tendrils of a pea plant (right)

1 toothed *gezähnt*
2 serrated *gesägt*
3 lobed *gebuchtet*
4 in whorls *wirtelförmig*
5 spine *(hier) Stachel*
6 cactus (pl. cacti) *Kaktus*
7 sweet pea *Wicke*

1.4 Transpiration takes place in the leaves

Leaves play an important role in the transport of water and minerals into the upper parts of the plant.
More than 90 per cent of the water which the plant takes up through its root hairs eventually leaves the plant. This occurs partly through **evaporation** from the surface of the leaves, but mainly in the form of **transpiration** through the **stomata**, tiny openings in the **epidermis** which cannot be seen with the naked eye. Under the microscope you can see that each stoma is flanked[1] by a pair of kidney-shaped **guard cells**, which can open and close the stoma by changing their shape. In most plants the stomata are distributed throughout the lower epidermis.
The evaporated water must be replaced to prevent the plant from drying out. Therefore the water which is taken up through the roots must be transported upwards in the xylem tubes of the roots, the stem and the leaves. The transpiration of water through the stomata is the main force that pulls the water upwards, but it is also important that a constant flow of water **molecules** is provided.
This happens because water molecules cling to each other and also onto the walls of the narrow and airtight[2] tubes of the xylem vessels. You may have experienced the unfavourable[3] effect of these forces if you have ever placed a narrow straw into a full soft drink[4] bottle without starting to suck at the straw right away. A force called **capillary action** drives the fluid up and out of the straw and onto the table or your clothes.
Transpiration has astonishing forces which can transport water and dissolved minerals upward to heights of a hundred and more metres in tall trees like the giant redwood[5] or eucalyptus[6] trees. But transpiration not only provides all parts of a plant with water and minerals but also has the beneficial effect of cooling the surface of a plant in hot weather.

Key terms

Transpiration
- capillary action *Kapillarwirkung*
- epidermis *Epidermis*
- evaporation *Verdunstung (über eine freie Oberfläche)*
- guard cell *Spaltöffnungszelle*
- molecule *Molekül*
- stoma (pl. stomata) *Spaltöffnung*
- transpiration *Verdunstung (über die Spaltöffnungen)*

Figure 14 Stomata and guard cells in the lower epidermis of a leaf, as seen through a light microscope

1 flank sb./sth *jdn./etwas flankieren*
2 airtight *luftdicht*
3 unfavourable *ungünstig, nachteilig*
4 soft drink *Limonade*
5 giant redwood *Mammutbaum*
6 eucalyptus tree *Eukalyptusbaum*
7 lettuce *Salat*

1 Why does the inside of a closed plastic bag become damp when you use it to carry fresh lettuce[7] home from the market?
2 What is the difference between evaporation and transpiration?
3 Why do certain plants such as the pine tree have leaves with a very small surface area?
4 Explain how a plant is able to provide all its organs with water and minerals.

1.5 Photosynthesis is a vital[1] process

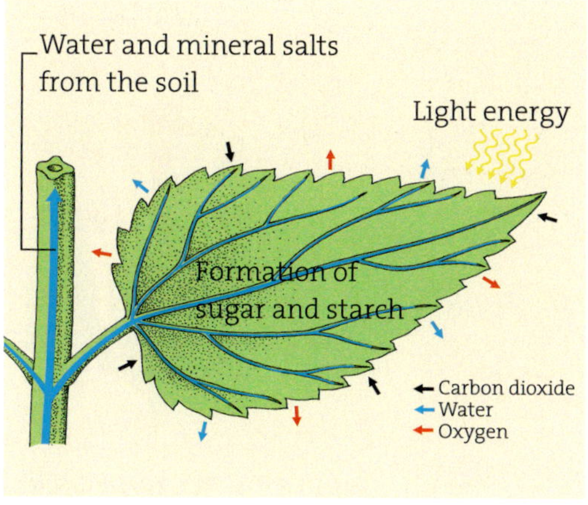

Figure 15 **Formation of nutrients in the leaf of a green plant**

Key terms

Photosynthesis
- carbon dioxide *Kohlenhydroxid*
- chlorophyll *Chlorophyll*
- glucose *Glukose*
- light energy *Lichtenergie*
- nitrogen *Stickstoff*
- oxygen *Sauerstoff*
- photosynthesis *Fotosynthese*
- pigment *Farbstoff*
- respiration *Atmung*
- starch *Stärke*

Photosynthesis is a unique[2] process that enables plants to produce their own food. It occurs in all green parts of a plant, particularly in the leaves.
Besides **nitrogen** and **oxygen**, the air in the atmosphere[3] contains a small amount of **carbon dioxide** (0.03%), which is produced by all organisms through **respiration**. The carbon dioxide enters the leaves of green plants through the stomata and reacts[4] with the water molecules which the plant has taken up from the soil, producing sugar and oxygen. The energy for this reaction comes from sunlight, which is trapped[5] by **chlorophyll**, the green **pigment** in the leaf cells. The **light energy** is stored in chemical form first in **glucose** and later in **starch** molecules, which are not water-soluble and can be stored in the plant cells. This is a complicated process which takes place in a series of small steps. Photosynthesis can be summed up with this basic[6] equation[7]:

$$\text{Carbon dioxide + Water} \xrightarrow[\text{Chlorophyll}]{\text{Light energy}} \text{Glucose + Oxygen}$$

Reactants[8] — *Products*[9]

Photosynthesis is the reason why plants do not have to eat. They produce sugar, which they later burn in respiration to gain the energy they need for their metabolic processes[10]. They can also use the sugar to produce a great variety of complex carbohydrates, proteins and fats.
Only green plants can produce these organic substances from simple inorganic ones (carbon dioxide and water). Therefore, all animals and humans depend on plants for food and oxygen. The concentration of oxygen in the atmosphere is 21 per cent, and two thirds of this is produced by the phytoplankton[11] in the sea and in fresh water.

1 vital *lebenswichtig*
2 unique *einzigartig*
3 atmosphere *Atmosphäre*
4 react *reagieren*
5 trap sb./sth. *jdn./etwas einfangen, fangen*
6 basic *grundlegend*
7 equation *Gleichung*
8 reactant *Reaktionspartner*
9 product *(hier) Reaktionsprodukt*
10 metabolic process *Stoffwechselprozess*
11 phytoplankton *pflanzliches Plankton, Phytoplankton*

1 Peas contain protein and the seeds of rape and sunflower contain fat. How do these organic substances get into the plants?
2 Why can we say that all the energy that is used by organisms comes originally from the sun?
3 Explain why it is that we have to eat, but plants do not. Have you ever seen or heard of plants that do eat? Discuss your answers with a partner, then share them with the rest of the class.
4 Collect information about the shedding of leaves in autumn. Write a short report and present your results to your class.

1.6 Flowers are the organs of reproduction

Key terms

Flowers
- anther *Staubbeutel*
- flower bud *Blütenknospe*
- carpel (pistil) *Fruchtblatt*
- corolla *Blütenkrone*
- filament *Staubfaden*
- flower diagram *Blütendiagramm*
- flower stalk *Blütenstiel*
- inflorescence *Blütenstand*
- nectar *Nektar*
- nectary *Nektarium*
- ovary *Fruchtknoten*
- ovule *Samenanlage*
- petal *Blütenblatt*
- pollen sac *Pollensack*
- receptacle *Blütenboden*
- sepal *Kelchblatt*
- stamen *Staubblatt*
- stigma *Stigma, Narbe*
- style *Griffel*

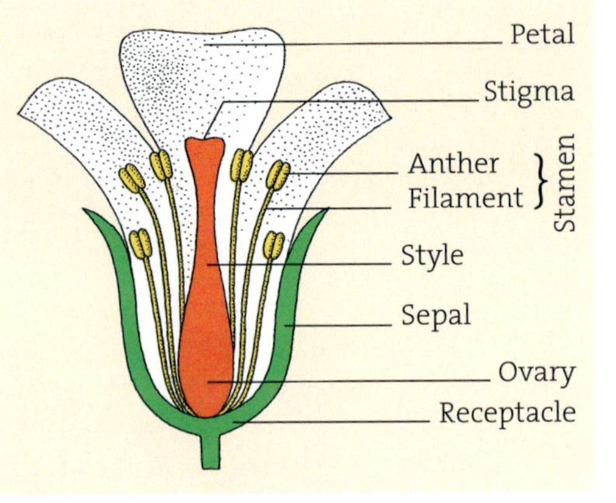

Figure 16 **The flower of a rape plant (longitudinal section[1])**

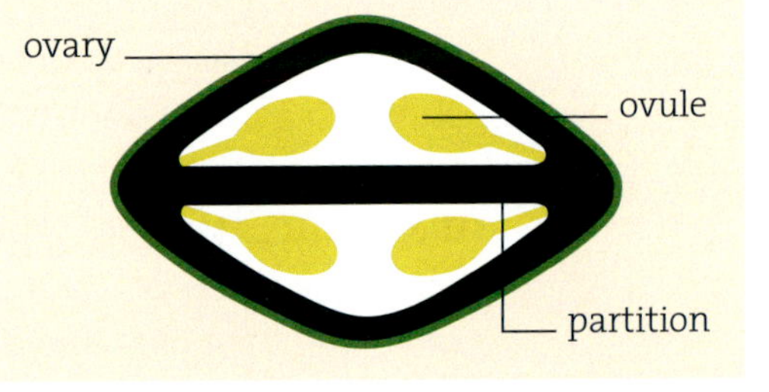

Figure 17 **Cross section of the rape ovary**

As everyone knows, flowers are often beautifully shaped and coloured. What most people do not know is that flowers are specialized leaves. In contrast to green leaves, they do not have the function of producing food, but instead they contain the male and female organs necessary for sexual reproduction. The attractive colours and different scents[2] help the plants to reproduce.

Flowers can look very different, but despite the variety of shapes and colours, you can identify the following parts on most flowers: **sepals**, **petals**, **stamens** and **carpels**. These parts sit on the **receptacle**, which can either be swollen and turned upwards or cup-shaped[3]. Sepals enclose and protect the **flower bud** and are mostly green in colour. In the fully developed flower they are either fused or are separate leaves, as in the rape or cherry flower (see Figure 16 and p. 205, Figures 37 and 38). The petals which follow on the inside are mostly coloured and scented[4], because their function is to attract insects or other creatures to visit the flower. Often there are **nectaries** at the base of the petals, which produce a sugary liquid called **nectar**. The nectar attracts visiting animals because it serves as their food. Like the sepals, the petals can either be fused, forming a tube, or remain separate and are called free petals. In some flowers, such as tulips[5], the sepals are missing. All the petals together form the **corolla**.

The arrangement of the sepals is an important characteristic which is used by botanists to assign[6] plants to a particular plant family.

Inside the corolla of the rape flower there are six stamens, two of which are shorter than the rest. Each stamen consists of a thin **filament** with an **anther** at its top. The anther contains four **pollen sacs**, in which pollen grains are formed. The anther is the male part of the flower and the pollen grains can be compared to an animal's sperms.

The female part of the flower, the carpel, lies in the centre of the flower. It consists of three parts: the forked **stigma** at the top, followed by the more slender[7] **style** and the swollen **ovary** at the bottom. The ovary contains one or more **ovules**, small bodies which contain the egg cell.

1 longitudinal section *Längsschnitt*
2 scent *Duft*
3 cup-shaped *kelchförmig*
4 scented *duftend*
5 tulip *Tulpe*
6 assign sb./sth. to sth. *jdm./etwas etwas zuordnen*
7 slender *schlank, schmal*

In the rape flower the ovary is made up of two fused carpels. It is separated into two halves by a thin partition[1] with several ovules attached to both sides (see Figure 17). In other plants, for example the cherry tree, the ovary contains one ovule with one egg cell.
When male and female parts are combined in one flower, it is called a hermaphrodite flower.

Figure 18 **A radial flower (dogrose)**

Figure 19 **A bilateral flower (sweet pea)**

Flowers can differ in symmetry[2], as well as in the number of sepals, petals and stamens. Regarding their symmetry, flowers are either radial or bilateral. In Figure 19 you can see that a bilaterally symmetric flower can be divided into two identical halves.
If you count the numbers of petals in the flowers of different species, you will find that some species have petals in multiples[3] of three, but most flowers with a corolla have petals in multiples of four and five.
Often several flowers are arranged in a specific way on the **flower stalk**. The botanical name for such a cluster of flowers is **inflorescence**. The most frequent types of inflorescence are shown in Figure 20.

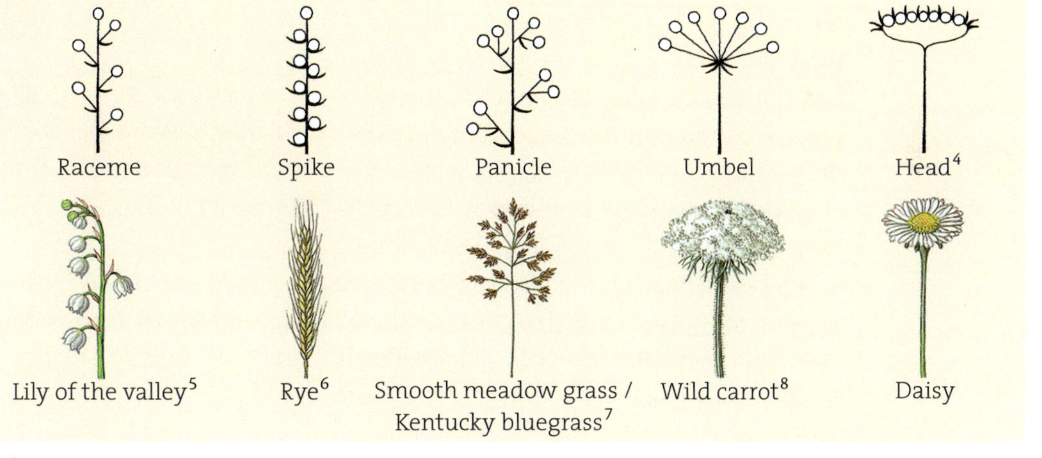

Figure 20 **Different kinds of inflorescence (schematic, with a corresponding species of flowering plant)**

All flowering plants are grouped into different families according to the characteristic differences in the structure of their flowers (the number and position of the sepals, petals and stamens, and the structure of the carpel). A **flower diagram** demonstrates the special arrangement of the main flower parts for all plants that belong to the same family (see Chapter 22, 'Activities', p. 205).

1 What is the role of stamens and carpels in the reproduction of flowering plants?
2 Why do animals visit flowers?
3 Which part of the flower is male and which is female? Make a list of all the male parts and all the female parts. Compare you results with those of a partner.

1 partition *Trennwand*
2 symmetry *Symmetrie*
3 multiples of sth. (a number) *ein Vielfaches von (einer Zahl)*
4 head *(hier) Kopf, Korb*
5 lily of the valley *Maiglöckchen*
6 rye *Roggen*
7 smooth meadow grass, Kentucky bluegrass *bläuliches Wiesen-Rispengras*
8 wild carrot *Wilde Möhre*

1.7 Sexual reproduction begins with pollination

Sexual reproduction in plants is only possible when the sex cell in the pollen grain fuses with the egg cell in the ovary. As the pollen grains in flowering plants cannot move by themselves, they have to be transferred somehow from an anther to a stigma. This can be done by insects or other animals, by wind and in some rare cases even by water. This process of transfer is called **pollination**.

If the pollen falls onto the stigma of the same flower we call this **self-pollination**. If the pollen is carried to the stigma of another plant, we talk about **cross-pollination**. It is much better for a plant species if the pollen is transferred to the flower of a different plant of the same species. Many plants avoid self-pollination by developing their anthers and stigma at different times.

Key terms

Pollination
- beehive *Bienenstock*
- catkin *Blütenkätzchen*
- cross-pollination *Fremdbestäubung*
- fruit bat *Fruchtfledermaus*
- hummingbird *Kolibri*
- pollen basket *Pollenkörbchen*
- self-pollination *Selbstbestäubung*

Pollination by insects

One common way of pollination is by insects. Plants such as rape, apple and foxglove[1] have colourful flowers with scented petals and nectar. These flowers attract insects, which eat or collect the nectar and the pollen grains.

Butterflies are adapted to drink nectar which collects at the base of large petals. Their long tube-like[2] mouth parts are normally coiled up[3], but can be extended to reach the nectaries at the base of the petals. As a butterfly drinks the nectar, the pollen sticks to its body.

Figure 21 Butterfly sucking nectar

Honeybees have special stomachs for transporting the nectar to the **beehive**. In these stomachs the nectar is changed into honey, which is used later to feed the bee larvae[4].

When a bee sticks its head into a flower, the pollen grains cling to its hairy body. Then it uses its front legs to brush the pollen into the **pollen baskets** on its third pair of legs. The pollen, which is rich in protein, is used as food for the growing bee larvae.

Figure 22 Honeybee collecting pollen

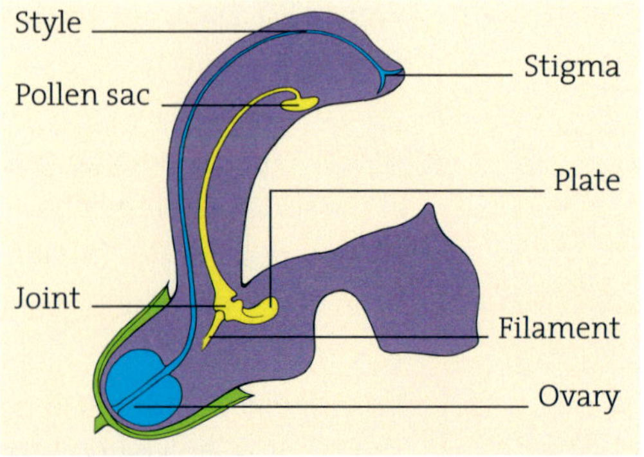

Figure 23 Inner structure of the meadow clary flower[5]

1 foxglove *Fingerhut*
2 tube-like *röhrenförmig*
3 coil sth. up *etwas aufrollen*
4 larva (pl. larvae) *Larve*
5 meadow clary *Wiesen-Salbei*

The meadow clary, a charming[1] herbaceous plant that grows wild in calcareous[2] meadows and on woodland borders, has been introduced into gardens because people enjoy its attractive blue flowers. This plant has developed a special mechanism to support its pollination. If a big honeybee or bumblebee lands on the lower lip of a flower and stretches its tongue into the flower tube to reach the nectar, its head pushes against a plate at the lower end of the two stamens. This works like a lever[3]: the two stamens immediately swing down and push the anthers onto the back of the insect. The impact[4] makes the stamens open, and the pollen grains are emptied onto the hairy body of the bee. If the insect then looks for nectar in an older meadow clary flower whose stamens have already dried and shrunk, its back rubs against the style and the flower is pollinated.

Figure 24 **Honeybee sucking nectar from a meadow clary flower**

The pollen grains often have a rough surface, which helps them to stick to an insect's body. When the insect visits another flower, some of the pollen is rubbed onto the stigmas. This process is called pollination.

In tropical countries pollination is also carried out by birds and bats. **Hummingbirds**, which drink nectar while hovering[5] with rotating[6] wings in front of flowers, transfer pollen from flower to flower on their thin beaks[7] or their foreheads[8]. **Fruit bats** also drink nectar and help to pollinate flowers.

Figure 25 **Hummingbird hovering in front of a tropical flower**

Figure 26 **Pussy willows, the male catkins of a willow tree**

Besides animals, wind plays an important role in the pollination of many plants. The flowers of grasses and the **catkins** of birch[9], alder[10] und poplar[11] have no coloured petals and no nectar. The flowers of beech[12] and oak[13] trees are also small and inconspicuous[14]. They are not visited by insects and instead rely[15] on the wind for pollination. Because of this, the flowers produce large amounts of tiny pollen grains, which are so light that they are easily carried away by the wind.

The stigmas of wind-pollinated flowers are often long and feathery, which helps them to catch the airborne pollen grains. Trees and bushes with catkins have separate male and female flowers. In the case of trees like hazel[16], birch and oak, both types of flowers are on one plant, whereas in poplars and willows[17] the male and female flowers are on different plants.

1 charming *hübsch*
2 calcareous *kalkhaltig*
3 lever *Hebel*
4 impact *Aufprall*
5 hover *schweben*
6 rotate *kreisen, rotieren*
7 beak *Schnabel eines Vogels*
8 forehead *Stirn*
9 birch *Birke*
10 alder *Erle*
11 poplar *Pappel*
12 beech *Buche*
13 oak *Eiche*
14 inconspicuous *unscheinbar*
15 rely on sth. *auf etwas angewiesen sein*
16 hazel *Hasel*
17 willow *Weide*

Figure 27 Left to right: separate female and male flowers on a hazel bush

1 Describe the different ways in which flowers are pollinated by animals.
2 Explain how wind-pollinated flowers differ from those that are pollinated by insects.
3 Discuss in class why it is better for a plant species to reproduce by cross-pollination than by self-pollination.
4 Find out: which of our common trees have hermaphrodite flowers and which have flowers with separate sexes; and which tree species have male and female flowers on the same plant, and which have them on different plants. Write a short report with pictures of the different flowers and fruits. Maybe you can even find the trees outside and take your own pictures. If this is not possible, draw your own pictures or download them from the Internet.

1.8 Fertilization, seed and fruit

Key terms

Fertilization
- core *Kerngehäuse*
- fertilization *Befruchtung*
- fruit *Frucht*
- nucleus *Zellkern*
- partition *Trennwand*
- pollen tube *Pollenschlauch*
- seed *Same*

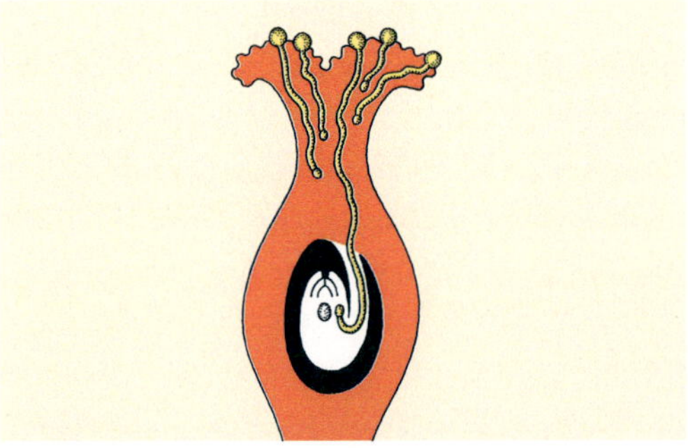

Figure 28 Pollen tubes growing through the style

After a pollen grain lands on a stigma, a **pollen tube** starts to grow into the stigma and down the style. The **nucleus** of the sex cell moves to the tip of the growing pollen tube. When the tube reaches an ovary, it pushes its way into the ovule and grows towards the egg cell. When the pollen tube bursts, the nucleus of the male sex cell fuses with the egg cell and **fertilization** occurs.
The fertilized egg cell starts to divide and develops into the plant embryo. The embryo and the surrounding ovule become the seed. The tissue which develops out of the ovule supplies the embryo with food as it grows.
As the seed develops, the sepals, petals and stamens die and drop off. The ovary gradually grows and changes its shape, and together with the seeds inside it becomes the **fruit**. The cherries you eat are actually the ovaries which have become thick and juicy[1]. An apple or a pear is formed by the swollen receptacle[2] of the flower, which becomes fleshy[3], enclosing the **core** with its brown **seeds**.

1 juicy *saftig*
2 receptacle *Blütenboden*
3 fleshy *fleischig*

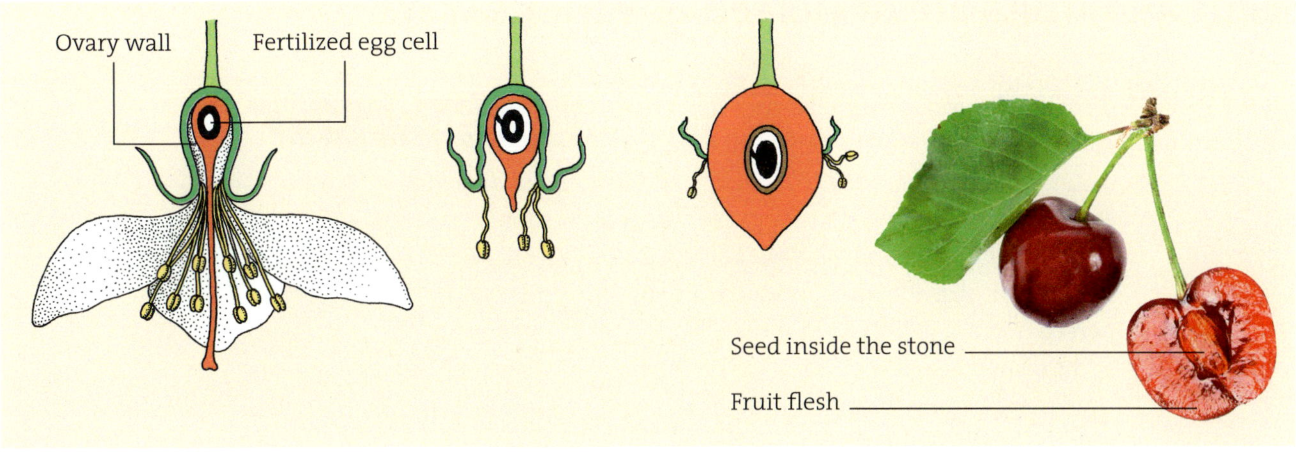

Figure 29 **Development of a fertilized cherry flower into a ripe fruit**

The fruit of the rape plant, which develops out of two fused carpels, dries out when it gets ripe. The two halves of the fruit split open and the transparent[1] **partition** becomes visible with the ripe seeds fastened on both sides.

Figure 30 **An apple cut in half**

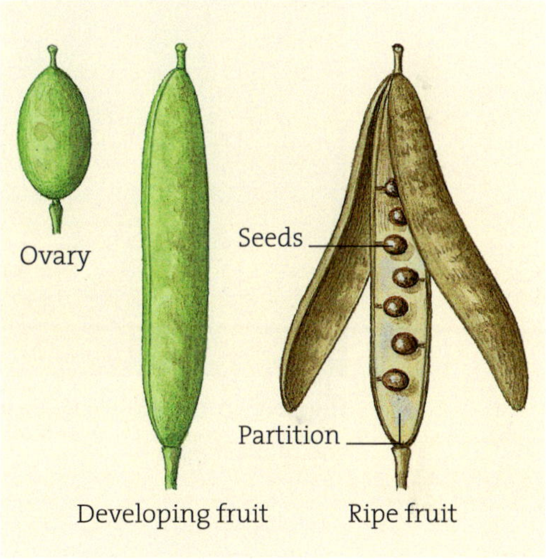

Figure 31 **Ovary and fruit of the rape plant**

1 transparent *durchsichtig*

1. Make a flow chart to show the process that takes place from the pollination of a flower to the formation of the seeds. For help see 'Working with flow charts' p. 84.
2. Use Figure 29 to explain how a cherry develops.
3. Why is it that some flowers do not develop into fruits?
4. Compare pollination and fertilization in flowering plants with the equivalent processes in humans.
5. Collect the following: a pod of a pea or bean, a ripe tomato or gooseberry, an apple or a pear, a kiwi fruit and a rosehip. Try to find the rest of the flower stalk, the sepals and petals on the outside. Open the pod of the pea and use a sharp knife or scalpel to slice the other fruits from top to bottom. (If this is not possible, look for pictures of the sliced fruits in books or on the Internet.) Then make drawings which show the inside structure of the different fruits. Label all the parts of the fruits in your drawings which you can recognize.
6. Collect the fruits and seeds of trees in your neighbourhood and glue them on a piece of cardboard together with descriptions and explanations of how these fruits are formed. You will find information on common trees and their fruits in guidebooks or on the Internet. Give a short presentation to your class about the fruits and seeds of trees using your poster to illustrate.

1.9 Seeds and fruits are dispersed in various ways

Key terms

Seed dispersal
- capsule *Kapsel*
- dispersal *Verbreitung*
- disperse sth. *etwas verbreiten, etwas verstreuen*
- scatter sth. *etwas verstreuen*

Most plants produce hundreds or even thousands of fruits and seeds. If they fell onto the ground next to the plant, they would have to compete with each other and with the parent plant for light, water and minerals. Therefore it is important that the fruits and seeds are **dispersed** over a large area, so that the offspring[1] of the plant reach new places with favourable conditions for growth and reproduction. This process of **dispersal** is done either by the plant itself, by the wind, by water, by animals or by humans. Some plants of the pea family (such as gorse[2], broom[3] and lupin[4]) burst open and fling out[5] their seeds when their fruits get dry. The seeds can travel for many metres. In geranium[6] fruits the sides curl[7] upwards[8] when they are touched and the tiny seeds are **scattered** in all directions.

Figure 32 Geranium seeds are scattered by the plant itself

Figure 33 Winged fruits of trees

You may have seen in springtime how the fruits of some trees have wings of various shapes which enable them to travel in the air. The wind carries them far away from the parent plant. Dandelion, clematis and thistle[9] seeds have feathery hairs protruding[10] from the fruits or seeds, which act as parachutes[11]. The wind can lift them high into the sky and scatter them over large distances. The seed **capsules** of the poppy[12] sit on high flower stems. Holes open up on all sides when the capsule dries out, and when the wind shakes the flower stalk, the tiny seeds are scattered like from a pepper shaker.

Figure 34 Poppy seeds are shed from the dried-out capsule

Figure 35 Dandelion plant and its downy[13] fruits

1 offspring *Nachkommen*
2 gorse *Stechginster*
3 broom *Besenginster*
4 lupin *Lupine*
5 fling sth. out *etwas hinauswerfen, herausschleudern*
6 geranium *Geranie*
7 curl *rollen, einrollen*
8 upwards *aufwärts*
9 thistle *Distel*
10 protrude *herausragen, hervorstehen*
11 parachute *Fallschirm*
12 poppy *Mohn*
13 downy *flaumig*

The fruits of many water plants or plants that grow near the water (such as water lily[1], iris[2] and alder) are very light because they contain air-filled tissues. They drift along the surface of the water until the air has escaped from the air chambers and the fruits sink to the bottom and start to grow. They may also be driven ashore[3] where they root in the moist soil and develop into herbaceous plants or trees.

The best-known example of a fruit that is dispersed by water is the coconut. The fruits of the coconut palm[4] are carried by the sea from the coast of the mainland to an island or from one island to another. Here, too, air is trapped inside the nut, which keeps it afloat[5].

Fruits and seeds are often dispersed by animals or humans. The seed coats or fruits of some plants have hooks which cling[6] to the fur of animals or the clothes of people who brush past them. Typical examples are the fruits of the woolly burdock[7] or goosegrass[8], whose stems and leaves stick to your clothes as well. After having been carried along for a while, they are rubbed off and fall to the ground.

Many juicy fruits like those of the blackberry[9] are eaten by birds or small mammals. The seeds inside the berries pass through their bodies unharmed and are excreted. If they fall on good soil, they will develop into new plants.

Figure 36 Burdock fruit

Figure 37 A dormouse[10] feeding on blackberries

Figure 38 Squirrel feeding on nuts

The nuts of hazel and beech or the acorns[11] of oak trees are the favourite food for squirrels[12] and many species of birds. Squirrels, mice and jays[13] hide fruits and seeds in the soil as food for the winter. They often do not find them again and the seeds start to germinate when the soil warms up in springtime.

1 Working with a partner, name the main forms of seed dispersal. Make a list of them in your exercise book.

2 Describe some forms of self-dispersal and give examples of plants that use them.

3 Explain in your own words why the dispersal of seeds is important for the survival of plant species. What would happen if the seeds landed right next to the parent plant?

4 What do seeds and fruits that are scattered by the wind look like? Make a sketch in your notebook of one you have observed outdoors.

5 In small groups, discuss your own experiences of seeds that fly in the wind or stick to your body and clothes. How will you see them differently in the future?

Did you know?

In 1941 a Swiss engineer named Georges de Mestral was given an idea by the burdock fruits that stuck to his clothes during a country walk. He studied the fruits under a microscope and went on to create a man-made version, using one strip of fabric covered in tiny hooks and another covered in tiny loops. Although many people laughed at his invention at the time, the Velcro™[14] fastener he invented is now a part of our everyday life, found on clothes, shoes, bags and countless other items[15].

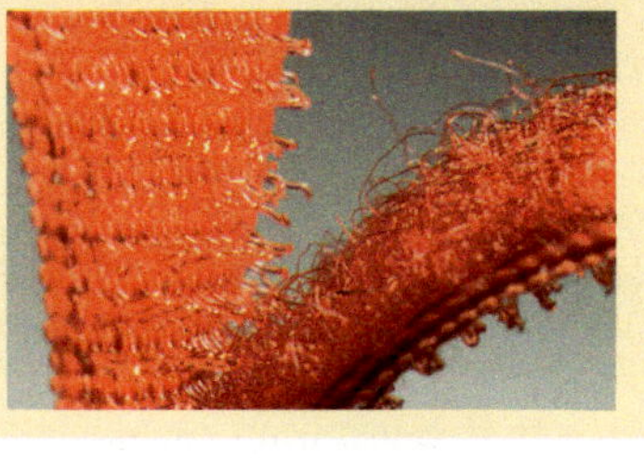

1 water lily *Wasserlilie*
2 iris *Schwertlilie, Iris*
3 ashore *an Land*
4 coconut palm *Kokusnuss-Palme*
5 keep sth. afloat *etwas schwimmfähig halten*
6 cling to sth. *an etwas haften, hängen*
7 woolly burdock *Filz-Klette*
8 goosegrass *Klebkraut*
9 blackberry *Brombeere*
10 dormouse *Haselmaus*
11 acorn *Eichel*
12 squirrel *Eichhörnchen*
13 jay *Eichelhäher*
14 Velcro™ fastener *Klettverschluss*
15 item *Gegenstand, Objekt*

1.10 Seeds can germinate and develop into new plants

When seeds are dispersed, many of them die because they do not find suitable places and conditions in which to **germinate** and grow into a new plant. Many seeds contain very little water, such as the dry seeds of a bean or pea plant. A hard outer covering, the **testa**, protects the main part of the seed, the tiny embryo. Though the seed appears to be dead, it is only in a state of **dormancy** and can remain **dormant** for a long time until the conditions around it are right for the seed to germinate.

Figure 39 shows the different structures inside a bean seed that has swollen up after being placed in water for several hours. The absorption of water by the dry seed is a purely physical process, which leads to a softening of the tissues. This is the signal for the embryo to wake up. Germination will begin if a second factor, namely the outer temperature, is suitable for this process.

When the water penetrates the testa, the **cotyledons**, which contain the food for the developing embryo, start to swell up and cause the testa to burst. At the same time, the soil particles which surround the seed are pushed aside. This provides space for the **radicle** to emerge from the seed and grow vertically[1] down into the soil. The radicle is covered with a thick fur of root hairs and absorbs water and minerals from the soil, which are transported to the rest of the seed. The radicle becomes the main root of the new plant, and lateral roots develop soon afterwards.

1 vertical *senkrecht, vertikal*

Key terms

Germination
- coleoptile *Keimscheide*
- cotyledon *Keimblatt*
- dormancy *Samenruhe, Keimruhe*
- dormant *ruhend*
- endosperm *Nährgewebe, Endosperm*
- germinate *keimen*
- plumule *Sprossknospe*
- radicle *Keimwurzel*
- seedling *Keimling*
- testa *Samenschale*

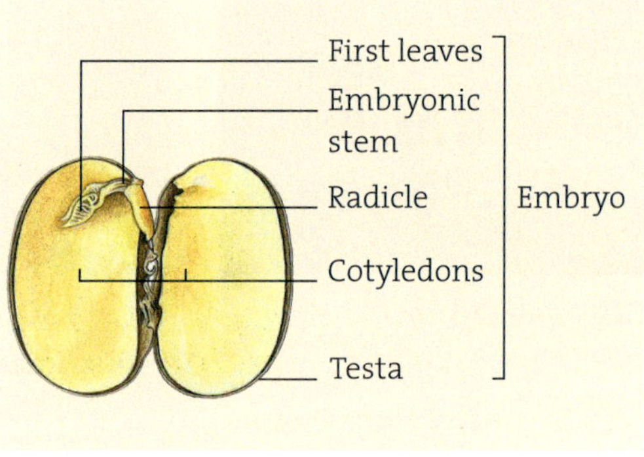

Figure 39 Bean embryo between the two cotyledons of an opened seed

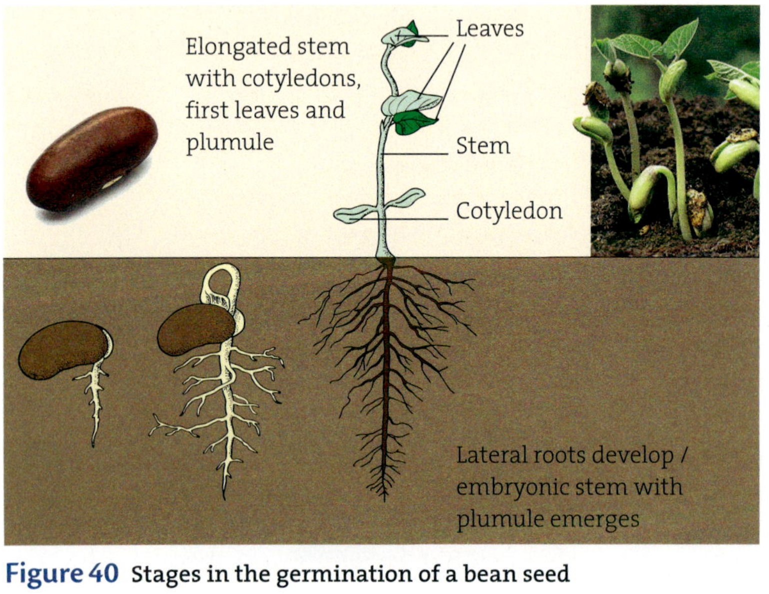

Figure 40 Stages in the germination of a bean seed

Did you know?

If wheat is properly stored under dry conditions, the embryo inside remains alive for up to 15 years and the seeds will germinate if suitable conditions are provided.
During the time of dormancy the seeds respire, as they need oxygen, and use some of the food stored for the embryo.

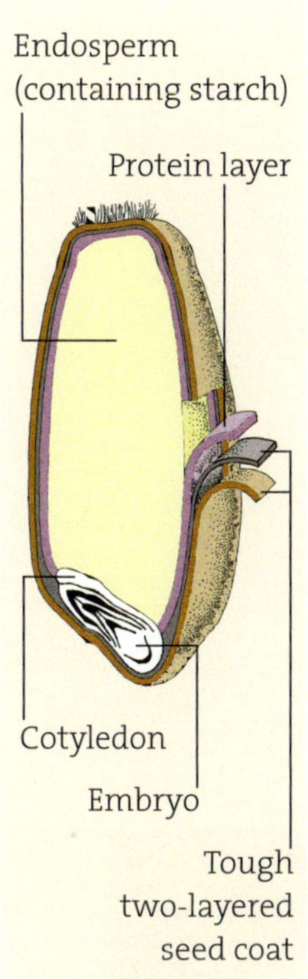

Figure 41 **The inside of a wheat grain (longitudinal section)**

Once the root is firmly anchored in the soil, the stem starts to grow. In the garden bean[1] the rapidly growing stem pulls the cotyledons out of the testa and lifts them upwards through the soil. As soon as the stem is above the soil, it straightens[2], the cotyledons separate and the **plumule** becomes visible.

As soon as they are exposed to light, the plumule and later also the embryonic stem turn green and start to photosynthesize (see Section 1.5). After a few days the cotyledons begin to shrink[3] and wither[4]. The stem grows longer and the first two leaves open up. The cotyledons, whose food storage has been used up, fall off. Now germination has come to an end and the embryo is an independent young plant, a **seedling**, which can develop into a mature[5] bean plant.

Not all seeds develop in exactly the same way as the bean seed, which belongs to the dicots. Plants of the grass family, to which all cereal plants belong, are monocots. The food in their seeds is not stored in the single cotyledon, but in a mass of tissue next to the cotyledon called **endosperm**.

The first leaves of a grass plant are protected by a hard sheath[6], the **coleoptile**, which splits open when the leaves fold out and start to photosynthesize.

1 garden bean *Gemüsebohne, Gartenbohne*
2 straighten *(sich) aufrichten*
3 shrink *schrumpfen*
4 wither *welken, verwelken*
5 mature *reif*
6 sheath *Hülle, Scheide*

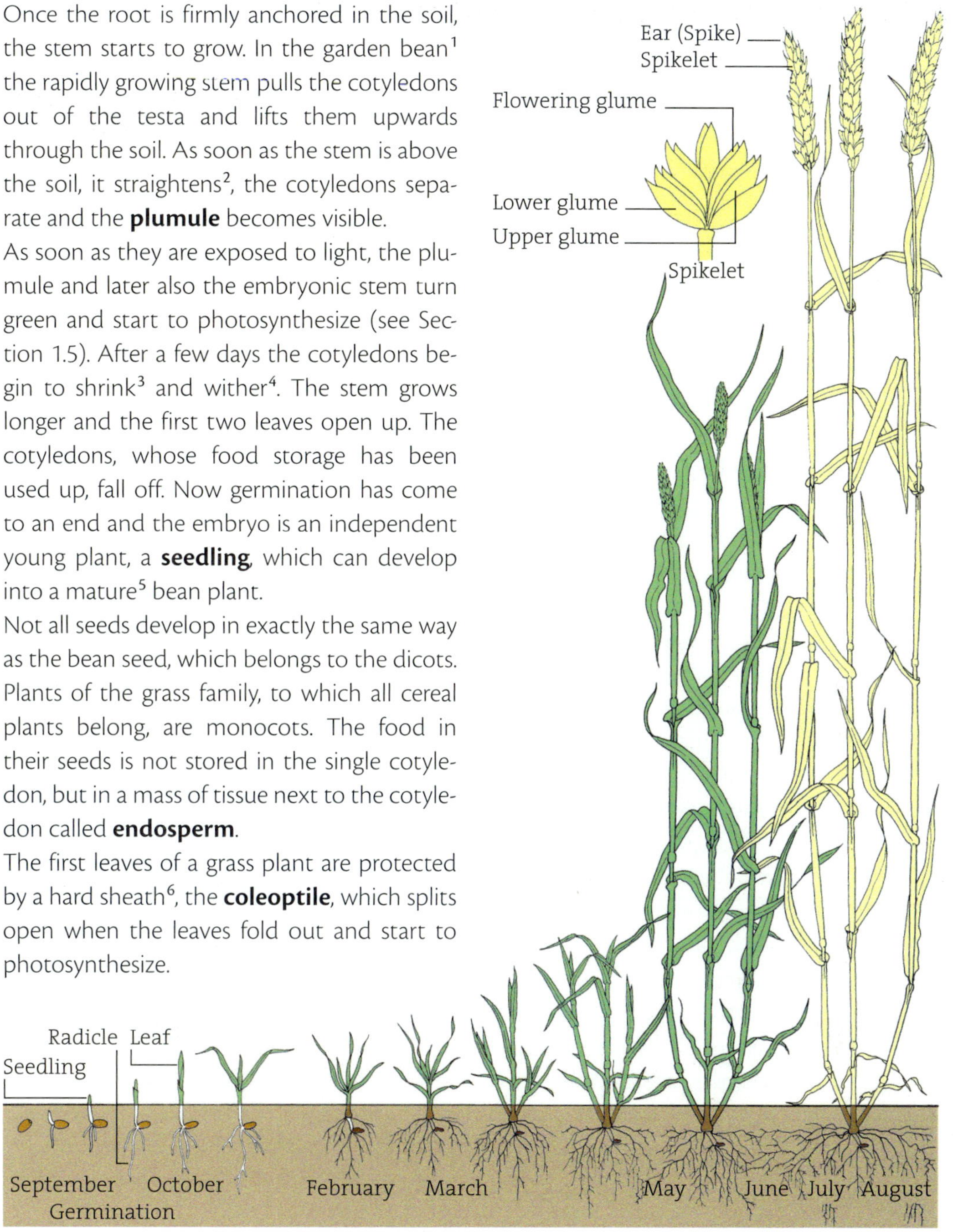

Figure 42 **Germination of a wheat grain and development into a mature plant**

1 Name the outer conditions which are necessary for a seed to develop into a plant.
2 Describe the functions of the cotyledons.
3 Use Figure 42 to describe the structure and the germination of a wheat grain. Compare it with the germination of the bean seed shown in Figure 40. What differences do you find? For language help see **Activate your English**, 'Comparing and contrasting', on the inside cover.
4 What are the advantages of seed dormancy for a plant species? What are the advantages for us? For example, how does it help gardeners who want to buy seeds by mail order?

Did you know?

The germinating seeds of cereals contain large amounts of protein, vitamin B and vitamin C. If you eat them together with muesli, lettuce or vegetables, you strengthen your body's defences.

1.11 Some plant organs allow for asexual reproduction

Many flowering plants can also reproduce **asexually** or **vegetatively** without pollination and fertilization. Strawberry plants, for instance, form long side-shoots in summer called **runners**. The buds at their ends grow into new plants whose roots anchor the daughter plant in the soil. For a short time, the new plants are provided with water and minerals through the runners, but then the roots of the daughter plant take over and the runners die away. Many weeds use this type of vegetative reproduction, which makes it difficult for gardeners to get rid of them.

Potato **tubers** are actually swollen underground stems of the potato plant which store food for the plant in the form of starch. Each potato 'eye' is a bud that can develop into a new plant. Anemones[1] and dahlias[2] also produce tubers.

The underground part of plants such as crocus and gladiolus is the swollen base of the flower stem, called the **corm**, which remains below ground all the time. Only the leaves and the flower stalk are above the ground. Being a stem the corm can produce buds, which grow into daughter corms. Although the original corm will wither and die at the end of the plant's lifecycle, the daughter corms will develop into new plants the following year.

Many grasses, and also plants including irises and water lilies, have **rhizomes**, which are long underground stems that are swollen with stored food. They grow horizontally and form branches under the soil which curve[3] upwards at their ends and produce new shoots with leaves and plants.

Key terms

Asexual reproduction
- asexual *vegetativ, ungeschlechtlich*
- bulb *Zwiebel*
- corm *Sprossknolle*
- foliage *Blätterdach, Blattwerk*
- rhizome *Rhizom*
- runner *Ausläufer*
- stage of dormancy *Ruhezustand*
- tuber *Knolle*
- vegetative *vegetativ*

Figure 43 Strawberry plant with runners

Figure 44 Potato plant with tubers

Figure 45 Corm of a gladiolus plant

Figure 46 Iris plants with rhizomes

1 anemone *Anemone, Windröschen*
2 dahlia *Dahlie*
3 curve *sich biegen*

The **bulbs** of onions, tulips and daffodils are also storage organs and organs of asexual reproduction. If you make a longitudinal cut through a tulip bulb in May or June, when the flower has already withered, you will find a thick mass of tissue at the bottom of the bulb. This is the very short and thick stem of the plant, from which the roots emerge. Most of the leaves of the tulip form the bulb. They are colourless, because they have no chlorophyll, and they are thick and fleshy with stored food. The glucose which is produced in the few green leaves at the flower stalk is transported into the bulb, where it is stored.

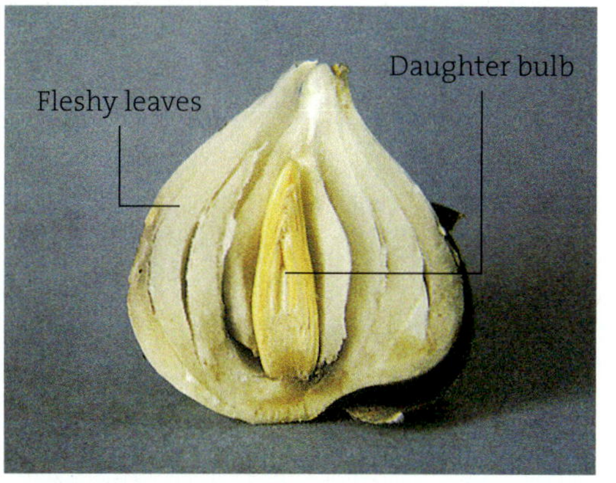

Figure 47 **Inside a tulip bulb (longitudinal section)**

When you pull a tulip plant out of the soil in May or June, you will find one or more smaller bulbs beside the one which produced this year's flower. They have developed asexually out of lateral buds from the short thick stem of the tulip plant.

These different forms of asexual reproduction in flowering plants usually do not lead to a quick and widespread distribution of the plants, in contrast to plants which disperse their seeds after sexual reproduction. However, the food storage which is often combined with the ability to reproduce vegetatively is a great advantage for herbaceous plants in temperate regions, whose shoots wither in autumn but whose bulbs, corms, rhizomes and tubers survive the winter in a **state of dormancy**. Dormant plants only expend[1] a minimal amount of energy and food, just enough to keep their cells and tissues alive. As a result, enough food reserves are left in the underground organs after the winter to enable these plants to grow rapidly in spring and early summer. This gives plants like snowdrop[2], crocus, daffodil[3], bluebell and lily of the valley an obvious advantage in competition with other plants for water, minerals and light. They can grow, spread their leaves to photosynthesize, and produce flowers and seeds before other plants have fully developed. This is especially important for plants that grow in broad-leafed woods, where in summer the **foliage** of trees and bushes is so dense that not much light reaches the forest floor.

Gardeners make use of the ability of plants to reproduce vegetatively when they divide tubers, rhizomes and corms or separate daughter bulbs and plant them in separate places. In the following year each divided part will grow into a new plant. When artificial methods are used to grow new plants by vegetative reproduction it is called **artificial propagation**.

Did you know?

There are plants that can reproduce asexually by producing tiny seedlings at the edge of their leaves. For example, bryophyllum[4] (see below), which is sometimes kept as a hardy pot plant, has embryonic tissue in the saw-shaped edges of its leaves. This tissue can develop tiny daughter plants, complete with stems, leaves and roots. They drop off and are ready to root in the soil.

Figure 48 **Bryophyllum leaf with daughter plants**

1 expend sth. *etwas ausgeben*
2 snowdrop *Schneeglöckchen*
3 daffodil *Narzisse, Osterglocke*
4 bryophyllum *Brutblatt*

You probably know already that **cuttings** from many herbaceous plants often develop roots when they are put in a glass of water. With many plants it is possible to grow a new plant simply by placing a leaf into water or wet compost. The process of rooting can be encouraged by dipping the lower end of the cutting into rooting powder[1], which contains plant hormones, before pushing it into the compost. This way of growing new plants is used by gardeners if they want to get more plants of one particular type and quality. The cuttings of a specific type of geranium will grow into plants with exactly the same colour of flowers or type of leaves as the original plant, because only the genetic information of the mother plant is present in the new plants grown from its tissue.

Key terms

Artificial propagation
- artificial propagation *künstliche Vermehrung*
- cutting *Steckling*
- graft *Pfropfung*
- grafting *pfropfen*
- rootstock *Unterlage*
- scion *Edelreis*

Figure 49 A stem cutting has grown roots

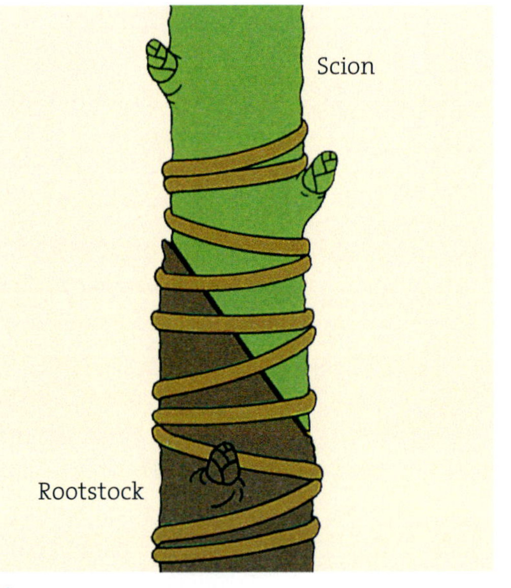

Figure 50 Grafting

Another method of artificial propagation is **grafting**, a method by which a piece of a plant, the **graft**, is fixed to another one, so that both parts fuse and become one plant. This method only works when the two plants are related. It is widely used in the breeding[2] of flowering bushes and fruit trees, because these plants are difficult to grow from seeds and often do not have the same qualities as the original plants. The freshly cut end of a thin stem, a twig or even a piece of bark with a bud of the **scion** is joined carefully to the cut[3] of the **rootstock**, the stem of the plant which remains rooted[4] in the soil. The two ends of the grafted plants must be joined in such a way that the cambium (see Section 1.2) of both plants is in contact. Then they are carefully bound together and sealed with wax on the outside to protect the wound against diseases.

1 rooting powder *Bewurzelungspulver*
2 breeding *Züchtung, Aufucht*
3 cut *(hier) Einschnitt*
4 root *sich bewurzeln, wurzeln*

1 What is the difference between sexual and asexual reproduction? Make a table with two columns comparing each aspect of reproduction.
2 Describe some ways in which plants can reproduce asexually, and give examples of plant species that do this.
3 Name two methods of artificial propagation and explain how and why they work.
4 Explain why gardeners prefer growing fruit trees from grafts rather than from seeds.
5 Why is it important in grafting that the tissue of the cambium of both plants is in contact?

Carrying out a long-term observation

Biological processes, especially growth processes, take time. This means you have to be patient if you want to study these processes in animals or plants. You also need to provide suitable conditions for the creatures you want to observe. If a plant starts to wither or an animal does not feel at ease in the observation cage, you cannot expect the plant or animal to show their normal reactions or behaviour. You can only expect to get true and useful results if you keep the animal or plant under conditions that are as natural as possible.

Figure 51 **Bean seedling with cotyledons and first leaves**

Observing the germination and growth of a bean plant

Your teacher will give you three dry seeds of a runner bean[1], for example scarlet runner[2], and a flowerpot with moist potting soil[3]. Observing their germination and growth is a long-term task and should be carried out for at least one month, or better still for two months.

At home:

1. Put the bean seeds into a flat dish with some water and let them soak for a few hours. The seeds should not be completely covered with water because they have to respire.
2. Plant the seeds in the soil, leaving a space of about 2 cm between them. Write your name on a nameplate and put it on your flowerpot.
3. Place your flowerpot in a sunny place on a window sill, on a balcony or in the garden.
4. Water your plants regularly, but not too often. The soil must never be completely soaked, because this stops the root cells respiring. You can add plant fertilizer to the water once a week, as soon as the seedlings appear above the soil.
5. Put one or more thin sticks into the soil to support the bean plants as they climb upwards.
6. Keep a diary of your observations, starting with 'Day 1' when you plant your beans. Do not be disappointed if nothing seems to happen in the first few days or if only one or two of your seeds develop into seedlings. Also make drawings if you see interesting changes in your plants.
7. Note down how often you add water or fertilizer.
8. Measure the height of your plants each week and record the figures in your diary.

At school:

Take your plants to school every second week to compare how well your classmates' plants and yours are developing. This is also a chance to swap ideas[4] on how to care for the plants.
After one month (and again after two months) *the winner of the month* is the student whose bean plant is the tallest. After two months a second quality can be added. The final winner is the person with the tallest plant with flowers or even with developing fruits.

1 runner bean *Stangenbohne*
2 scarlet runner *rotblühende Stangenbohne*
3 potting soil *Blumenerde*
4 swap ideas *Ideen austauschen*

Conclusions

1 Compare the development of your runner beans to the germination of the French bean which is shown in Figure 40 and described in the text.

2 Explain the different results experienced by you and your classmates. Why were some of the plants higher or stronger? Why did some develop flowers or fruits while others did not?

Activity 1 Studying the root and the stem

A Study the structure of root hairs

Materials
- cress[1] seeds (or mustard[2] seeds, radish[3] seeds or grains of wheat)
- Petri dish[4]
- hand lens[5] (magnification 4x) or binocular microscope[6]

Procedure
1. Put some cress seeds on wet tissue paper or on a layer of wet cotton wool in a Petri dish and let them germinate and grow into seedlings. This will take about two to three days.
2. Using a pair of pincers, place a single seedling on a microscope slide.

1 Observe the radicle against a dark background (e.g. a small piece of black cardboard) using a hand lens or a binocular microscope. If you use a microscope, put the radicle into a drop of water and cover it with a cover slip.

2 Make a labelled drawing of the radicle.

3 What does a single root hair look like? How many cells does it consist of?

B Demonstrate the transport of water and minerals through a stem

Materials
- 2 stems of a daisa or a busy Lizzie[7] plant with white flowers or another herbaceous plant (e.g. a tulip, ideally with white or yellow petals, a touch-me-not[8] or a coleus[9], ideally with white petals)
- sharp knife
- 2 beakers[10] (about 250 ml)
- water
- eosin[11] or red ink

Figure 52 Observing changes in the plants

Procedure
1. Fill the two beakers half-full with water and colour the water in one of them dark red using the eosin or ink.
2. Take two stems of the same plant species and hold them under water while you make a fresh horizontal cut through the stems. Place one stem into each beaker.
3. Wait for one hour and observe any differences in the colour of the two stems, the leaves and the flowers. Note down your observations.
4. Then cut a 2 cm long piece from the lower end of each stem. Rinse[12] the pieces of stem briefly in clear water.
5. Now observe the fresh cuts with a hand lens or make a thin cross section through both pieces of stem and observe them under the microscope.
6. Make a labelled drawing of the cut surface or the cross section of the stem that stood in the beaker of coloured water.

1 Write a short report about the results of this experiment.

1 cress *Kresse*
2 mustard *Senf*
3 radish *Radieschen*
4 Petri dish *Petrischale*
5 hand lens *Lupe*
6 binocular microscope *binokulares Mikroskop*
7 busy Lizzie *Fleißiges Lieschen*
8 touch-me-not *Springkraut*
9 coleus *Buntnessel*
10 beaker *Becherglas*
11 eosin *ungiftiger roter Farbstoff*
12 rinse sth. *etwas abspülen*

Activity 2 Studying the functions of leaves

A Demonstrate the importance of light and chlorophyll for the formation of starch in green leaves

The presence of starch can be proved by the iodine test. When iodine solution[1] is added to parts of a leaf which contains starch, those parts take on a violet or dark blue colour.
Before the experiment can begin, the leaves have to be completely destarched[2]. This can be done by keeping the plant in a dark place for a few days. During this time the starch is broken down into sugar and transported to other parts of the plant, especially into the root. In the root, the sugar is changed back into starch and stored. To make sure that the leaves of the plant used in the experiment are free of starch, you can carry out the iodine test with one of its leaves.

Materials
- geranium or other potted plant, bean plant or nasturtium[3]
- hotplate[4] or Bunsen burner[5] with a tripod[6]
- 250 ml beaker
- test tube[7]
- Petri dish
- aluminium foil
- tweezers[8]
- water
- methylated spirit[9] (alcohol)
- iodine solution
- safety goggles[10]

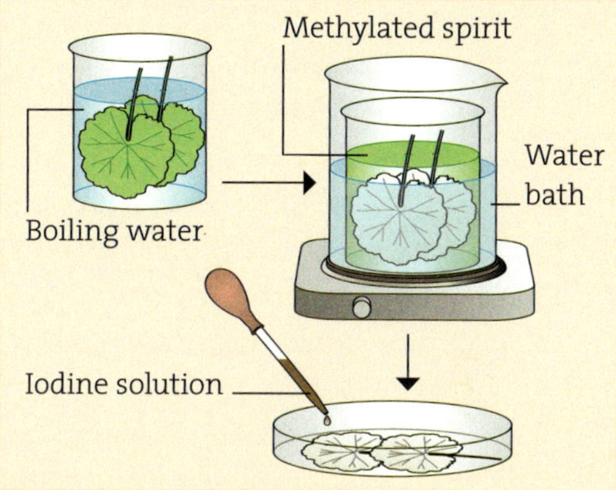

Figure 53 Experiment with two destarched leaves

1 iodine solution *Jodlösung*
2 destarch sth. *die Stärke von etwas entfernen*
3 nasturtium *Kapuzinerkresse*
4 hotplate *Heizplatte*
5 Bunsen burner *Bunsenbrenner*
6 tripod *Dreifuß*
7 test tube *Reagenzglas*
8 tweezers *Pinzette*
9 methylated spirit *Brennspiritus*
10 safety goggles *Sicherheitsbrille*
11 variegated *panaschiert*
12 ornamental plant *Zierpflanze*

Procedure
1. Cover one of the two leaves you are using completely with aluminium foil. Then place the plant in bright light, for example on a sunny window sill, for several hours.
2. Cut off the two leaves, remove the aluminium foil from one of them and put them in a beaker of boiling water. This will kill the leaf cells.
3. Use the tweezers to lift the leaves out of the water and put them in a small beaker or a test tube with methylated spirit.
4. Place the beaker or test tube in a hot water bath. The alcohol will remove the chlorophyll from the leaf cells.
5. Use the tweezers to take the leaves out of the methylated spirit and rinse them under cold water.
6. Place the leaves in a Petri dish and cover them with iodine solution (for about 15 minutes). Compare the colour of the leaves.

Note: It is very important to wear safety goggles for this experiment.

1 Why does the plant have to be destarched before the experiment?
2 What were the results of the iodine test?
3 Why do plants store starch and not sugar in their leaves? (Remember what happens to sugar when you put it into water.)

Repeat the above experiment with a variegated leaf. This is a leaf which is not completely green, but has white or yellow margins or patches. Variegated leaves are found in ivy, geraniums and many other ornamental plants. For this experiment no aluminium foil is used.

1 Describe what the leaf looks like when you take it out of the iodine solution.
2 Interpret the results.

B Demonstrate that oxygen is released in photosynthesis

The water plant elodea[1], also known as Canadian pondweed, is a classical object in experiments with photosynthesis. You can find this hardy plant in streams and ponds or you can buy it in a pet shop.

Materials
- 5 shoots of pond weed, each 15 cm long
- knife or scalpel[2]
- 20 ml test tube
- 250 ml beaker
- glass funnel[3] with a tap[4]
- wooden splint[5]
- matches[6]

Figure 54 **Experiment with a glowing splint**

Procedure
1. Fill the beaker half-full with water.
2. Remove some leaves from the cut ends of the pond weed and make fresh cuts at a 25° angle. Place the five shoots into the beaker.
2. Open the tap of the funnel, turn the funnel upside down and lower it into the beaker so that all the plants are underneath it and covered with water. Now you can close the tap.
3. Either shine a light on to the beaker from a distance of about 10 cm or place the beaker on the glass plate of an overhead projector.
4. The gas bubbles released from the cut ends of the shoots will be collected in the part of the funnel below the closed tap.
5. Light the end of the wooden splint. Then extinguish[7] the flame and hold the glowing[8] splint above the opening of the funnel. Now open the tap and observe what happens.

1 Describe your observations to a partner, and then write a report on the experiment in your exercise book.

2 Explain the result of the experiment.

Activity 3 Demonstrating where food is stored in a plant seedling

Materials
- bean seeds (runner beans or French beans)
- beaker or large Petri dish
- some wet tissue paper or cotton wool
- pipette
- iodine solution

Procedure
1. Place the bean seeds on the wet filter paper or cotton wool and leave them to germinate and grow until the radicle, the stem and the cotyledons are well developed. This will take a few days.
2. Then place a drop of iodine solution on each of the three organs of the seedlings. Observe what happens.

1 Where is a colour reaction visible?

2 Interpret the results.

1 elodea, Canadian pondweed *(Kanadische) Wasserpest*
2 scalpel *Skalpel*
3 funnel *Trichter*
4 tap *Absperrhahn*
5 wooden splint *Holzspan*
6 match *Streichholz*
7 extinguish sth. *etwas auslöschen*
8 glowing *glühend*

Activity 4 Studying asexual reproduction

A Observe the growth of a plant from a bulb

Materials
- bulb , e.g. from a tulip, a hyacinth[1] or an onion
- jar with a neck narrow enough for the bulb to be placed on top of it without falling through
- water

Procedure
1. Fill the jar with water and place the bulb on the neck of the jar, so that the lower end of the bulb just touches the water.
2. Place the jar on a window sill.

1 Observe the plant and note how long it takes for it to develop roots.

B Grow a new plant from a stem or leaf cutting

Every pupil grows a plant from the cutting of a herbaceous plant of their choice. You can grow your cutting either in water or in potting compost, which you can get in a gardening shop.

Materials
- 10 cm-long cutting from a herbaceous plant (e.g. a geranium, busy Lizzie or other herbaceous plant)
- sharp knife
- jam jar, small vase with water or flowerpot with potting compost and a small saucer for collecting the water
- rooting powder (if available)

Procedure
1. Cut the stem of the plant diagonally just below one of the leaves.
2. Remove the lower leaves on the shoot.
3. Put the cutting in water or push it firmly into the potting compost.
4. Keep the compost moist but not soaking wet.
5. Put the glass or flowerpot with your cutting on a warm window sill, but avoid direct sunlight.

Figure 55 Stem cuttings of tradescantia[2] have grown roots

1 hyacinth *Hyazinthe*
2 tradescantia *Ampelkraut, Wasserranke*

1 Observe your cuttings every day for at least two weeks.
2 Keep a diary in which you write your observations each day.
3 If you have rooting powder, test its effect on two separate cuttings. Treat one cutting with the rooting powder and then plant both cuttings into flowerpots at the same time.
4 Compare the growth of your cuttings in class after two weeks and decide which plants can be grown from cuttings more easily than others.

1.12 Plants produce more than just sugar and starch

Proteins

Legumes[1], like peas and beans, contain high amounts of protein. The different proteins are important building materials for cells. Soya beans[2] are especially rich in protein. They are used for making tofu[3] products, among other things.

Figure 56 Soya beans and tofu products are rich in protein

Fats

The seeds of many plants contain a lot of fat. By grinding[4] and pressing the seeds you can obtain[5] different oils. Oils from sunflower seeds and olives are used for cooking, while rape oil is used both for food and for the production of biodiesel.

Figure 57 Olive oil is extracted[10] from the fruits of the olive tree

Fibres

Hemp[6] and cotton[7] plants are excellent sources[8] of fibres for making fabrics[9]. Clothes made from these plants are comfortable to wear because the fabrics absorb moisture very well and dry quickly.

Figure 58 The fibres from the stem of the hemp plant are used to make fabrics

Figure 59 The flower of the blueweed[12] contains blue and red dyes

Dyes[11]

Fabrics used to be dyed exclusively with chemicals obtained from plants. People used juices from fruits or other parts of a plant, or extract dyes from wood or tree bark. Today, artificially made dyes are cheaper and produce more intensive colours.

1 legume *Hülsenfrucht*
2 soya bean *Sojabohne*
3 tofu *Tofu*
4 grind sth. *etwas mahlen*
5 obtain sth. *(hier) etwas gewinnen*
6 hemp *Hanf*
7 cotton *Baumwolle*
8 source *Quelle*
9 fabric *Stoff*
10 extract sth. *etwas extrahieren, (hier) etwas gewinnen*
11 dye *Farbstoff*
12 blueweed *Natterkopf*

Spices

We use some plants as spices[1] because of their essential oils[2]. Many of these plants, such as nutmeg[3] and clove[4], are found in tropical or subtropical areas.

Scents[5]

The essential oils of some plants are used as scents for perfumes or as an additional flavouring in foods. Some of these scents can also be imitated artificially.

Substances that heal

Camomile[6] is just one of the many plants which have healing effects: its petals disinfect and support the healing of wounds, they soothe[7] itching[8] and they are antispasmodic[9], i.e. they release cramps[10].

Figure 60 Cloves are used as a spice, for example in pumpkin[11] pie

Figure 61 Rose flowers contain essential oils

Figure 62 The camomile flower contains many healthy substances

1 spice *Gewürz*
2 essential oil *ätherisches Öl*
3 nutmeg *Muskatnuss*
4 clove *Gewürznelke*
5 scent *Duftstoff*
6 camomile *Kamille*
7 soothe sth. *etwas beruhigen, lindern*
8 itch *jucken*
9 antispasmodic *krampflösend*
10 cramp *Krampf*
11 pumpkin pie *Kürbiskuchen*
12 gummy bear, jelly bear *Gummibärchen*
13 package *Verpackung*
14 plant-based *auf pflanzlicher Basis, aus Pflanzen gewonnen*
15 remedy *Heilmittel, Arzneimittel*

1 Make a table that includes the names of different plants, the substances they contain and what they are used for. Besides the ones on these pages, try to find three other examples.

2 Tofu is an essential part of many vegetarian diets. Give reasons for this and identify other plants which can fulfil the same function in a vegetarian diet.

3 On a packet of gummy bears[12] you can see that chlorophyll is used as one of the dyes. Which colour of gummy bear do you think contains this dye? Look at other food packages[13] and write down which of these foods contain dyes from plants. Then share and discuss your conclusions in small groups.

4 Find out what the last plant-based[14] medicine or remedy[15] was that you used or received. Your parents may be able to help you remember. Gather information about the plant, where it is found and how it is prepared and used for healing purposes. Present your results to the class.

Summary

There is an immense[1] variety among the 250,000 species of flowering plants known today, but they all have roots, stems, leaves and flowers as their main organs. The roots provide the plants with water and minerals, which are transported through the stems to the leaves and flowers. Transpiration and photosynthesis take place in the green leaves. These two essential processes are the basis for the existence of all plants, animals and humans on earth. Flowers, which are specialized leaves, are the organs of sexual reproduction; their sepals and petals protect the sexual organs and attract organisms to the flowers. The stamens contain the male gametes while the ovary inside the pistil carries the female gametes. In pollination the ripe pollen is transferred onto a stigma, either by the wind, by insects, by animals or by people. When male and female gametes fuse in fertilization, an embryo develops inside the ovule, which is now called the seed. The fruit contains the seed with cotyledons or endosperm as storage tissue and other parts of the flower necessary for dispersal. The germination of the seed and the development of the embryo into a mature plant depend on many factors. The different forms of asexual reproduction in which new plants develop out of one adult plant all result in offspring which have the genetic material and therefore the same qualities as the mother plant. In artificial propagation the ability of plants to reproduce asexually is used for plant cultivation.

Key terms

See the boxes:

Plants and their parts .... p. 160
Roots .... p. 161
Stems .... p. 162
Parts of stems and leaves .... p. 164
Transpiration .... p. 165
Photosynthesis .... p. 167
Flowers .... p. 168
Pollination .... p. 170
Fertilization .... p. 172
Seed dispersal .... p. 174
Germination .... p. 176
Asexual reproduction .... p. 178
Artificial propagation .... p. 180

A Study tasks

1. Develop a mind map based on the information in the summary above (see 'Improve your skills', p. 22).
2. Write down all the words from one or two Key terms boxes (depending on the number of words in each box) on cards. Then, working with a partner, take a card and ask your partner to explain what it means. Swap roles until you have used all the cards. If you are not sure how to express something in English, check the text for words that can help you and copy them into your vocabulary notebook. When you have explained all the words in one Key terms box, move on to another one and repeat the activity.

B Practice your English

Connect the sentence parts correctly.

1. Many plants that flower in early spring ...
2. Xylem vessels have the task ...
3. Sugar solution is transported in a system of tubes ...
4. The stomata can open and close ...
5. Photosynthesis can occur in all parts of a plant ...
6. After a plant seedling has used up the stored food in its cotyledons ...
7. While a bee is sucking nectar from a flower ...
8. A fibrous root system allows plants ...
9. Flowers that belong to the same family ...
10. Hummingbirds drink nectar from a flower ...

1 immense *unermesslich, ungeheuer*

a) ... they wither and fall off.
b) ... the pollen grains cling to its hairy body.
c) ... that contain chlorophyll.
d) ... of transporting water to the highest branches of a tree.
e) ... have a similar arrangement of flower parts.
f) ... by changing their shape.
i) ... while hovering in front of it.
g) ... have bulbs as storage organs.
h) ... which is called phloem.
j) ... to take up a lot of water in a short time.

C The organs of a flowering plant

1. Name the different parts of a cherry blossom as shown in Figure 63 below.
2. Explain what must happen in a flower before a fruit can develop.
3. Make a sketch of Figure 64 in your notebook and label the numbered parts of the plant.
4. Give a short explanation of the functions of the organs you have labelled in Figure 64.

Figure 63 Longitudinal section of a cherry blossom

Figure 64 Structure of a flowering plant

D Complete the sentences

1. A seed only starts to germinate if ...
2. The function of the xylem is to ..., whereas the phloem ...
3. A woody plant develops growth rings because ...
4. Transpiration has a cooling effect on the surface of plants, but its main function is to ...
5. Photosynthesis is the most important chemical process on earth, because ...
6. In order to be able to carry out photosynthesis, plants need ...
7. The plant uses some of the ... it produces in photosynthesis for its own ..., but most of it ...
8. Each stamen is made up of ...
9. The female part of a hermaphrodite flower consists of ...
10. Pollination is carried out by ...

E Say it in German

Your school newspaper is doing a special issue[1] on the environment and the need to protect it. Write a short article for the newspaper on the topic of photosynthesis. Explain what it is, how it works and why it is so important for life on earth.

F Critical thinking

1 Explain why the stomata are on the lower side of the leaves in most plants. Find an example of a plant which has leaves with the stomata on the upper side and explain why this is.

2 Explain why it is more appropriate to compare the development of a plant embryo and a young plant to the development of a bird or reptile embryo than to the embryo of a mammal.

3 Why is it that a biologist calls tomatoes and cucumbers[2] fruits, but not rhubarb[3] and carrots?

1 issue *Ausgabe*
2 cucumber *Gurke*
3 rhubarb *Rhabarber*

Plants of Economic Importance

Chapter objectives

- To **underline** the economic importance of cultivated grasses
- To **describe** the characteristics of the grass family and the cabbage family
- To **explain** how cultivated grass and cabbage plants are useful to us
- To **learn** how to collect and sort plants
- To **investigate** the structure and function of flowers and seeds
- To **understand** that plants and animals must respire

For thousands of years, people have cultivated plant species from the grass family and the cabbage or mustard[1] family. These plants provide essential foods, important forage crops[2] and products such as spices, dyes[3], fuel and building materials. In this chapter we will study the structures typical for the two plant families, identify important species and look at their wide range[4] of uses. You will also learn how to collect and prepare plants for a herbarium[5], so you can build your own library of plants.

2.1 Grasses have flowers, too

Figure 1 **Structural features of a grass plant**

Key terms

The structure of grasses
- awn *Granne*
- glume *Hüllspelze, Hochblatt*
- husk *Spelze*
- ligule *Blatthäutchen*
- inner/outer scale *Vor-/Deckspelze*
- sheath *Blattscheide*
- spikelet *Ährchen*
- stalk *Halm*

The family of 'true grasses' (Latin name: *Poaceae*) is one of the the biggest and most widespread families of flowering plants, containing about 10,000 species.
Grass flowers look different from other flowers because they have no brightly coloured petals, but only stamens and carpels. Their small and inconspicuous[6] flowers are arranged in **spikelets**. A spikelet is a special type of inflorescence, which can contain from one to 20 single flowers. As in the more familiar flowers of other plants, each tiny flower has stamens and an ovary with feathery stigmas for catching the pollen. Instead of petals there are two **scales** which often have **awns**, or hard bristles[7]. The scales later form the **husk**, which encloses the ripe grain. In the different species of grasses the spikelets are arranged in different ways, as you can see in the cultivated grasses in Section 2.7, p. 200.
All grasses share certain structural characteristics. They have slender[8] and mostly hollow stems with regularly spaced nodes[9]. The leaves are flat and sword-shaped with parallel veins running from the stem to the tip of the leaf. The leaves fold in a **sheath** around the stem, often forming a **ligule** which can be shaped differently depending on the species.

1 mustard *Senf, Ackersenf*
2 forage crop *Futterpflanze*
3 dye *Färbemittel, Farbstoff*
4 range *Angebot, Auswahl*
5 herbarium *Herbar, Sammlung getrockneter Pflanzen*
6 inconspicuous *unauffällig*
7 bristle *Borste*
8 slender *schlank, schmal*
9 node *Knoten*

1 Name some of the differences between the rape flower (see Chapter 1, p. 168) and the grass flower. What do they have in common? For language help see **Activate your English**, 'Comparing and contrasting', on the inside cover.
2 Describe the structural features which all grasses have.
3 From what you have learned so far, how do you think the pollen of grasses might be dispersed?

2.2 Grasses provide food for humans and animals

People use grasses in many ways. **Meadows** with their great variety of wild grasses provide food for **grazing** animals. Dried grass, or **hay**, is stored in barns or in **silos**, which are big, airtight containers. In a silo the grasses **ferment**, just as cabbage ferments in the making of pickled cabbage[1]. The resulting grass **silage** is less dependent on good weather than the production of hay, and so it is ideal for farmers in temperate[2] climates. Silage is a juicy animal feed, rich in nutrients, which can be used throughout the year. It is especially needed during the winter months, when no fresh grass is available.

Today **maize** is becoming more and more important as **forage** for pigs and cattle. The whole maize plant – cob[3], leaves and stem – is used for feeding farm animals. Many farmers specialize in raising one type of animal, keeping large numbers of cattle, pigs or poultry[4]. In these farms, great quantities of **fodder** are needed continuously and have to be kept in store[5].

Straw, the dried stems of **cereal plants**, is commonly used as bedding[6] for cattle and horses. It can also be used as packing material, as material for thatched roofs[7], which are becoming increasingly popular, and in combination with clay[8] or in the form of bales[9] as a natural building material for houses, which offers high-quality insulation[10]. It is also used as raw material in the paper industry.

Shredded[11] material from cereal plants and other energy-rich plants, but also biological waste from plants, animals and humans, is an increasingly important source of **renewable energy**. The chemical energy stored in biomass[12] can be released by burning the material directly or by feeding it to micro-organisms that are kept in big containers (fermenters[13]) without oxygen, which can be seen on more and more farms today (see Figure 3). The microorganisms break down the organic material and release a mixture of gases. **Biogas** can be used for heating, cooking and for running engines, for example in cars and lorries.

Key terms

Uses of grasses (1)

- biogas *Biogas*
- cereal plant *Getreidepflanze*
- ferment *fermentieren*
- fodder *Viehfutter*
- forage *Viehfutter*
- graze *grasen*
- maize *Mais*
- hay *Heu*
- meadow *Wiese*
- renewable energy *erneuerbare/ nachwachsende Energie*
- silage *Silage, fermentiertes Tierfutter*
- silo *Silo, Getreidespeicher*
- straw *Stroh*

1 pickled cabbage *Sauerkraut*
2 temperate *gemäßigt*
3 (maize) cob *Maiskolben*
4 poultry *Geflügel*
5 keep sth. in store *etwas bevorraten*
6 bedding *Lagerstreu*
7 thatched roof *reedgedecktes Dach*
8 clay *Lehm*
9 bale *Ballen*
10 insulation *Isolierung*
11 shred sth. *etwas hechseln, zerkleinern*
12 biomass *Biomasse*
13 fermenter *Fermentierer, Gärbehälter*

Figure 2 Traditional silo for making and storing silage on a farm in New England

Figure 3 Grazing cattle with fermenters for biogas in the background

Tropical grasses like **bamboo** are also used as construction materials and for making pieces of furniture. By far the most important way of using grasses, however, is for the **cultivation** of cereal plants whose nutritious[1] grains play a significant role in our diet. For about 11,000 years different cereals have been grown and used for food in many parts of the world. Throughout this long history of cultivation, plants have been selected that produce more and bigger grains, are easier to harvest and can be stored for long periods. Among the many cultivated grasses that are used as human food, the most frequent are **wheat**, **rye**, **maize**, **oats** and **barley**, as well as **rice**, **millet** and **sorghum**.

The fruits of cereal plants – the grains – are rich in carbohydrates and also contain protein and minerals. Whole grains are good sources of **dietary fibre**, **essential fatty acids**, and other important nutrients. In many countries cereal grains form the basis of the diet; such foods are called **staple foods**.

Most cereals are grown in temperate environments with fertile soil and moderate rainfall, but maize and rice need warm and humid temperatures, and rice grows best in flooded[2] fields.

The most important cereals economically[3] are wheat, rice and maize. Together these three cereals account[4] for more than 80 per cent of the grain production worldwide.

Key terms

Uses of grasses (2)
- cultivation *Anbau, Bebauung*
- dietary fibre *Ballaststoff, Faserstoff*
- essential fatty acid *essentielle Fettsäure*
- staple food *Grundnahrungsmittel*

Important cereal plants
- bamboo *Bambus*
- barley *Gerste*
- maize *Mais*
- millet *Rispenhirse*
- oats *Hafer*
- rice *Reis*
- rye *Roggen*
- sorghum *Mohrenhirse*
- wheat *Weizen*

1 In what ways do people make use of cultivated grasses? Make a table showing the different types of cultivated grasses and the ways they are used. Compare your table with a partner's.

2 Explain the advantage of making silage for animal feed, in comparison to making hay.

3 The practice of using cultivated plants to produce biogas, a renewable energy source, has become very popular. However, it is also controversial, since it means less food is available for the native people, who depend on the plants for food. Find out more about this issue and discuss your findings in class.

2.3 Wheat is of great economic importance

Every year more than 600 million tons of wheat are produced. This annual[5] grass plant has a hollow stem and an upright head. There are no awns or very short ones at the spikelets, which contain between three and six flowers and are arranged alternately[6] on opposite sides of the flower stem. The **ear** of the plant remains in an upright position when the grains are ripe, instead of bending under its own weight. The ripe grains are big and smooth and separate easily from the surrounding glumes.

Key terms

Wheat
- bran *äußere Schicht des Getreidekorn, Kleie*
- ear *Ähre, Kornähre*
- endosperm *Nährgewebe, Endosperm*
- flour *Mehl*
- gluten *Kleber*
- mill *mahlen*

1 nutritious *nahrhaft*
2 flooded *überflutet, überschwemmt*
3 economically *wirtschaftlich*
4 account for sth. *(hier) etwas ausmachen*
5 annual *einjährig*
6 alternately *abwechselnd*

Wheat needs a heavy and fertile soil and a lot of moisture during the growth period. Through selective breeding many varieties of wheat have been developed, which are resistant not only to low temperatures, drought[1] and wind, but also to various diseases.

Figure 4 **Field of wheat**

Figure 5 **Ears and grains of wheat**

Wheat is the most important cereal for making bread and pasta[2]. For white **flour** the **bran** and the embryo are removed, leaving only the **endosperm**, which consists mainly of starch, to be used in the **milling** process. For wholemeal[3] flour the full grain, including the bran, is milled. This gives the flour its darker colour and coarser[4] structure. Bread and pasta are healthier when they are made from wholemeal instead of white flour. Wholemeal flour contains not only starch but also protein, minerals and the cellulose of the bran, which cannot be digested but supports the digestive process (see Unit 1, Chapter 5, p. 81).

Today some of the nutrients which get lost in the production of white flour are added after milling, but white bread does not contain as much fibre as wholemeal bread.

Wheat is the only cereal that contains **gluten**, which makes wheat flour sticky when it comes into contact with water. This is important for baking bread, because the gluten holds the loaf of bread together and keeps the carbon dioxide which is produced by the yeast[5] cells inside the rising dough[6].

1 drought *Dürre*
2 pasta *Nudeln, Teigwaren*
3 wholemeal *Vollkorn-*
4 coarse *grob*
5 yeast *Hefe*
6 dough *Teig*

1 Use Figure 6 to name as many different products as you can that are made from wheat flour. How many other wheat products can you think of?
2 Why is wholemeal bread healthier than white bread?
3 Collect information from books and the Internet and write a short report on the history of the cultivation of grasses. Help your teacher to display the reports around the class, then read your classmates' reports to see what different facts each of you found.

Figure 6 **Bakery products made of wheat**

2.4 The cabbage flower has the form of a cross

Key terms

Cabbage plants (1)
- cabbage family *Kreuzblütler*

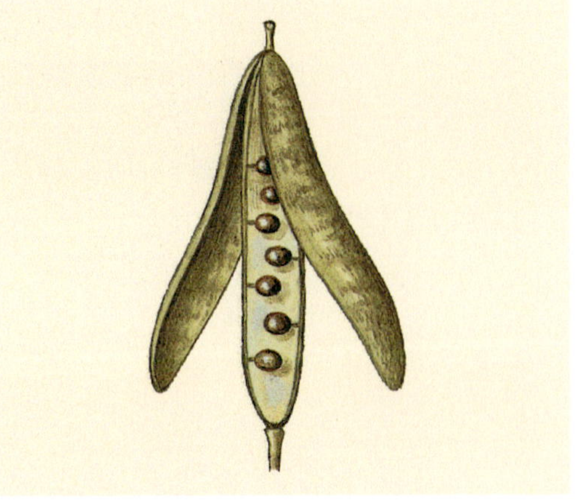

Figure 7 Ripe fruit of a plant of the cabbage family

The plants shown in Figures 8 to 19 differ in the size, shape and colour of their flowers, but they all belong to the **cabbage family** (Latin name: *Brassicaceae*). The members of this family have flower parts in multiples of four and petals that are arranged in a cross-shape. Their seed pods are divided by a wall with many seeds on both sides. Among the 3700 species of this family are ornamental plants[1] with attractive and often fragrant[2] flowers like the yellow or brown wallflower[3], the stock[4] with its red or pink blossoms and the garden silverleaf[5], whose fruits look like silver coins. In many rockeries[6] the pink or purple aubretia[7] and white candytuft[8] form beautiful flower cushions.

Ornamental species of the cabbage family

Figure 8 Wallflower

Figure 9 Aubretia

Figure 10 Stock

Figure 11 Candytuft

1 ornamental plant *Zierpflanze*
2 fragrant *duftend*
3 wallflower *Goldlack*
4 stock *Levkoje*
5 garden silverleaf *Garten-Silberblatt*
6 rockery *Steingarten*
7 aubretia *Blaukissen*
8 candytuft *Schleifenblume*

Figure 12 Garden silverleaf

Figure 13 Fruits of the garden silverleaf

There are numerous wild-growing species which can be found at the edges of fields, in meadows and along country lanes. These include plants with yellow flowers like charlock[1] and others with white or mauve[2] blossoms like wild radish[3], garlic mustard[4], shepherd's purse[5], field penny-cress[6] and the cuckoo flower[7].

Wild-growing species of the cabbage family

Figure 14 Charlock

Figure 15 Wild radish

Figure 16 Garlic mustard

Figure 17 Shepherd's purse

1 charlock *Ackersenf*
2 mauve *hellviolett*
3 wild radish *Hederich*
4 garlic mustard *Knoblauchhederich, Knoblauchsrauke*
5 shepherd's purse *Hirtentäschelkraut*
6 field penny-cress *Acker-Hellerkraut*
7 cuckoo flower *Wiesen-Schaumkraut*

Figure 18 Field penny-cress

Figure 19 Cuckoo flower

1 Draw the flower of one member of the cabbage family.
2 This section names a number of species from the cabbage family.
 a Find three of them close to where you live using the pictures here or a guidebook to identify them.
 b Either make a drawing of each plant or prepare and label them as described in 'Improve your skills', p. 203.

2.5 Cabbages are the result of artificial selection

Key terms

Cabbage plants (2)
- artificial selection *künstliche Auslese*
- cabbage variety *Kohlsorte*

Some species of the cabbage family have been cultivated for many centuries and now belong to the most important food crops worldwide. Cabbages are an important vegetable group which were developed from a wild-growing common ancestor[1] using the method of **artificial selection**. The wild cabbage, native to coastal areas of southern and western Europe, is thought to be the ancestor of all types of cabbage which are used for food today. With every generation, the early farmers only chose certain individuals among the offspring[2] of the descendants[3] of the wild cabbage plants for further reproduction, namely those that possessed certain desirable[4] qualities. Such qualities were thicker edible[5] roots, larger leaves, fleshy flower stalks, shorter or swollen internodes or bigger side buds on the stem. On the following page, Figure 20 shows the wild cabbage plant and a number of well-known **cabbage varieties**.

The same type of selective procedures have lead to the development of the small red radish[6], the larger white or brown radish[7] varieties, the turnip[8] and even the popular Chinese cabbage[9], which looks more like a cabbage than a radish. In spite of the great structural differences, these are all descendants of the wild radish (see Figure 15), a small annual plant with white or yellow flowers that grows on waste ground[10].

Radishes, turnips and the various cabbage varieties belong to the same family because the flowers they produce all have the typical cross-shaped petals. Other common vegetables like tomatoes, red and green peppers, carrots, cucumbers, beans, peas and asparagus are members of other plant families.

1 ancestor *Vorfahre*
2 offspring *Nachkommenschaft*
3 descendant *Nachkomme*
4 desirable *wünschenswert*
5 edible *essbar, verzehrbar*
6 small red radish *Radieschen*
7 white/brown radish *Rettich*
8 turnip *weiße Rübe*
9 Chinese cabbage *Chinakohl*
10 waste ground *Ödland*

1 Explain which parts of the wild cabbage plant were changed by artificial selection to get the different cabbage varieties shown in Figure 20.
2 Make a mind map showing how the edible cabbage varieties shown in Figure 20 are related to radishes and turnips.

Figure 20 Various cabbage varieties and their common ancestor

2.6 Cultivated cabbage plants have many uses

Figure 21 Horseradish plant

Key terms

Uses of cultivated cabbage plants
- antiallergic *nicht allergisch*
- biofuel *Biobrennstoff*
- dyer's woad *Färberwaid*
- lubricate sth. *etwas schmieren, einfetten*
- rapeseed oil *Rapsöl*
- rape plant, rapeseed *Raps*
- spice *Gewürz*

The foods of the cabbage family are not only nice to eat, but also good for our health. For more than 30 years scientists have been exploring the link between the human diet and certain types of cancer. The results of several studies in the USA, China and Japan show that vegetables from the cabbage family contain substances which can help to prevent cancers of the digestive system and also reduce the risk of cancer of different other organs.

The cabbage family provides not only many important vegetables but also **spices** that are used to give flavour to the food we eat, such as cress[1], horseradish[2], and black and white mustard[3]. Mustard seeds are the basis for all the different types of mustard that are used for flavouring meat and other dishes.

Dyer's woad, another member of the cabbage family, has been cultivated in Germany since the ninth century for the blue colour contained in its leaves. The colour can be extracted[4] from the dried leaves and used for dyeing linen[5] textiles and for painting. In monasteries it was used for the beautiful illustrations in the handwritten gospel[6] books. Woad was the only source of blue colour in Europe until the end of the 16th century, when it became possible to import indigo from India. Since 1880, when the chemical industry began to produce blue colours, there has been a steady decline[7] in the cultivation of woad, but in recent decades production has increased once again. Today woad is used in inkjet printers[8] and in natural dyes for textiles. In contrast to many synthetic colours, it is **antiallergic** and safe for the environment.

Figure 22 Dyer's woad

One plant of the cabbage family with a very long history is of special importance today: the **rape plant**, whose fruit is called **rapeseed**. For about 4000 years this plant has been cultivated for its oil-containing seeds. Rape is a heavy nectar producer, and the honey which honey bees make from it has a light colour and a slightly peppery taste. The oil produced from rapeseed was at first only used for lighting lamps, and later to a small degree also for cooking, for animal feed and for **lubricating** machines. Over the last 100 years biologists and chemists have found new varieties of rape with oil that tastes much better: this oil is used for the production of vegetable oil and margarine. In addition to this, scientists have also developed a method of refining[9] rapeseed oil for use as **biofuel** in engines.

1 cress *Kresse*
2 horseradish *Meerrettich*
3 black/white mustard *schwarzer/weißer Senf*
4 extract sth. from sth. *etwas aus etwas extrahieren*
5 linen *Leinen, Leinwand*
6 gospel *Evangelium*
7 decline *Rückgang*
8 inkjet-printer *Tintenstrahldrucker*
9 refine sth. *etwas veredeln, aufbereiten, raffinieren*

Figure 23 Flowering rape field with an oil-processing[1] plant in the background

Figure 24 Service station offering biofuel for sale

The leading producers of rapeseed oil are the Euopean Union, the USA, Canada, India, China and Australia. With the rise in the price of petroleum products, biofuels from rapeseed have become more attractive economically. The German biofuel industry has expanded to keep pace with the growing demand. At the end of 2008 there were about 1900 service stations in Germany that offered biodiesel as a renewable energy alternative to the limited resources[2] of fossil fuels[3].

1 processing *Verarbeitung*
2 resource *Rohstoffquelle, Bodenschatz*
3 fossil fuel *fossiler Brennstoff*
4 display *Ausstellung*

1 Choose a vegetable that does not belong to the cabbage family and collect information about it. Make a poster which illustrates and describes the plant, its history, and how it is grown, harvested and used. Then organize a display[4] in the classroom of all the posters. For help see 'Improve your skills', p. 133.

2 Petroleum products, natural gas and coal are limited resources. This is why they are often called non-renewable sources of energy. In the future, renewable energy sources like biofuels will become more and more important.

a What other renewable energy sources are there? Find out what their advantages are and what difficulties there are in using them efficiently and on a large scale.

b Divide into teams and prepare arguments for a discussion on the advantages and disadvantages of nonrenewable and renewable energy sources. One team takes the position of nonrenewable sources, the other team argues for renewable sources. The words and phrases in the **Activate your English** box will help you formulate your arguments.

3 **Say it in German:** Imagine you and a friend are on your way to the sports grounds and are walking along a country lane beside large fields of rapeseed, which are in full flower. Looking across the beautiful yellow sea of flowers your friend asks you if you know what this mass of rapeseed plants might possibly be used for. Writing in your notebook, tell them what you know about the different uses of the rape plant today. Alternatively, do a role-play with a partner.

Activate your English

Discussing sources of energy

One (big/clear) (dis)advantage of ... is that it is (not) ... / can (not) be ... / (doesn't) ...
Using natural gas / biofuel / solar energy / ... makes / does not make sense because ...
One of the strengths/weaknesses of ... is that ...
The problem with ... is that ...
That may be true, but ... / That's not true
I agree (with you), but ... / I disagree (with you).
You may have a point, but ...
I would just like to point out that ...
I would argue that ...
On the one hand ..., on the other hand ...
Please don't interrupt me.
Would you please let me finish?

2.7 A survey of common cereals

Maize

With a height of up to 2.5 metres the maize plant grows taller than most other cereal plants. Its stem is filled with a spongy[1] tissue, or marrow[2]. Male and female flowers grow on the same plant but are separate: a large number of male flowers sit at the top of the plant, while the female flowers form a cone-shaped inflorescence halfway down the stem in the axil of the leaves. Maize is an annual plant which originates from southern tropical regions of North America, where it was originally cultivated by native Indians as an important food source. It was introduced to Europe in the 15th century.

Today maize is also grown in subtropical[3] and temperate regions of Europe and America, where it is mostly used for forage.

Figure 25 Flowering maize plants

In South America and in southern Europe the ripe maize is ground into flour and used for baking or making polenta[4]. Global annual production of maize is more than 700 million tons.

Rice

Rice is an important staple food in many parts of the world, especially in south, east and south-east Asia. It can be grown practically anywhere. While it is normally grown as an annual plant, in tropical areas it can be cultivated throughout the year. Rice is traditionally grown by flooding the fields during the entire growth period. Other methods of cultivation are also used, depending on the climate and terrain, but in all cases large amounts of water are necessary. In contrast to most other cereals, the inflorescence of the rice plant is a panicle[5] – a loose, irregularly branched cluster[6] of spikelets.

By removing the outer husk of the harvested grain you get brown rice, which contains protein, B vitamins and the minerals iron and calcium. White rice, which is popular in many cultures, is produced by removing the bran layer, i.e. the rest of the husk and germ[7]. Important nutrients are lost, however, in this process. In cultures where only white rice is eaten as a staple food, people can suffer from a deficiency disease[8] called beriberi[9], which causes muscular paralysis[10] and can even lead to death. In countries where rice is mainly used, people often eat brown rice.

Parboiling[11] is a method of making white rice which retains[12] about 80 per cent of the vitamins and minerals which normally get lost in the peeling[13] and polishing[14] process. The brown grains are cooked and pressed to make the nutrients migrate from the outer brown layer into the starch-containing endosperm. Today there are hundreds of rice varieties which differ in size and cooking qualities. Every year about 350 million tons of rice are harvested worldwide.

Figure 26 Rice panicles

1 spongy *schwammig*
2 marrow *Mark*
3 subtropical *subtropisch*
4 polenta *Polenta, Maisgericht*
5 panicle *Rispe*
6 cluster *Büschel*
7 germ *Keim*
8 deficiency disease *Mangelkrankheit*
9 beriberi *Beriberi*
10 paralysis *Lähmung*
11 parboiling *Herstellung von Parboiled-Reis*
12 retain sth. *etwas beibehalten*
13 peel sth. *etwas schälen*
14 polish sth. *etwas polieren, abschleifen*

Wheat

Together with rice and maize, wheat is one of the most important cereal plants. Section 2.3 looks more closely at this cereal plant, which is highly valued in Europe for its use in making bread and other baked products, as well as pasta.

Rye

This annual grass plant has spikelets with two or more flowers and clearly visible awns. As the grains become ripe, the seed head[1] bends down. The grains are greyish or greenish in colour and longer and more slender than wheat grains.

In contrast to wheat, rye is frost-resistant and can endure temperatures as low as −35 °Celsius. It can grow on poorer soils and also tolerates droughts better than other cereal grains. Since the Middle Ages, rye has been widely cultivated in Central and Eastern Europe. It is the main bread cereal in areas east of the French border with Germany and north of Hungary. Rye is also grown in temperate[2] regions of Asia as well as North and South America.

Figure 27 Ears of rye

Bread made from rye flour is denser[3] in structure than bread made from wheat. It is higher in fibre than other types of bread and often has a darker colour and stronger flavour. Today rye crops[4] are used more for feeding livestock[5] than for human consumption[6]. The yearly production worldwide is about 30 million tons.

Barley

Barley is the grain of an annual grass plant which does not make great demands on the quality of the soil and the climatic conditions. It also has the shortest growth period of all cereals, which means that it can be grown and harvested in mountainous areas and in regions with short summers, even near the Arctic Circle, where it was used for making bread in the past. As bread made entirely from barley flour is often crumbly[7], it is usually baked in combination with other types of flour.

The ear of the barley plant is easily distinguished from wheat and rye by the length of its awns, which are nearly as long as the seed head. Each spikelet contains three flowers. When the grains are ripe, they are spindle-shaped[8] and are fixed to the glumes and the awn.

Figure 28 Ears of barley

Barley was used in ancient Egypt for making bread and beer. Today about half of the world's barley production of about 170 million tons is used for feeding animals and a large part of the rest is used for the production of beer and malt whisky[9].

1 seed head *Fruchtstand*
2 temperate *gemäßigt*
3 dense *dicht*
4 crop *Ernte, Ernteertrag*
5 livestock *Vieh, Viehbestand*
6 human consumption *menschlicher Verzehr*
7 crumbly *krümelig, brüchig*
8 spindle-shaped *spindelförmig*
9 malt whisky *Malzwhisky*

Oats

The seed head of oats is a panicle. Every spikelet contains two flowers with a very long glume and short awns. Oats grow in low-quality soil throughout the temperate zones. They require low summer heat and have a greater tolerance to rain than other cereals. Oats are used for livestock feed and in human consumption mainly for porridge[1] and muesli. Bread cannot be made from oat flour. In Britain oats are traditionally used in beer production. Every year about 40 million tons are harvested wordwide.

Figure 29 Field of oats

Millet

This ancient food plant has been cultivated in China since 3000 BC. Its different varieties grow to between 30 cm and 1.3 metres high. As the grains develop, the loose seed head bends down. The seeds remain inside the husk after threshing[2].

Today millet is an important staple food in many parts of Asia, in Russia and in western Africa. The flour obtained from millet is mainly used for porridge or flatbread[3], but the grains are also cooked and eaten like rice. In western Europe and the USA millet is mainly used as forage. The yearly production worldwide is over 30 million tons.

Figure 30 Millet panicles

1 porridge *Haferbrei*
2 thresh sth. *etwas dreschen*
3 flatbread *Fladenbrot*
4 drought *Dürre*
5 beverage *Getränk*
6 account for sth. *verantwortlich sein für etwas, etwas bedingen*

Figure 31 Sorghum

Sorghum

This strong grass plant grows to heights of up to 5 metres. Its structure is similar to that of maize and, like maize, its stems are filled with marrow. The inflorescence is a loose but upright panicle 10–60 cm long, with male and female flowers. The grains are white, yellow or red and are smaller than wheat grains but bigger than millet grains. Sorghum grows easily on poor soil and is resistant to drought[4] and heat. It is Africa's most important cereal, but is also grown on the Indian subcontinent, in China, Argentina, Australia and in southern Europe. The crop is used mainly for boiled porridge, flatbread, as well as for beer and other alcoholic beverages[5]. The annual worldwide crop of sorghum is more than 50 million tons .

1 Which of the main cereal plants have an ear and which have a panicle? Make a table giving key information for each plant in this section, including wheat (see Section 2.3, p. 192). Use the following headings: Type of seed head, Main growth areas, Use by humans, Yearly production. For further help see 'Working with tables and line graphs', p. 53.
2 Which three cereal plants are the most important globally? What percentage of the total yearly production of cereal grains do these three plant species account[6] for?
3 Find out what polenta is and how it is eaten. How does it differ from porridge or muesli?
4 Explain why parboiled rice is healthier than normal white rice. For language help see **Activate your English**, 'Comparing and contrasting', on the inside cover.

Collecting and sorting plant species

A **herbarium** is a collection of plant species which have been identified, dried and preserved[1], together with their names and information about the place where they were found. It is a well-known method used to find out which species of plants exist in a particular area, e.g. a meadow, forest or river bank. You can also make a herbarium which contains only **mosses** or **ferns** or specimens of a special family of flowering plants. It is an interesting hobby and a fun way of learning about plants growing in your area or country.

If you collect plants in a meadow in springtime, your herbarium will be a kind of snapshot[2] of the state of plants at that specific time of year. If you go on collecting plants in the same area at other times of the year, you will get a record[3] of the plant succession in that ecosystem[4].

To make a herbarium, you first need to decide what kinds of plants you want to collect. The next step is to get the necessary materials for collecting, identifying and preserving the plants.

Key terms

Collecting plant species
- fern *Farn*
- herbarium *Herbar, Sammlung getrockneter Pflanzen*
- moss *Moos*
- mount sth. *etwas befestigen, (auf)montieren*
- plant press *Pflanzenpresse*
- plant succession *Sukzession; Abfolge der Pflanzen, das nacheinander Auftreten von Pflanzenarten (in einem bestimmten Gebiet)*

Materials
- small paper or plastic bags
- bigger bag to hold the small bags
- pruning shears[5] or a pocket knife
- tags[6] or sticky labels
- pencil or pen
- field notebook
- field guide or flower identification book
- plain white paper, slightly heavy (A4, approximately 100 g/m²)
- strips of paper and glue or sticky tape

For a plant press[7]:
- some old newspapers
- blotting paper[8] (optional)
- several strong sheets of cardboard (A4)
- long piece of string
- some books or heavy objects

Figure 32 Collecting and identifying plants in a meadow

Collecting the plants
1. Only collect plants that you already know and that you are sure are not protected species.
2. Only collect healthy, undamaged plants and try to get as many different parts of the plant as possible (leaves, flowers, fruits).
3. If you want to show the root of a plant in your collection, clean the soil off it before putting the plant into one of the small bags.
4. If you already know the name of a species or at least the family, write it on a name tag and put that into the bag with the plant. Handle the plants carefully and do not crush them. For each plant write down the name, where you found it and the date on a tag or better in a field notebook, together with any other information on the plant and the surrounding vegetation.

1 preserve sth. *etwas aufbewahren, konservieren*
2 snapshot *Momentaufnahme*
3 record *Nachweis, Beleg*
4 ecosystem *Ökosystem*
5 pruning shears (pl.) *Gartenschere, Baumschere*
6 tag *Anhänger*
7 plant press *Pflanzenpresse*
8 blotting paper *Löschpapier*

Drying, pressing and mounting the plants

Figure 33 **Drying the plant**

Figure 34 **Pressing the plant**

1. Do not wait too long before drying and pressing the plants once you get home. First make sure, however, that you have exactly identified your plants, as the drying process may change the natural colour of the flowers and leaves.
2. Start by laying out each specimen[1] on half a sheet of newspaper. Arrange the different parts of the plant in their natural position, but make sure that at least one leaf is bottom side up. Many leaves look different on their top and bottom sides.
3. Add the filled-out name tag to the specimen, then fold the other half of the newspaper over the plant and place it between further layers of newspaper. You can speed up the drying process by placing a piece of blotting paper on the plant before folding over the newspaper.
5. When each plant has been prepared in this way, they can be placed on top of each other, though it is essential that you put a sheet of cardboard between each specimen. Then add two stronger pieces of cardboard to each side of the packet and tie it all together with a piece of string. Place some books, a brick or any other heavy object on top of the packet.
6. About every two days, exchange the inner layers of newspaper for fresh ones to keep the plants from growing mould[2].

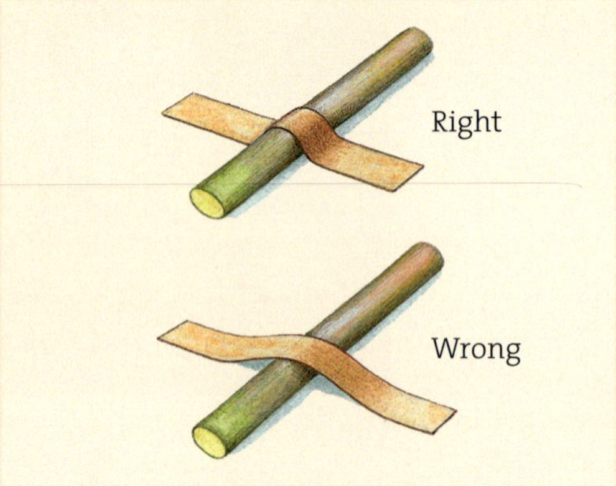

Figure 35 **Fastening the plant on the herbarium sheet**

Figure 36 **A finished herbarium sheet**

7. When the specimens are dried and pressed, they can be **mounted** onto sheets of paper. To hold the plants in place, use either glue and thin strips of paper or pieces of sticky tape.
8. Then glue on a final name plate in the lower right-hand corner. This should indicate: the name of the plant family, the common name of the species in both English and German (plus, if possible, the scientific name in Latin), the place where the plant was growing, the date of collection and the name of the collector.
9. The herbarium sheets can be stored in a box, folder or binder[3]. They should be kept in a safe, dry place where they will not become dusty[4] or mouldy[5].

1 specimen *Einzelprobe, Exemplar*
2 mould *Schimmel*
3 binder *Ordner*
4 dusty *staubig*
5 mouldy *schimmelig*

Activity 1 Studying flowers

A Make a drawing of the longitudinal section of a flower

Materials
- cherry flower
- small glass plate
- tweezers
- scalpel or razor blade with a safety handle
- binocular microscope or hand lens

Procedure
1. Place the cherry flower on its side and press it lightly onto the glass plate, holding it in place using your fingers or the tweezers. Then slice[1] the flower in two from top to bottom.
2. Place the longitudinal section under the microscope and observe and identify the different parts of the flower.
3. Make a labelled drawing. Your drawing should be big enough to recognize the details inside the flower.

Figure 37 Cherry flower sliced longitudinally

B Develop a flower diagram

A flower diagram is the schematic picture of the arrangement of the main parts of the flower. In order to make a flower diagram it is necessary to dissect[2] the flower.

Materials
- cherry flower
- tweezers
- hand lens
- 2 sheets of plain white paper
- sticky tape

Procedure
1. Using the tweezers, carefully remove the sepals, petals, stamens and carpels from the flower stalk.
2. Place the flower parts on the sheet of paper and arrange them in circles from the centre to the outside of the flower. Fix them onto the paper with sticky tape (see Figure 38). Count the different parts.
3. Now draw concentric[3] circles on the second sheet of paper and complete the flower diagram (see Figure 39).

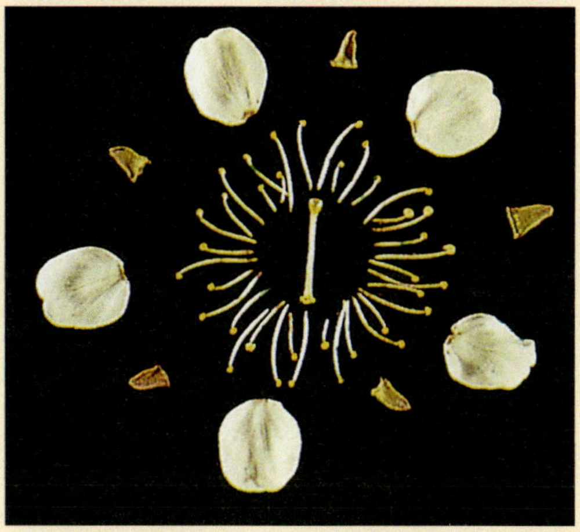

Figure 38 Dissected cherry flower

1 slice sth. *etwas (durch)schneiden*
2 dissect sth. *etwas zerlegen*
3 concentric *konzentrisch*

Information: Plants with the same flower structure as the cherry tree belong to the rose family[1]. Those with the same flower diagram as the rape plant are members of the cabbage or mustard family.

1 Identify other plants with the same flower structure as the cherry tree.
2 Make a flower diagram of the rape plant using the same procedure as for the cherry flower.
3 Compare your diagram of the rape flower with that of the cherry flower. What are the differences? For language help, see **Activate your English**, 'Comparing and contrasting', on the inside cover.

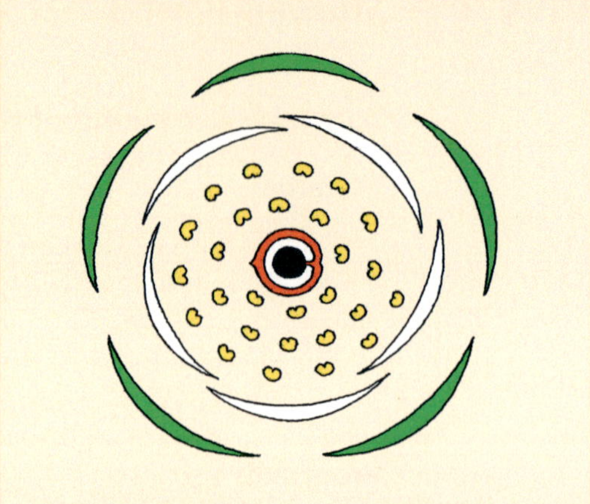

Figure 39 Diagram of a cherry flower

Activity 2 Studying the structure and the germination of seeds

A Investigate the structure of a bean seed

Materials
- dry bean seeds of the scarlet runner[2] or the broad bean[3] (from a plant nursery or a shop that sells garden materials)
- kitchen knife
- dissection needle[4]
- pincers
- hand lens or binocular microscope
- pipette
- iodine solution

Figure 40 Seeds of the scarlet runner bean

Procedure
1. Before starting your investigation, soak the bean seeds in water for several hours. When the seeds have swollen to about twice their original size, the seed cover is soft enough to be opened.
2. Remove the seed cover with the kitchen knife.
3. Split the bean into two halves by using your fingernails to cut off one of the two cotyledons.
4. Observe the cotyledons under the microscope or with your hand lens.

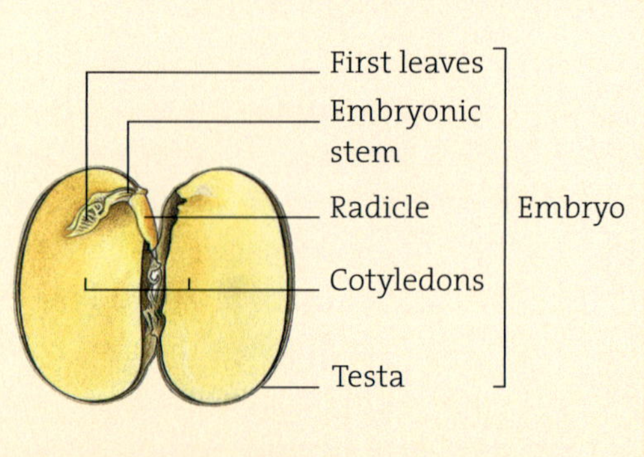

Figure 41 Bean embryo between the two cotyledons

1 Working with a partner, compare the inner surface of the two cotyledons. Make labelled drawings of what you can see.
2 Scrape off[5] some material from the surface of one cotyledon and add a drop of iodine solution.
3 Describe and explain the effect of the iodine solution on the cotyledon.

1 rose family *rosenartige Gewächse*
2 scarlet runner *Feuerbohne*
3 broad bean *Dicke Bohne, Saubohne*
4 dissection needle *Präpariernadel*
5 scrape sth. off *etwas abkratzen*

B Demonstrate the forces which develop when seeds take up water

Materials
- dry bean seeds of the scarlet runner or the broad bean
- gypsum[1] (from a do-it-yourself store or a building centre)
- water
- 2 cheese boxes (camembert) or any other small cardboard box (approx. 10 cm x 10 cm)

Procedure
1. Mix the gypsum with water to make a plaster mixture. Then fill the cardboard boxes completely with the plaster.
2. Press about ten dry bean seeds into the soft plaster in one of the boxes so that they are completely covered. Allow the plaster to get hard and dry.
3. Take both plaster blocks out of the cardboard boxes and place them into a large Petri dish with about 1 cm of water and leave them there overnight.

1 Describe what the plaster blocks look like the following day.
2 Explain what has happened.
3 Explain why this process is important for a germinating plant under natural conditions.
4 Why were two plaster blocks used in the experiment?

C Determine the conditions for seed germination

Decide in a class discussion which factors could influence seed germination. Working in groups of three, choose one factor to test, e.g. by using dry or wet filter paper or cotton wool, by storing the seeds in a dark or light place, by keeping the seeds at different temperatures, or by covering them completely with water. You can use different kinds of seeds for the experiments, e.g. peas or beans, mustard seeds and grains of wheat or maize.

Materials
- Petri dishes with lids
- jam jars or baby food jars
- paper towels or cotton wool
- water
- potting soil
- seeds of different plant species

Procedure
Each group uses three seeds of one kind for testing their factor of influence. Only the factor under investigation is variable; all other factors must remain constant when you repeat the test with different seeds. In your group, carry out a control experiment in which the factor you are investigating is either present or missing.

1 gypsum *Gips*

1 Note down your expectations before you start and record the results of each experiment. Why are three seeds used in each experiment and why is a control necessary?
2 Make a table in which you fill in the condition to be tested, a short description of the procedure, your expectations and your observation.
3 Compare the results with your expectations and comment on the results.

2.8 Producers and consumers must respire

Green plants are called **producers** because by means of photosynthesis (see Chapter 1, p. 167) they use the sun's energy along with carbon dioxide and water to produce energy-rich glucose. Oxygen, a by-product[1] of this process, is released into the air. Plants use the glucose to produce complex carbohydrates, but also combine it with different minerals to make fats, proteins and nucleic acids.

As the cells of animals and humans have no chlorophyll, they cannot carry out photosynthesis and therefore depend on green plants for food. By eating plants, **herbivorous** animals benefit[2] directly from the energy-rich[3] substances produced by photosynthesis, while **carnivorous** animals and **omnivorous** organisms such as humans use them indirectly when they eat meat. For this reason they all belong to the group of **consumers**.

Key terms

Producers and consumers
- carnivorous *Fleisch fressend*
- cellular respiration *Zellatmung*
- consumer *Verbraucher, Konsument*
- food chain *Nahrungskette*
- herbivorous *Pflanzen fressend*
- omnivorous *alles fressend*
- producer *Produzent*
- reactant *Reaktionspartner*
- respire *atmen*

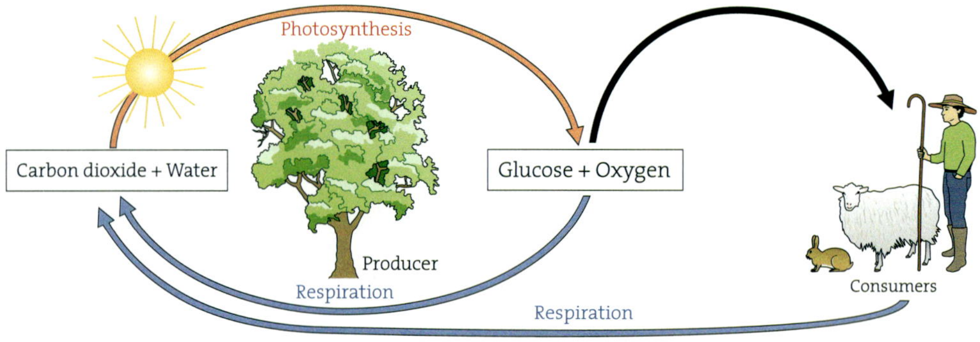

Figure 42 **Cycle of exchange between producers and consumers**

Not only the organic matter of the plants but also the oxygen is essential for all organisms that **respire**. As green plants cannot rely[4] on the sun's energy directly for all their life processes, they need to respire just as animals and humans do. Respiration releases energy by breaking down energy-rich substances, most importantly the carbohydrate glucose. In this process plants use about one-fifth of the glucose they have gained through photosynthesis. This process is known as **cellular respiration** and takes place inside all living cells. It is summarized in the formula:

Glucose + Oxygen	----------→	Carbon dioxide + Water + Energy
Reactants		*Products*

At night, when no light is available for photosynthesis, green plants respire and use oxygen from the large store[5] in the atmosphere, which is continuously filled by photosynthesizing plants on parts of the globe[6] where sunlight is available.

1 by-product *Nebenprodukt*
2 benefit from sth. *profitieren von etwas, Nutzen aus etwas ziehen*
3 energy-rich *energiereich*
4 rely on sb./sth. *sich auf jdn./etwas verlassen*
5 store *Vorrat*
6 globe *Erdkugel, Globus*

1 Make a table comparing the processes of photosynthesis and cellular respiration.
2 A **food chain** shows which organism is consumed by another organism. Set up a food chain by connecting the names of the organisms by arrows. Begin with a species of green plant and finish the chain with a carnivorous animal or a human. Develop a similar chain for marine organisms. Find out who in your class can set up the longest chain.
3 **Say it in German:** Explain the cycle shown in Figure 42.

Activity 3 Demonstrating that plants respire

The apparatus which you will use in this experiment is similar to the one used to prove that the air you breathe out contains carbon dioxide (see Unit I, Chapter 2, p. 42). The indicator for carbon dioxide is again lime water[1], which becomes milky white when carbon dioxide dissolves in it. The sodium hydroxide solution[2] binds the atmospheric carbon dioxide.

Materials
- 3 small Erlenmeyer flasks (250 cm^3) with tight glass stoppers fused with glass tubes that let air pass through in only one direction
- lime water
- sodium hydroxide solution
- dried peas soaked in water for several hours until swollen
- 2 short plastic or rubber tubes
- suction pump[3]

Figure 43 Germinating peas respire

Procedure
1. Fill one flask half-full with sodium hydroxide solution, the second one with clear lime water and the third one with the swollen peas.
2. Connect the three bottles as shown in Figure 43 and put the apparatus in a warm place.
3. Connect a suction pump to the free glass tube of the bottle containing lime water.
4. Observe the changes after one or two days.

1 lime water *Kalkwasser*
2 sodium hydroxide solution *Natronlauge*
3 suction pump *Saugpumpe*

1 What changes can you detect? Make notes in your exercise book and compare your notes with those of a partner.
2 Draw conclusions from your observations.
3 Why was it necessary to put sodium hydroxide into the first bottle?

Related activity Investigating if germinating seeds release energy in the form of heat

Use a thermos flask half-filled with seeds of peas, beans or cereal grains that have been soaked in water for 24 hours. Fit a thermometer through the hole in a rubber bung that closes the top of the thermos flask. The thermometer should reach into the centre of the germinating seeds. Read the temperature after about 10 minutes and then several times during the next two days.

1 Use your notes on temperature to develop a diagram (x-axis = time, y-axis = temperature).
2 Comment on the results and explain how this and the above experiment are related.

Summary

Grasses have slender stems, long thin leaves with parallel veins, and small flowers arranged in spikelets. Wild and cultivated grasses are used in different ways, but most importantly as food for humans and for livestock. Through selective breeding many important cereal plants have been developed, whose nutritious grains serve as staple foods in many countries. The most important cereals worldwide are rice, wheat and maize. Wheat is most commonly used for breadmaking. Wholemeal or wholewheat flour contains more minerals, protein and fibre than processed white flour. In the same way, brown rice is healthier than white rice.

The cabbage family contains a large variety of wild-growing and ornamental species of plants, whose often attractive flowers all have cross-shaped petals. As with the grass family, humans have cultivated wild species of the cabbage family. Through artificial selection, different parts of the wild cabbage plant have been changed to develop a large variety of cabbage plants. These belong to the most important food crops, but their products are also used as spices (mustard), for dyeing (dyer's woad), for producing paper (straw) and as a source of renewable energy .(rapeseed oil)

Key terms

See the boxes:

The structure of grasses .... p. 190
Uses of grasses 1, 2 .... pp. 191,192
Important cereal plants .... p. 192
Wheat .... p. 192
Cabbage plants 1, 2 .... pp. 194,196
Uses of cultivated cabbage plants .. p. 198
Collecting plant species .... p. 203
Producers and consumers .... p. 208

A Further projects

1. Make a herbarium of cereal plants collected from fields in your neighbourhood. Add any cultivated grasses you find in other places on excursions or during your holidays.
2. Join with some classmates to set up an exhibition of plants of the cabbage family. Arrange the species in groups such as wild-growing plants, ornamental plants, vegetables, plants used as spices, plants cultivated for natural dyes and plants cultivated for oil.

B True or false?

Are these statements true or false? Correct the false statements.

1. Cereal grains are staple foods in many countries.
2. Grass flowers have brightly coloured petals.
3. The different types of cabbages have developed without human influence.
4. Silage is used in winter when no fresh food is available for farm animals.
5. Grasses are typical dicots with broad leaves and a network of veins.
6. Plants of the cabbage family can be identified by their three petals, which are always blue.
7. Oats are the most important cereal plant for making bread.
8. The seeds of the rape plant sit on both sides of a thin wall, which partitions the fruit.
9. As plants carry out photosynthesis, they do not need to respire.
10. White flour is healthier than the grey wholemeal because it is cleaner.

C Critical thinking

1 Comment on the following statement: 'The energy which you use to contract the muscles in your arms and legs comes originally from the sun.'

2 In industrialized countries, but also in some developing countries like Brazil, plants that can be used as food like maize, rape and sugar cane are used instead for making engine fuel. Hold a panel discussion[1] and discuss the pros and cons[2] of this practice. For language help, see **Activate your English**, 'Holding a panel discussion', on the inside cover.

1 panel discussion *Podiumsdiskussion*
2 the pros and cons *das Pro und Kontra*

Unit III

Cells, Tissues and Organs

Chapter 1 Investigating Cells

Chapter objectives

- To **underline** the central role of the cell in all organisms
- To **name** the parts of the compound microscope and **describe** their functions
- To **learn** how to use a compound microscope correctly
  To **compare** the functions of a light microscope with those of an electron microscope
- To **explain** how a wet mount is made

Key terms

Cells

- amoeba (pl. amoebae) *Amöbe, Urtierchen*
- bacterium (pl. bacteria) *Bakterie,*
- egg cell *Eizelle*
- nerve cell *Nervenzelle*
- paramecium (pl. paramecia) *Pantoffeltierchen, Paramaecium*
- seed fibre *Samenfaser*

Humans, just like animals, plants, fungi[1] and **bacteria**, are made of cells. The average[2] size of a cell is 0.01 mm, which is unimaginably[3] small, but cells can be much smaller than this, too. The smallest bacteria, for example, measure just 0.0001 mm across. Cells can also be a lot bigger. Some of the largest animal cells are **egg cells**. While the human egg cell is only 0.1–0.2 mm across, just large enough to be seen with the naked eye, the egg cell of an ostrich[4] can be more than 7 cm in diameter[5]. Among plant cells the **seed fibres** of cotton[6] are 3 to 5 cm long. The longest cells in the human body are the **nerve cells**, which have a length of one metre or more.

1 fungus (pl. fungi) *Pilz*
2 average *durchschnittlich*
3 unimaginably *unvorstellbar*
4 ostrich *Strauß*
5 diameter *Durchmesser*
6 cotton *Baumwolle*

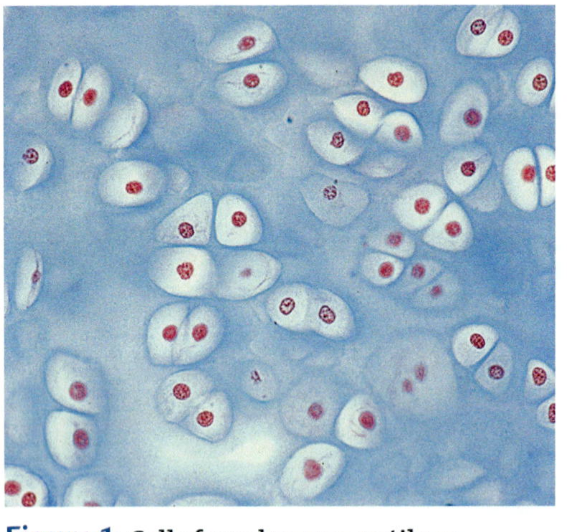

Figure 1 Cells from human cartilage

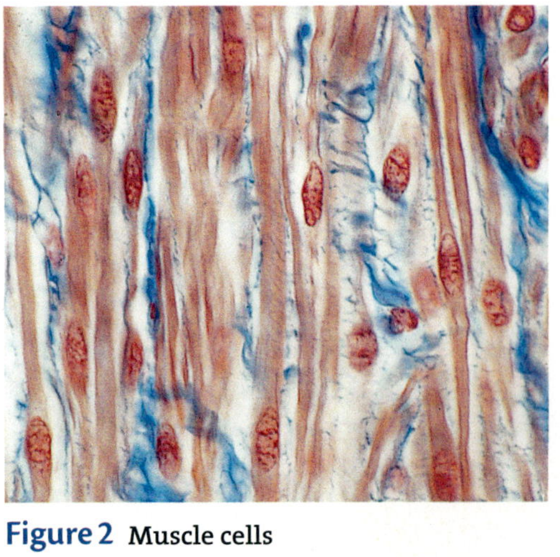

Figure 2 Muscle cells

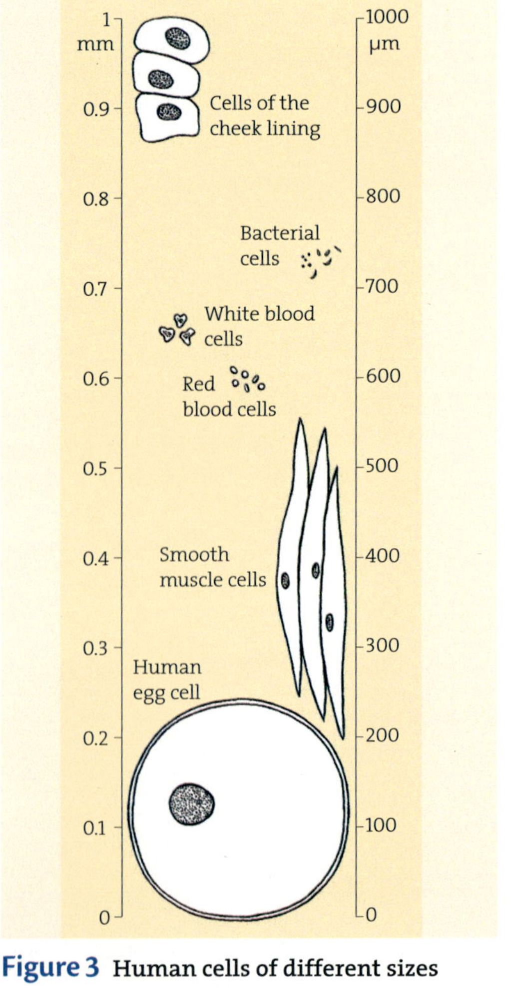

Figure 3 Human cells of different sizes compared to bacterial cells

While the human body is estimated[1] to have more than 10,000 billion cells, the bodies of other organisms like **amoeba** or **paramecium** consist of only one cell. These single cells can carry out all vital[2] processes of a living organism from movement and feeding to reproduction.

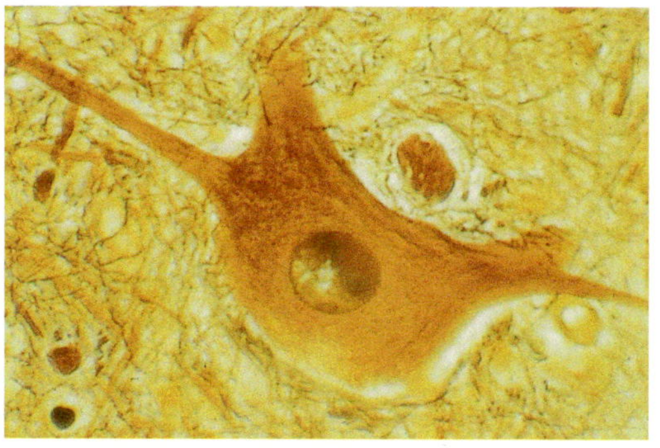

Figure 4 A single nerve cell

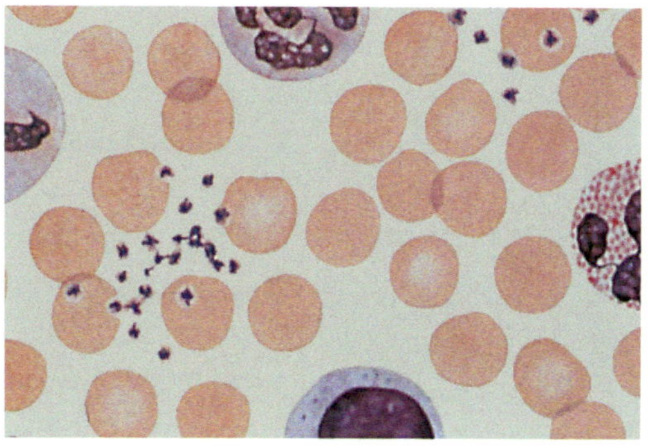

Figure 5 Human blood cells

1.1 Cells cannot be studied with the naked eye

Key terms

Magnifying objects (1)
- compound microscope *zusammengesetztes Mikroskop*
- field of view Gesichtsfeld, Blickfeld
- hand lens *Lupe*
- light microscope *Lichtmikroskop*
- magnification *Vergrößerung*
- magnify sth. *etwas vergrößern*
- magnifying power *Vergrößerungskraft*
- specimen *Untersuchungsobjekt*

Figure 6 A scientist at work

For about 170 years scientists have known that the bodies of organisms consist of cells. As most cells are too small to be seen with the naked eye[3], the development of the **compound microscope** gave biologists the necessary tool for studying cells. Compared to the **hand lens**, whose single lens is placed between the eye and the object that you are investigating[4], a compound microscope has two or more lenses. The rays[5] of visible[6] light first pass through the **specimen** and then through the lenses, which magnify the image[7] of the object several times. Modern **light microscopes** used in biological and medical laboratories can **magnify** an object up to 1500 times under optimal conditions. This means that small animal and plant cells can be identified and counted, for example the white and red blood cells in a blood sample and even tiny bacteria can be seen.

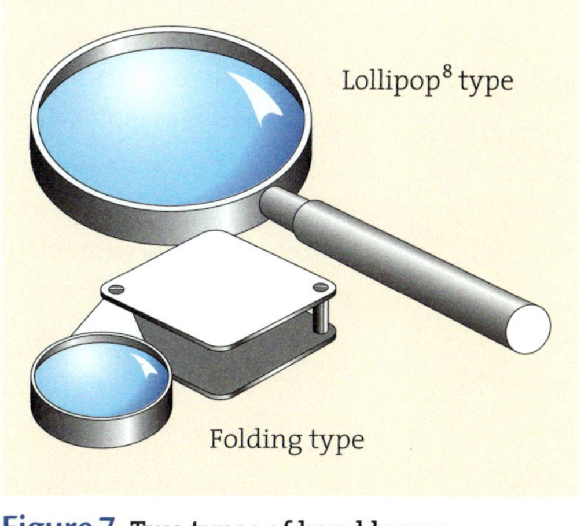

Figure 7 Two types of hand lenses

1 estimate sth. *etwas schätzen, abschätzen*
2 vital *lebenswichtig, vital*
3 with the naked eye *mit dem bloßen Auge*
4 investigate sth. *etwas untersuchen*
5 ray *Strahl*
6 visible *sichtbar*
7 image *Bild, Abbild*
8 lollipop *(wie ein) Lutscher*

At a distance of less than about 25 cm, the picture you see becomes blurred[1]. This is the so-called **focusing distance**. At this distance the average human eye can identify two black dots that are 0.1 mm apart[2] as being separate. This ability is called the **resolving power**. If an object is smaller than 0.1 mm, it is beyond the resolving power of the human eye and you will see the two points as one. The only way to increase the resolving power of the human eye is by using a magnifying tool.

A hand lens is a single glass lens which is **convex**, which means it collects the light rays and **focuses** them in such a way that we can move closer to the object than 25 cm and still get a clear image of the enlarged object. With a **magnifying power** of 5 to 20 times, hand lenses are useful for seeing the surface structures of objects better. The lollipop type of hand lens does not magnify the object as much as the smaller folding type (see both in Figure 7), but it offers a larger field of view.

The whole plant seen with the naked eye

A single leaf seen through a hand lens

Cells from the inside of a leaf seen through a compound light microscope

You can see that the cells are arranged like the bricks in a wall. The cells are separated from each other by **cell walls** and there are large numbers of chloroplasts in the cells inside the leaf.
The resolving power of a good light microscope is 0.25 µm [3], which is about half the average wavelength[4] of visible light.

A chloroplast, an organelle, seen through an electron microscope

At this high **magnification** you can see in the cell's **microscopic structure** that there are layers of thin membranes inside the chloroplast, which are darker in colour in places where the **chorophyll** molecules are embedded[5] in the membranes.
A transmission electron microscope (TEM) has a resolving power of 0.2 nm [6] and can magnify objects up to 240,000 times.

Figure 8 The Canadian pondweed[7] ***(Elodia canadensis)*** **seen with the naked eye and with different optical instruments**

1 blurred *verschwommen*
2 apart *getrennt, auseinander*
3 µm = micron, micro metre *Tausendstel Millimeter* (10^{-3} *mm*)
4 wavelength *Wellenlänge*
5 embedded *eingebettet*
6 nm = nanometre *Milliardstel Millimeter* (10^{-9} *mm*)
7 Canadian pond weed *Kanadische Wasserpest*

Key terms

Magnifying objects (2)
- convex lens *Sammellinse*
- electron microscope *Elektronenmikroskop*
- focus sth. *etwas scharf stellen* focusing distance *Bezugssehweite*
- image *Bild, Abbild*
- resolving power *Auflösungsvermögen*

Cells and organelles
- cell wall *Zellwand*
- chlorophyll *grüner Blattfarbstoff, Chlorophyll*
- chloroplast *Blattgrünkörper, Chloroplast*
- microscopic structure *Feinstruktur*
- organelle *Zellorganell*
- unicellular organism *Einzeller*

Standard light microscopes like those you will work with in school can magnify objects up to 400 times their normal size. This is sufficient[1] for studying **unicellular organisms** and the cells of plants, animals and your own body in the biology lab[2]. These microscopes also enable you to see the larger **organelles**, like the **chloroplasts**, which carry out specific functions within the cell (organelles are covered in greater detail in the next chapter).
Viruses[3], molecules[4] and atoms[5] are too small to be viewed with today's light microscopes. They can only be observed with an **electron microscope**, which is an instrument too complicated and expensive to be used in school. Here, electrons are used instead of light to create a magnified image of a specimen.

1 sufficient *ausreichend*
2 lab *Praktikumsraum, Labor*
3 virus *Virus*
4 molecule *Molekül*
5 atom *Atom*

Did you know?

If the 100 billion nerve cells of a single human brain were laid end to end, they would reach to the moon and back.

1 Explain why the study of cells is important.
2 Why are different optical instruments useful for investigating cells?
3 Compare the magnifying power of a hand lens, a compound microscope and an electron microscope. For language help see **Activate your English**, 'Making comparisons', on the inside cover.
4 Which observations can be carried out with the naked eye and for which investigations are special optical instruments needed?
5 **Say it in German:** Prepare a short report on the different sizes and numbers of cells in living organisms.

1.2 The invention of the microscope has led to important discoveries in cytology

Figure 9 Robert Hooke's compound microscope

Key terms

Under the microscope
- Cell theory *Zellentheorie*
- cytology *Zelllehre, Zytologie*
- Antoni van Leeuwenhoek (1632–1723)
- Robert Hooke (1635–1703)
- Robert Brown (1773–1858)
- Matthias Schleiden (1804–1881)
- Theodor Schwann (1810–1882)
- Rudolph Virchow (1821–1902)

The first microscopes were invented towards the middle of the 15th century. These simple devices[1] were actually hand lenses. Scientists used them to magnify insects, plants and other natural objects. In about 1590 two Dutch spectacle[2] makers, Hans and Zacharias Janssen, constructed the first compound light microscope, which had lenses at both ends of a hollow tube, an ocular and an objective lens.

These early compound microscopes provided greater magnification, but when you see what the construction was like (see Figure 9) and where the light came from, it is no wonder that they only produced rather blurred images. This was the reason why most scientists in the early days of the compound microscope preferred to use the traditional simple microscope with a single high-power lens.

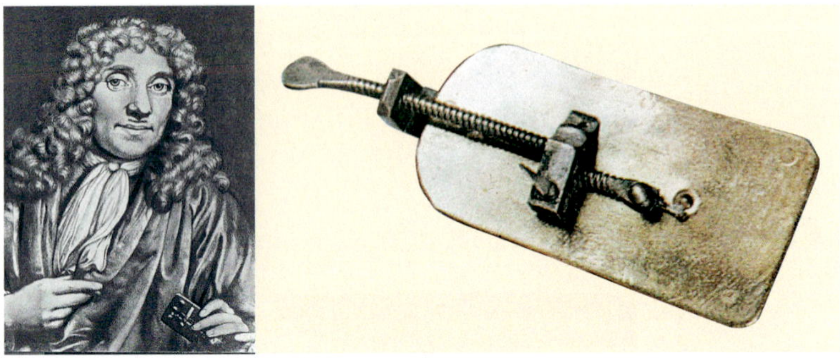

Figure 10 Van Leeuwenhoek and his simple microscope

Even today, experts are astonished by the discoveries that the Dutchman **Antoni van Leeuwenhoek** made in the years 1670–1680 using a rather primitive-looking instrument (see Figure 10). He looked at a wide range of objects, although he did not always know exactly what it was that he saw. From the drawings he made of the objects, we know that he saw blood cells, sperm cells, unicellular organisms in pond water and even bacteria.

About a decade before van Leeuwenhoek made his observations, the Englishman **Robert Hooke** was using one of the early compound microscopes to study many different biological objects, from insects to birds' feathers. He also looked at a thin slice of cork and called the box-like structures he saw 'cells' (see Figure 11). He used the word incorrectly in the modern sense, because what he saw were the empty cavities of former cells, of which only the dry cell walls were left. But his word 'cell' became the universal term for the building blocks of all living things.

In 1678, after van Leeuwenhoek had written a report about his discoveries to the Royal Society, Hooke was able to confirm his findings, thus paving the way[3] for the wide acceptance of van Leeuwenhoek's microscopic work.

The gradual improvements made to the light microscope during the 19th century led to important discoveries about the anatomy of cells. In 1831 the Scotsman **Robert Brown** announced that cells contain a large central body, the nucleus[4], and a few years later scientists in different parts of Europe reported that cells contain a clear, jellylike fluid, which we now call cytoplasm[5].

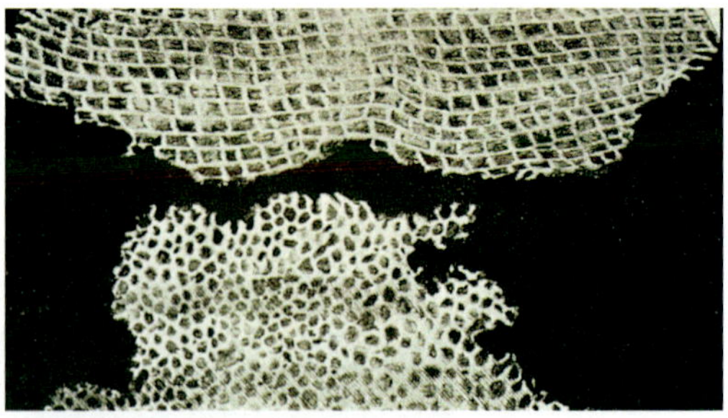

Figure 11 Hooke's drawings of cork 'cells' from his book *Micrographia* (1667)

After they had observed thousands of plant and animal cells under the microscope, the German scientists **Matthias Schleiden** and **Theodor Schwann** reached the conclusion in 1838 that all living things are composed of one or more cells. In 1858 the German physician **Rudolf Virchow** concluded from his observations of growing and dividing cells that cells can only develop out of cells that already exist. The observations of these three scientists were the basis for the **cell theory**, which is one of the universally accepted theories of biology. It states that:

1. All organisms consist of cells.
2. The cell is the basic unit of structure and function in organisms.
3. All cells come from other living cells.

1 device *Gerät*
2 spectacles (pl.) *Brille*
3 pave the way for sth. *etwas den Weg ebnen*
4 nucleus *Zellkern*
5 cytoplasm *Zellplasma*

Electron microscopes

Key terms

Electron microscopy
- beam of electrons *Elektronenstrahl*
- electromagnet *Elektromagnet*
- scanning electron microscope (SEM) *Rasterelektronenmikroskop*
- transmission electron microscope (TEM) *Transmissions-Elektronenmikroskop*
- micrograph *mikroskopisches Foto, mikroskopisches Bild*

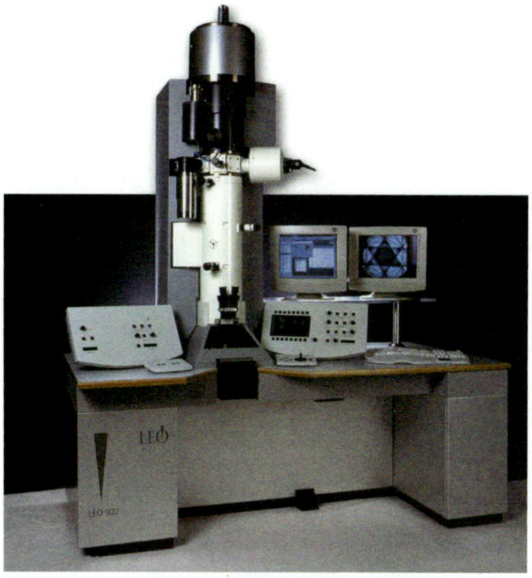

Figure 12 Modern electron microscopes are controlled by computers

The first electron microscopes were built by German engineers in the early 1930s. The **transmission electron microscope (TEM)**, which is an important tool of biological research[1], functions in principle like a light microscope, but uses a focused **beam of electrons** instead of light, and **electromagnets** instead of glass lenses. The electrons pass through the specimen, which must be sliced[2] very thinly, and interact[3] with it. This interaction of the electrons with the specimen forms an image which is magnified and focused onto a fluorescent screen[4] called a **micrograph**. As the electrons have a much shorter wavelength compared to visible light, and because a high vacuum in the body-tube of the TEM prevents the electrons from colliding[5] with gas molecules, the TEM can achieve magnifications of up to 240,000 times. However, as specimens must be extremely thin, only dead cells and tissues can be observed in the TEM after a complicated preparation, which includes the danger of damaging the delicate inner structures of the cells.

A later development (since about 1965) is the **scanning electron microscope (SEM)**, which takes detailed pictures of the surfaces of cells, tissues and even of whole organisms. The specimen is sprayed with a very thin layer of metal and then scanned from side to side with a beam of electrons. The electrons bounce off[6] the metal surface of the specimen and form an image on the fluorescent screen. A television camera transfers the pictures onto a monitor. The SEM produces impressive three-dimensional pictures of dead objects, but can only magnify up to 80,000 times. In both types of electron microscope the pictures are generated in black and white but can be coloured afterwards to make certain structures stand out[7] more clearly.

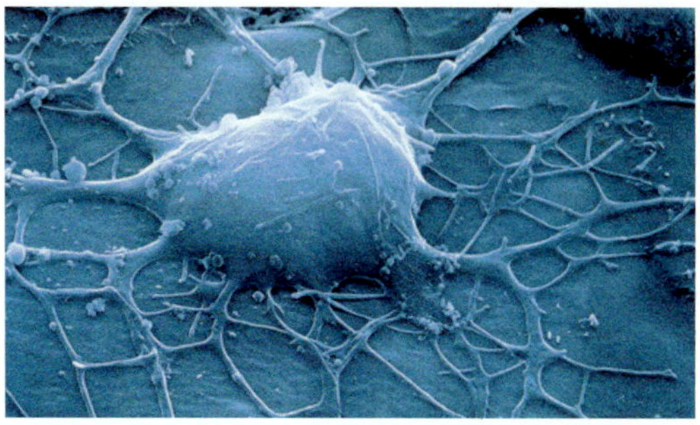

Figure 13 Photo of a nerve cell taken by an SEM

1 biological research *biologische Forschung*
2 slice sth. *etwas in Scheiben schneiden*
3 interact with sth. *(hier) durch etwas beeinflusst werden*
4 flourescent screen *Leuchtschirm*
5 collide with sb./sth. *mit jdm./etwas zusammenstoßen*
6 bounce off sth. *von/an etwas abprallen*
7 stand out *hervortreten*

1 Describe the first investigations into cells and who carried them out.
2 Explain why van Leeuwenhoek's discoveries were so astonishing.
3 List the differences between Hooke's microscope and a modern light microscope like those used in your school.
4 Why are there no glass lenses in an electron microscope and how does the microscope focus instead?
5 Explain the advantages and disadvantages of using an electron microscope instead of a light microscope.

1.3 The compound microscope is an essential tool[1]

Construction of a compound microscope

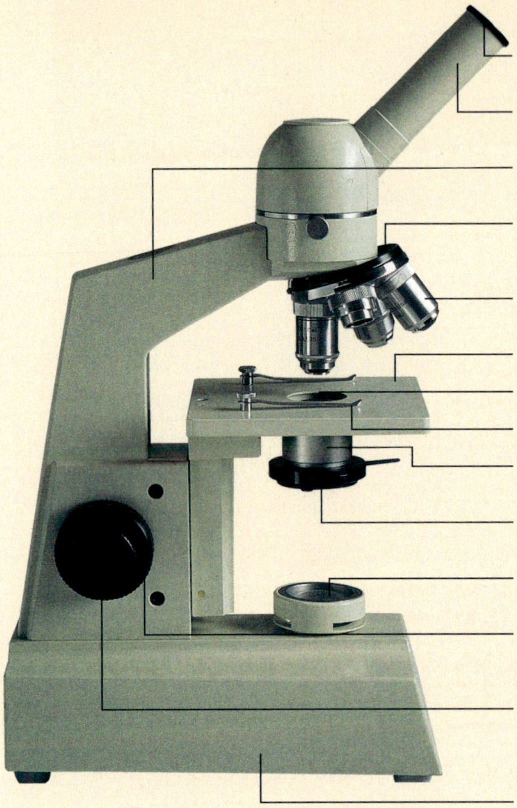

Name	Function
Eyepiece (ocular)[3]	Magnifies like a hand lens (5x, 10x, 15x).
Body tube[4]	Keeps ocular and objective lenses in the right position.
Arm[5]	Supports the microscope.
Revolving nosepiece[6]	Allows the user to switch between different objectives.
Objectives[7] (low, medium, high power)	Lenses for magnifying the specimen (2.5x–45x).
Stage[8]	Supports the glass slide
Stage opening	Allows light to pass through the stage.
Stage clip[9]	Holds the slide and specimen in place.
Condenser[10]	Lens that focuses light onto the specimen.
Iris diaphragm[11]	Regulates the amount of light entering the microscope.
Lamp	Makes the specimen easier to see.
Coarse adjustment[12]	Brings objects into rapid but coarse focus.
Fine adjustment[13]	Brings objects slowly into fine focus.
Base (stand)[14]	Keeps the microscope in a safe upright position.

Figure 14 The parts of the compound microscope

The compound microscope is one of the biologist's most important tools. A microscope is an expensive and sensitive instrument, so it is vital to learn the basic skills[2] needed to use it and the names of the parts of the microscope.

An important note: when the terms 'microscope' and 'light microscope' are used in this book, they always refer to the compound microscope as described in this section.

Determining the magnifying power of the compound microscope: The numbers which are printed on the eyepiece and the objective tell you how much each of these lens systems magnify the specimen. By multiplying the number on the eyepiece by the number on the objective you will get the total magnification of the microscope.

Rules for proper use of the microscope

1. When you carry the microscope, always hold it in both hands, one beneath the base and one holding the arm.
2. Place the microscope at least 5 cm from the edge of the work table.
3. Keep the lamp cable out of the way.

1 tool *Werkzeug, Gerät*
2 basic skill *grundlegende Fähigkeit, Fertigkeit*
3 eyepiece (ocular) *Okular*
4 body tube *Tubus*
5 arm *Arm, Stativ*
6 revolving nosepiece *Objektivrevolver*
7 objective *Objektiv*
8 stage *Objektivtisch*
9 stage clip *Objektklammer*
10 condenser *Kondensor*
11 iris diaphragm *Irisblende*
12 coarse adjustment *Grobtrieb*
13 fine adjustment *Feintrieb*
14 base (stand) *Fuß*

4. First lower the stage all the way down.
5. Rotate the nosepiece so that the low-power objective is locked in.
6. Only then put the slide[1] containing the specimen on the stage. Make sure the specimen is in the centre of the stage opening and then fix the slide with the stage clips.
7. Now, while looking at the stage from the side, raise it until it almost touches the objective. (In some types of microscopes you have to lower the revolving nosepiece.)
   If you look through the objective while raising the stage, you can accidentally crash the slide into the objective and destroy the lens.
8. Look down the eyepiece and slowly lower the stage until the specimen is in focus. To make sure you see it, you can move the slide back and forth just a little bit. It is easier to see a moving object than a stationary[2] one.
9. Get an overview of your specimen with the low-power objective. When you have found an interesting area that you want to look at in greater detail, simply turn the nosepiece to medium power and then to high power without changing the coarse adjustment knob[3]. Refocusing with the fine adjustment will usually be enough.
10. Never use the coarse adjustment knob for high-power focusing. You may damage not only your specimen but also the high-power objective.
11. When looking at the specimen under high power, it may be necessary to open the diaphragm a little in order to increase the lighting.
12. Dry the microscope completely when you have finished working. Check that the low-power objective is directly over the stage and that the body tube is as far down as possible. This protects the eyepiece and the objectives from possible damage.

Figure 15 Fixing the slide with the stage clips (see step 6)

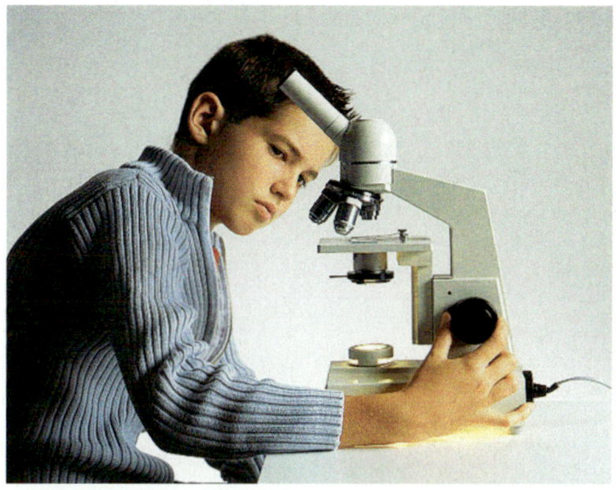

Figure 16 Raising the stage (see step 7)

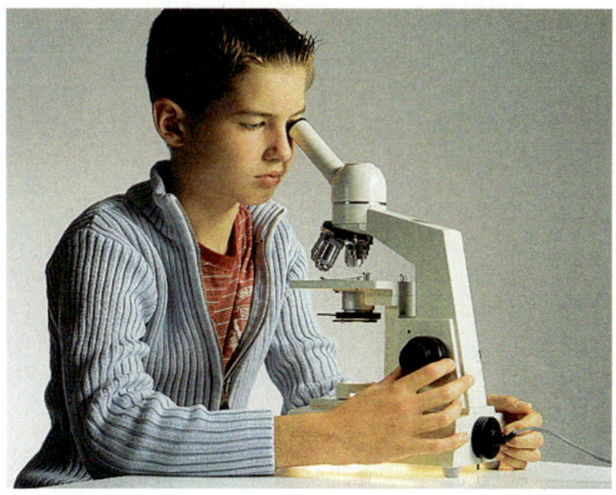

Figure 17 Bringing the specimen into focus (see step 8)

1 slide *Objektträger*
2 stationary *stationär, unbeweglich*
3 knob *Knopf*

Do not forget to clean the lenses of your microscope regularly with a soft cloth.

1 Why is it useful always to start your observation with the low-power objective?
2 **Say it in German:** Describe the steps you must follow in order to observe an object under the compound microscope.
3 Explain why you should not turn the coarse adjustment knob while looking through the eyepiece.

Preparing a wet mount

Key terms

Preparing microscopic objects
- cover slip *Deckglas*
- dissecting needle Präpariernadel
- hand-microtome *Handmikrotom*
- irrigation method *(in der Mikroskopie) Durchzugstechnik*
- lancet *Lanzettmesser, Lanzettnadel*
- microscopic slide *Objektträger*
- mount a specimen *ein Präparat auf den Objektträger bringen*
- permanent mount *Dauerpräparat*
- pipette *Pipette*
- scalpel *Skalpell*
- sectioning razor *Rasiermesser für mikroskopische Schnitte*
- single-edge razor blade *Rasierklinge mit einer scharfen Seite*
- temporary mount *Frischpräparat, zeitlich begrenzt haltbares Präparat*
- tweezers, forceps (pl.) *Pinzette*
- wet mount *Feuchtpräparat*

Before you can look at a specimen under a light microscope, you need to prepare it carefully. The object to be studied needs to be so thin that light from the lamp under the stage can penetrate[1] it. Therefore, a thicker object must be cut into very thin slices.

You can do the slicing by hand using a **single-edge razor blade** or a **sectioning razor** (see Figure 18). If you have to make a series of thin slices, a **hand-microtome** is quite useful (see Figure 19).

Once you have a thin slice, place the specimen into a drop of water in the centre of the **microscopic slide**. Then put a **cover slip** over the specimen to keep it in place and to stop the objective lens from coming into contact with the water (see Figures 20, 21 and 22).

It is important when adding the cover slip that no water gets onto its upper surface and no air remains under it. Hold the cover slip between your thumb and forefinger and touch the edge of the drop of water with one side of the cover slip. Hold it at a 45-degree angle while you use a pencil or **dissecting needle** to lower the cover slip carefully onto the specimen. This ensures that there are no air bubbles between cover slip and slide. If there are air pockets under the cover slip, these will appear as dark rings in your field of view. You can get them out by tapping[2] softly on the cover slip with the blunt[3] end of your pencil. The water should just fill the space between the cover slip and the slide. If there is too much water on the slide, hold a strip of filter paper (5 cm x 1 cm) or the edge of a paper towel to one side of the cover slip and remove the excess[4] water. Now the **wet mount** is ready for observation.

Take care that the specimen does not dry out while you are observing it. If any space under the cover slip suddenly becomes dry, quickly place a drop of water next to the cover slip.

1 penetrate sth. *etwas durchdringen*
2 tap sth. *etwas klopfen*
3 blunt *stumpf*
4 excess *überschüssig, überflüssig*

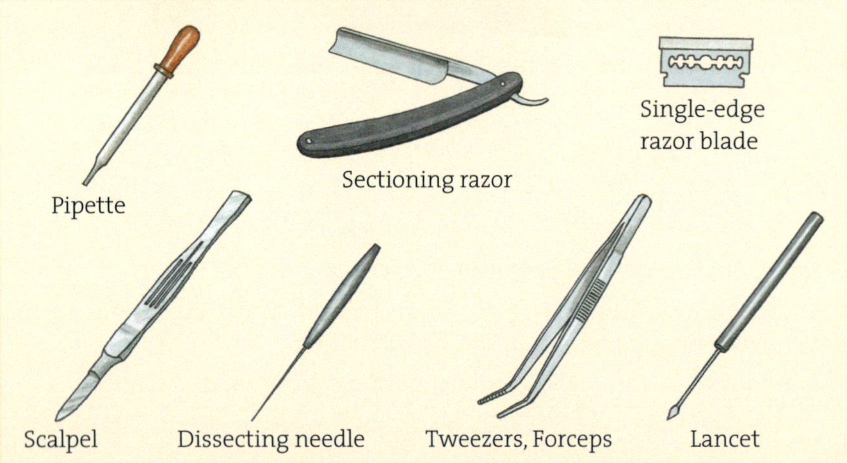

Figure 18 Tools needed for microscope work

Figure 19 Using a hand-microtome

Figure 20 Putting a drop of water on the slide

Figure 21 Placing the specimen on the slide

Figure 22 Covering the specimen with a cover slip

The use of **selective stains** (also known as **selective dyes**), which colour only certain structures in your specimen, makes it possible to see details much more clearly. **Vital stains** (e.g. **methylene blue** or **neutral red solution**) do not harm living cells. Other stains such as **iodine solution** kill the cells but do not destroy their structures.

You can add a stain to your wet mount by using the **irrigation method**. Follow these steps:

1. Remove the slide from the stage.
2. Place a drop of stain at the edge of the cover slip.
3. Hold a piece of paper towel or filter paper to the opposite edge of the cover slip.
4. Draw the stain through the sample.

Key terms

Microscopic stains
- iodine solution *Jodlösung*
- irrigation method *Durchzugstechnik*
- methylene blue solution *Methylenblau-Lösung*
- neutral red solution *Neutralrot Lösung*
- selective stain, selective dye *selektive Färbung*
- vital stain *Vitalfärbung, Färbung eines lebenden Objekts*

Carry out the method very carefully so that you do not disturb the structures in your specimen. If you want to observe unicellular organisms to see how they move, feed, etc., you should leave enough space for the living animal under the cover slip. This can be done by adding tiny pieces of wax[1] or modelling clay[2] to the bottom four corners of the cover slip before you place it on the specimen. If you want to make sure that the little creatures cannot move away too quickly once you have found them in the drop of water, you can put some fibres of cotton wool into the drop of water on your slide before you place the cover slip on it. These will act as walls, stopping the creatures from moving very far.

Besides wet mounts and **temporary mounts**, in which specimens can be observed while they are still alive, **permanent** (or prepared) **mounts** are also important in microscopic work. Many important discoveries about plant and animal structures were made by observing tissues which had been stained and permanently mounted. In permanent mounts very thin sections of stained tissue are mounted on a glass slide in a medium of fluid, transparent plastic or resin[3] (Canada balsam), which hardens quickly. This keeps air out and preserves the specimen under the cover slip for an indefinite[4] period[5], so that it can be studied again and again.

1 wax *Wachs*
2 modelling clay *Knetmasse*
3 resin *Harz*
4 indefinite *unbegrenzt, unbestimmt*
5 period *Zeitraum*

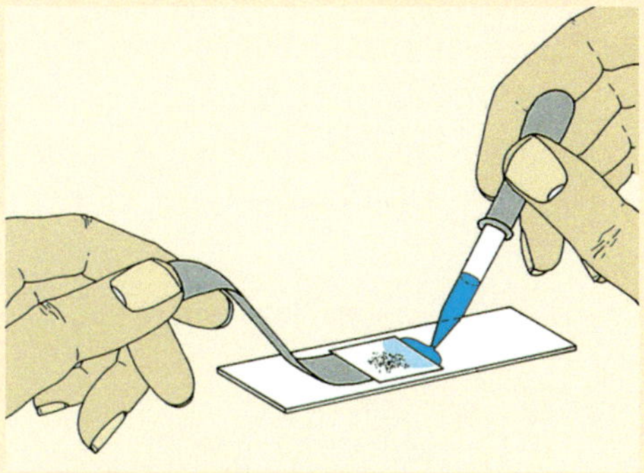

Figure 23 Irrigation method

1 Explain why a cover slip is placed over the specimen before you observe it under the microscope.
2 Why are selective stains used in microscopic investigations?
3 What are the advantages and disadvantages of wet mounts and permanent mounts?

Activity 1 Practise working with the microscope

Materials
- compound microscope
- glass slide
- cover slip
- pipette
- beaker of water
- the letter 'e' cut from a newspaper
- scissors

Procedure
1. Use a pipette to put a small drop of water on a clean glass slide.
2. Cut a small piece out of a newspaper with the letter 'e' on it.
3. Carefully place the letter 'e' into the drop of water using your tweezers and push it down to completely cover it with water.
4. Place a clean cover slip on the drop using the technique[1] described in 'Preparing a wet mount', p. 220.
5. Put the slide on the stage and view the letter under the low-power objective (the shortest objective). Then carry out the tasks below.

1 Describe what happens to the object when you move the slide …
a to the right.
b to the left.
c towards you.
d away from you.
e clockwise[2].
f anticlockwise[3].

2 View the letter under medium power and repeat Task 1.

3 Remove the slide from the stage and use a small piece of millimetre paper to measure the width of your field of view when you use different objectives. Note down the results and compare them with your classmates' results.

Activity 2 Determine the total magnification

1. Locate the numbers on the eyepiece and on the low-power objective, the medium-power objective and the high-power objective. Copy and fill in the chart below:

	Low-power	***Medium-power***	***High-power***
Eyepiece			
Objective			
Total magnification			

2. What is the rule for determining the total magnification of a compound microscope?

1 technique *Untersuchungsmethode*
2 clockwise *im Uhrzeigersinn*
3 anticlockwise *entgegen dem Uhrzeigersinn*

Summary

Cells vary in size and shape according to the different tasks they have to fulfil. Some organisms have billions of cells while others consist of just a single cell. Cells are the basic building blocks of living things. A cell contains all the qualities of a living organism and can only come from existing cells. This knowledge, known as cell theory, is based on observations scientists made with light microscopes in the 19th century. Electron microscopes are used today for detailed investigations of the inner and outer structure of cells. The compound light microscope, used in schools, is a sensitive tool, and guidelines for its proper use must be followed. Cells and tissue samples need to be prepared before they can be observed under the light microscope. There are special techniques for the preparation and staining of temporary or permanent mounts.

Key terms

See the boxes:

A Study questions

1. Find out which professions involve the use of microscopes and what the microscopes are used for in each case. Write a report about your findings and present it in class.
2. Write down all the words from the Key terms boxes on separate cards. With a partner, take turns drawing cards and explaining the terms. If you are not sure how to express something in English, check the text for words that can help you and copy them into your notebook.
3. Collect information on Antoni van Leeuwenhoek's microscope, how it was used and what he discovered with this rather simple instrument. Give a short talk about it in class.

B Complete the following sentence

1. The biggest animal cells are the ... cells of ...
2. The modern light microscope is also called a compound microscope, because ...
3. While the light microscope works with lenses, the electron microscope uses ...
4. When observing a specimen with the microscope, always start with the ... objective.
5. Objects to be studied under the light microscope must be very thin, because ...
6. The irrigation method can be used if ...
7. The advantage of using selective stains is that ...
8. When carrying the microscope, you should always ...
9. For investigating the microscopic structure of cells, scientists use ... today.
10. If you want to colour cells and tissues without killing them, you can ...

C Say it in German

Prepare a short presentation in German about the early microscopes and the discoveries that were made with them. Include the scientists who made important discoveries about cells.

D Critical thinking

1 More than 150 years passed between Leeuwenhoek's discoveries of living cells and the developments which led to the basis of the cell theory in the 19th century. Why do you think it took so long for the cell theory to emerge?

2 Microscopes led to the discovery that bacteria can cause diseases. How do you think this discovery changed the way that doctors treated infectious diseases?

Chapter 2

Cells: Their Structure, Function and Development

Chapter objectives

- To **compare** the structure and functions of animal and plant cells
- To **learn** how to make microscopic drawings of living cells
- To **explain** the importance of cellular respiration
- To **describe** cell division and differentiation in plant and animal cells
- To **describe** and **explain** the characteristics of cells, tissues, organs, organ systems and organisms
- To **learn** to carry out microscopic observations of animal and plant cells and to **identify** their characteristic structures

If you observe the cells of plants, animals or humans under the microscope, you will find that there are certain features which all cells have in common. At the same time, you will see that there are also characteristic differences between animal and plant cells.

Under the microscope cells often seem to be flat and two-dimensional[1], because you can only ever focus on one horizontal level, but in reality all cells have a three-dimensional shape and are often shaped like balls or bricks[2].

1 two-dimensional *flächig, zweidimensional*
2 brick *Ziegel*
3 pith *Mark*
4 elongated *länglich, lang gestreckt*
5 schematic drawing *schematische Zeichnung*

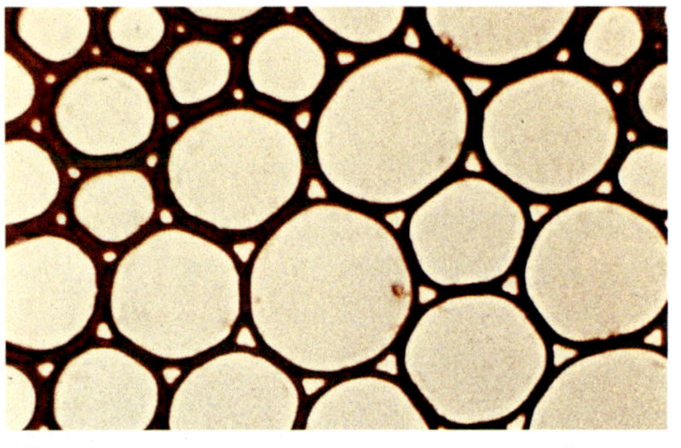

Figure 1 Ball-shaped cells in the pith[3] of the elder plant (magnification 100x)

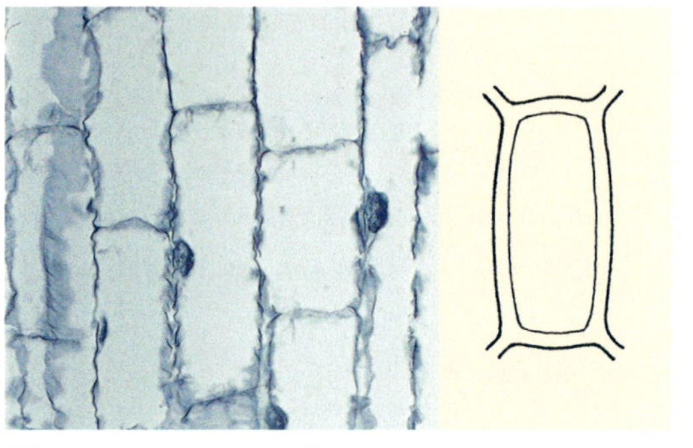

Figure 2 Elongated[4] cells in the epidermis of an onion scale (magnification 100x)

2.1 All plant cells have characteristic organelles

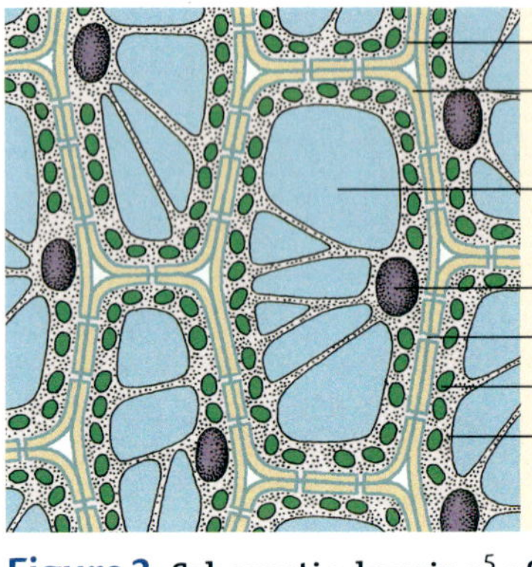

Figure 3 Schematic drawing[5] of a plant cell as seen through the light microscope

The cells of a leaf differ greatly from those in the stem or the root of a plant, but in spite of these differences there is one characteristic which all plant cells have in common: the **cell wall**. It lies next to the outer **cell membrane**,

Key terms

Structures and organelles of plant cells

- cell membrane *Zellmembran*
- cell sap *Zellsaft*
- cell sap vacuole *Zellsaftraum*
- cell wall *Zellwand*
- cellulose *Zellulose*
- chloroplast *Chloroplast*
- chromoplast *Chromoplast*
- cytoplasm *Zellplasma*
- leucoplast *Leukoplast*
- mitochondrion (pl. mitochondria) *Mitochondrium*
- nucleus *Zellkern*
- organelle *Zellorganell*
- starch grain *Stärkekorn*

which encloses the cytoplasm and the cell contents. It is thicker and tougher than the cell membrane and consists mainly of **cellulose**. Tiny pores in the cell wall allow the exchange of water and substances between neighbouring cells. When additional substances are embedded between the cellulose fibres, the wall becomes impermeable[1] to water and gases.

The cell membrane, which looks like a very thin line when viewed through the compound microscope, is not a barrier to the exchange of water and gases, but regulates[2] the exchange of other substances.

The **cytoplasm** is a semi-fluid[3] mixture[4] of substances containing food, storage materials (e.g. **starch grains**) and special **organelles**. The organelles are compartments within the cell in which metabolic processes[5] can take place without disturbing one another (See 'Further facts', p. 227). With a light microscope you can easily identify the **nucleus**, the **chloroplasts** and the **chromoplasts** because they contain pigments that are not soluble[6] in water. The pigment chlorophyll in chloroplasts is important for photosynthesis. The red and yellow pigments of chromoplasts give colour to plant organs. The smaller **leucoplasts**, found in cells which are not exposed to light, are colourless. The continuous[7] movement of the cytoplasm transports substances from one part of the cell to the other.

The nucleus, which contains the cell's genetic information, is the most important of all organelles. It is essential[8] for the life of the cell and controls the various processes that happen in the cytoplasm. The nucleus is located either near the cell wall or in the centre of the cell, connected by thin strands[9] of circulating cytoplasm to the thicker layer along the cell walls. **Mitochondria**, the places where energy is released from food substances, are difficult to see because they are extremely small. Under the light microscope they simply look like black dots.

Most plant cells have a large vacuole, a cavity[10] within the cytoplasm, filled with **cell sap**, a solution containing sugars, salts and often pigments. It is separated from the cytoplasm by a membrane known as the tonoplast. The cell wall and the cell sap vacuole play important roles in maintaining the shape of the plant cell.

Did you know?

In autumn, when the days become shorter and cooler, the leaves of most broadleaved trees and bushes change their colour from green to yellow, red or brown. The colour change is partly caused by the breaking down of chlorophyll. The yellow and orange pigments which were covered up by the green chorophyll during spring and summer gradually become visible. The bright red and purple colours are newly produced in autumn, while the brown colours are caused by accumulated[11] waste products[12] in the leaf cells.

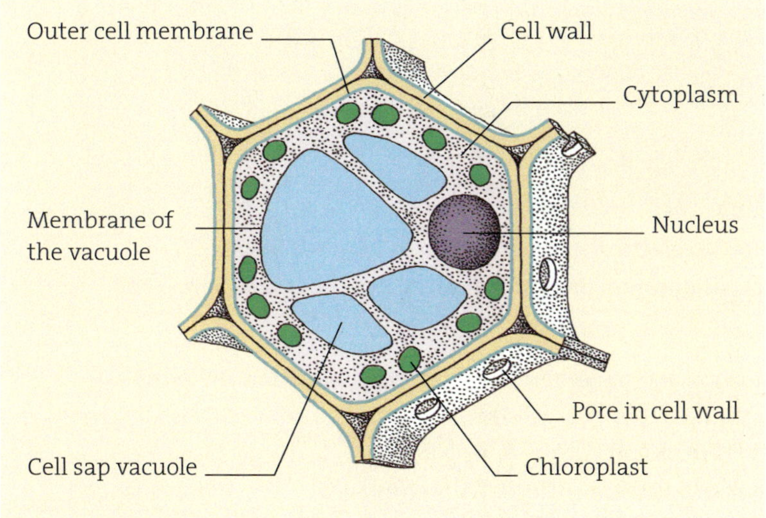

Figure 4 Three-dimensional picture of a plant cell

1 impermeable *undurchlässig*
2 regulate sth. *etwas regulieren*
3 semi-fluid *halbflüssig*
4 mixture *Mischung*
5 metabolic process *Stoffwechselprozess*
6 soluble *löslich*
7 continuous *andauernd, ununterbrochen*
8 essential *unerlässlich*
9 strand *Strang, Faden*
10 cavity *Hohlraum*
11 accumulated *angesammelt*
12 waste product *Abfallprodukt*

2.2 Animal cells come in many different shapes

> **Key terms**
>
> **Different animal cells**
> - nerve cell *Nervenzelle*
> - paramecium *Pantoffeltierchen*
> - sperm cell *Spermium, männliche Samenzelle*
> - red blood cell *rotes Blutkörperchen*
> - white blood cell *weißes Blutkörperchen*

The cells of humans and animals display[1] an even greater variety than those of plants. For example **nerve cells** (see Figure 5a), which have to transfer and process information, often look like stars or comets[2]. Electrical impulses reach the cell body through many short projections[3] called dendrites[4], then travel through one long projection, an axon[5], to other cells in the muscles or glands. In contrast to nerve cells, the skin cells of animals and humans look like bricks or cubes[6], and human **sperm cells** (see Figure 5d) have an oval-shaped head and a long tail, which propels the cell forward[7]. As animal cells have no cell wall, their shapes can change. You can see this in unicellular organisms like the **paramecium** (see Figure 5c), but also in **white blood cells** that move around freely inside the organism. This also applies to **red blood cells** (see Figure 5b) that are normally disc-shaped[8] when they are pushed along the blood vessels by the pressure of the heart. They can look like sausages when they squeeze through narrow capillaries.

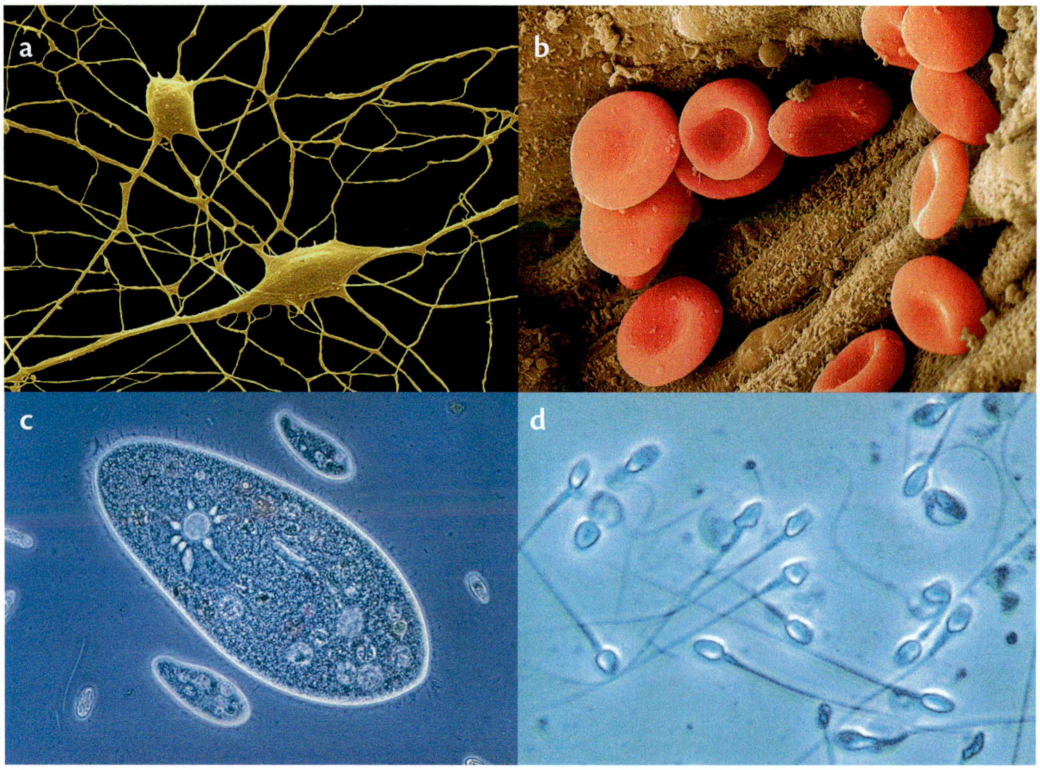

Figures 5a–d **a) Nerve cells (coloured electron micrograph); b) human red blood cells (coloured electron micrograph); c) paramecium (magnification 200x); d) human sperm cells (magnification 1000x)**

Three structures that can be found in all animal cells are the outer cell membrane, the cytoplasm and mitochondria. In contrast to plant cells, animal cells have no large cell sap vacuoles and the cytoplasm completely fills the cell. Food and storage substances like fat or glycogen[9] drift[10] through the cytoplasm as tiny vacuoles (droplets enclosed by membranes). The two star-shaped structures visible in paramecium are contractile vacuoles[11], responsible for transporting water out of the cell.

1 display sth. *etwas zeigen, zur Schau stellen*
2 comet *Komet*
3 projection *Ausstülpung, Vorwölbung*
4 dendrite *Dendrit*
5 axon *Axon*
6 cube *Würfel*
7 propel sth. forward *etwas vorwärtstreiben*
8 disc-shaped *diskusförmig*
9 glycogen *Glycogen, 'tierische Stärke'*
10 drift *treiben*
11 contractive vacuole *pulsierende Vakuole*

Further facts

Organelles

Cells contain organelles, which are structures that carry out specific tasks within the cell. The name 'organelle' means 'little organ'. This name shows that the relationship between cells and organelles is similar to the relationship between an organism and its organs. Organelles are often enclosed by membranes. Examples of organelles are the nucleus, plastids[1], mitochondria and vacuoles of different sizes.

- The nucleus contains all the information required to control the functions of the cell.
- Chloroplasts, chromoplasts and leucoplasts are plastids that are found in plant cells. The chloroplasts are the places where photosynthesis takes place in leaves and stems, whereas the pigments in chromoplasts give colour to petals, fruits and roots. The red and orange colours in tomatoes and carrots are typical examples of the effect of chromoplasts. The colourless leucoplasts, which are located mainly in root and seed tissues of plants, are specialized in the storage of starch, oil or protein.
- Mitochondria are very important because they are the 'power stations' of all living cells (see the next page).
- Vacuoles of different sizes carry out functions of storage, transport and excretion.

1 Compare the four cells shown in Figure 5. Write down which organelles or structures of animal cells you can recognize in any of them and describe their functions.
2 Explain which structures in a plant cell are responsible for the stability of its shape..
3 Use the information in Sections 2.1 and 2.2 to make a table showing the different structures and organelles to be found in the cells of humans, animals and plants. Your table should look like this:

Organelle/Structure	Plant cell	Human and animal cell	Function of the organelle/structure
Cell wall	Yes	No	Gives the cell shape and stability
Outer cell membrane	...	...	...
...	...	...	...

2.3 Mitochondria are the site of cellular respiration

The cells of all organisms need energy for their life processes, e.g. for growth, movement, sensitivity and division. **Cellular respiration** is the process by which energy is released from food substances high in energy such as glucose[2], starch and fats with the help of oxygen[3]. The end products of this process are carbon dioxide[4] and water.

This equation[5] gives a simplified illustration of cellular respiration:

Glucose + Oxygen ➡ Carbon dioxide + Water + ENERGY

or

$$C_6H_{12}O_6 + O_2 \rightarrow CO_2 + H_2O + \text{ENERGY}$$

1 plastid *Plastid, Farbstoffträger*
2 glucose *Traubenzucker*
3 oxygen *Sauerstoff*
4 carbon dioxide *Kohlendioxid*
5 equation *Gleichung*

The process of cellular respiration can be compared to a burning candle. If you light a candle, the **wax**, a highly energetic substance, starts to burn with the help of oxygen. This burning process releases energy in the form of light and heat, as well as water and carbon dioxide.

The following equation shows the chemical process that takes place when wax burns:

Wax + Oxygen → Carbon dioxide + Water + ENERGY

When glucose is burned in cells, it happens without fire. A sudden release of large quantities of energy, as in the burning of organic substances like wax, wood or coal, would destroy the heat-sensitive[1] proteins in cells. Therefore the energy is released in small portions in a chain of steps controlled by a large number of **enzymes**. Some of it is released in the form of heat and can be used to maintain the body temperature, but most of it is stored in the form of **ATP**. ATP is a chemical compound that feeds energy into all types of metabolic[2] processes, e.g. movement, growth and **active transport** between cells.

Key terms

Cellular respiration
- active transport *aktiver (energieverbrauchender) Transport*
- ATP = adenosine triphosphate *Adenosintriphosphat*
- cellular respiration *Zellatmung*
- enzyme *Enzym*
- wax *Wachs*

Mitochondria are the organelles of cell respiration in the cells of humans, animals and plants. Every cell contains between 10 and 1000 of these tiny oval or rod-shaped[3] bodies, depending on the intensity of its metabolism[4]. Muscle cells, liver cells and nerve cells contain a particularly large number of mitochondria.

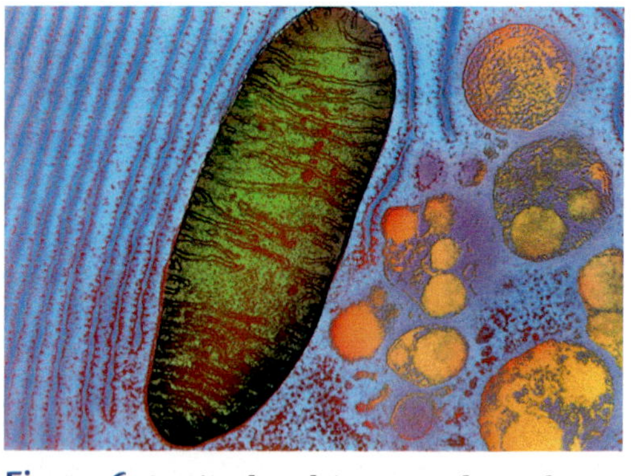

Figure 6 **A mitochondrion seen through an electron microscope**

Mitochondria are too small to be seen through the type of light microscope used in schools, but their fine inner structures can be seen through an electron microscope. As you can see in Figure 6, the inside membrane of the mitochondrion is folded many times, providing a large inner surface with room for many enzymes and compartments in which different reactions can take place without disturbing each other.

The rate of cellular respiration in body cells is regulated by the amount of ATP which is needed for metabolic processes and the level of ATP available in the cell. During phases of physical activity when the muscle cells need a lot of energy, the level of ATP decreases, which causes the mitochondria to increase the ATP production. In contrast, when you rest, the ATP level is high, which signals the mitochondria to decrease or stop the ATP production.

1 heat-sensitive *hitzeempfindlich*
2 metabolic *den Stoffwechsel betreffend*
3 rod-shaped *stabförmig*
4 metabolism *Stoffwechsel*

1 Name some of the energy-consuming processes which take place in the flower of a plant.

2 In mitochondria, layers of membranes provide a large inner surface. Name other organelles or organs which also have a large inner surface. Why are these inner surfaces necessary? Your notes from other units of this book may help you.

3 **Say it in German:** Working with a partner, describe to each other how the energy-releasing processes in our cells can be compared to the burning of a candle. Explain how these two processes resemble each other and where there are differences and why. Your partner should ask you questions if something is not sufficiently explained.

4 Working with a partner, plan an experiment to prove that plant cells respire. Present your plan to the class.

2.4 Body cells divide by mitosis

The lives of plants, animals and humans begin with a single **fertilized** egg cell. Soon after **fertilization** the cell starts to divide, first into two identical cells, then four, eight, 16 and so on. All these daughter cells must have the same **genetic information** as the parent cell, because the genetic information determines[1] the kind of species and the individual characteristics. Here 'daughter' does not mean that the new cells are female, but that they are both produced by cell division from the parent cell.

The genetic information of a cell is contained in the **chromosomes**, which look like long thin threads[2] in the nucleus. They can only be seen under the light microscope during cell division. Every species on earth possesses a definite number of chromosomes in each body cell (see Table 1 below).

Key terms

Cell division and differentiation

- cell division *Zellteilung*
- chromatid *Chromatid*
- chromosome *Chromosom*
- differentiate *differenzieren*
- differentiation *Differenzierung*
- division of the nucleus *Kernteilung*
- embryonic cell *embryonale Zelle*
- fertilization *Befruchtung*
- fertilize sth. *etwas befruchten*
- genetic information *Erbinformation*
- mitosis *Mitose*
- root hair *Wurzelhaar*
- root tip *Wurzelspitze*
- shoot tip *Sprossspitze*

Species	Number of chromosomes	Species	Number of chromosomes	Species	Number of chromosomes
Garden pea	14	Mouse	40	Cow	70
Tomato	24	Rabbit	44	Dog	78
Oak	24	Human	46	Pigeon[3]	80
Tobacco	48	Horse	64	Goldfish	94

Table 1 Chromosome numbers in the body cells of humans and some plants and animals

Before **cell division** starts, each chromosome produces an exact copy of itself. These copies are called **chromatids**, and the two chromatids of one chromosome connect in one place, the centromere[4], which makes it look like an X. A series of complex[5] processes then take place which

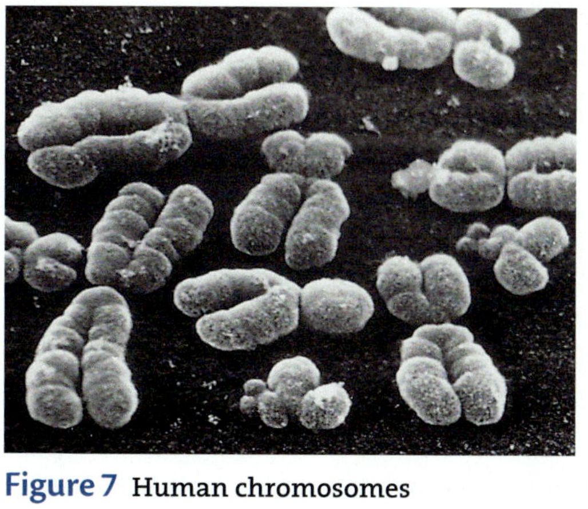

Figure 7 Human chromosomes

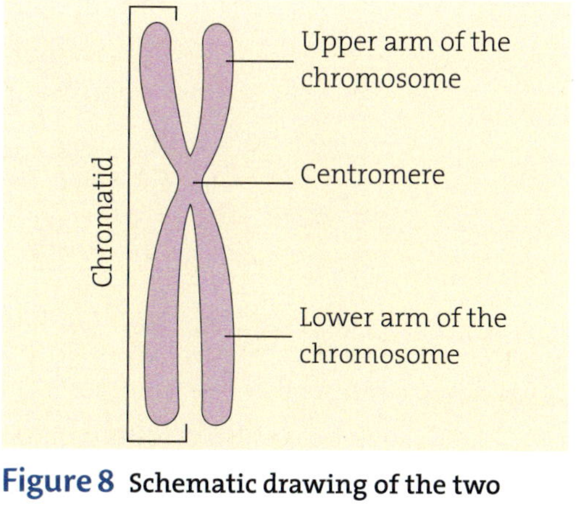

Figure 8 Schematic drawing of the two chromatids of a chromosome

1 determine sth. *etwas bestimmen*
2 thread *Faden*
3 pigeon *Taube*
4 centromere *Zentromer*
5 complex *komplex, kompliziert*

lead to the **division of the nucleus**. During these processes, called **mitosis**, the chromatids of the chromosomes split (see Figure 8) and one half moves into each new daughter cell. When mitosis is completed, the cytoplasm divides and two new cells are formed, identical to each other and with the same chromosomes as the parent cell.

Cell division in plants

In plant cells a thin cell membrane forms in the middle of the dividing cell and grows to the opposite sides, separating the new cells. Then new cell walls develop alongside the outer plasma membranes.

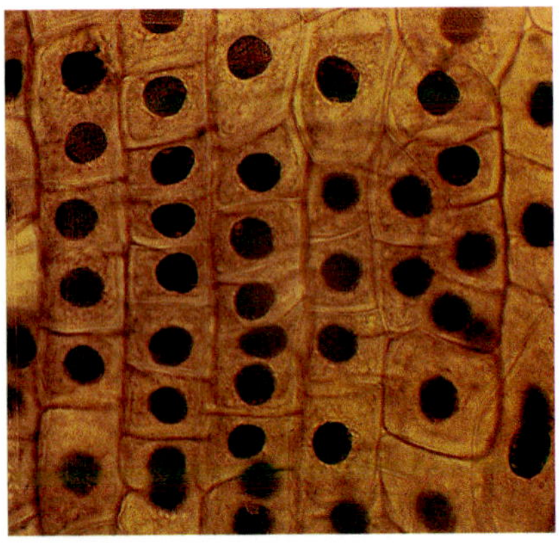

Figure 9 Young plant cells (magnification 300x)

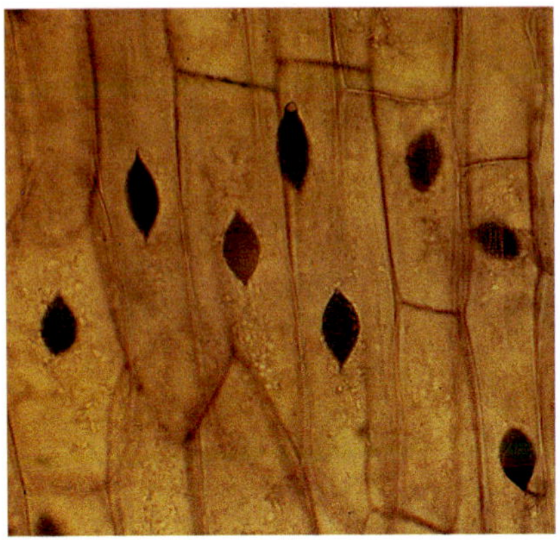

Figure 10 Mature[1] plant cells (magnification 300x)

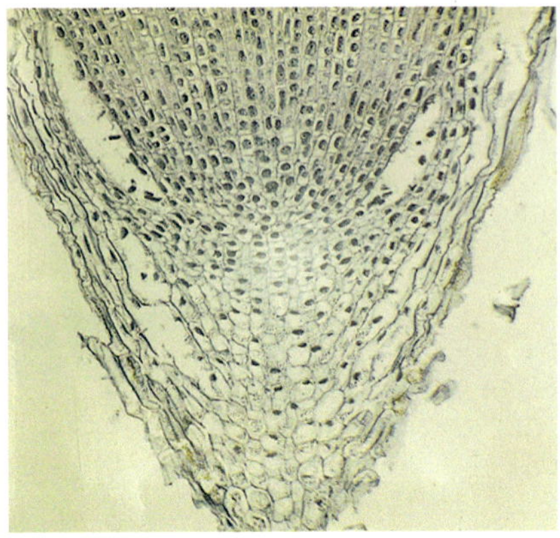

Figure 11 Cells in the root tip and in the adjacent[2] root tissue of a flowering plant

Cell division in animals

In the cells of humans and animals the cell membrane of the original cell 'pinches in[3]' to separate the chromatids into two cells. Following this, two new plasma membranes form. The original cell has become two daughter cells, which at first are half the size of the parent cell but grow quickly by taking up substances from outside.

When you observe the process of cell division in the **shoot tip** or the **root tip** of a flowering plant[4] under a compound microscope (see Figures 9 and 12), you will see that **embryonic cells** all look the same. They are square-shaped, without the large cell sap vacuole of mature plant cells, and they have thin and elastic cell walls. But when you look at areas of the root that lie further away from the root tip, you will see that the cells have grown in size due to an increase in the amount of cytoplasm, and they have started to **differentiate** according to the tasks they will have to carry out for the rest of their lives. The **root hair** cells with their thin cell walls and large vacuoles take up water and minerals. Other cells in the centre of the root transport water

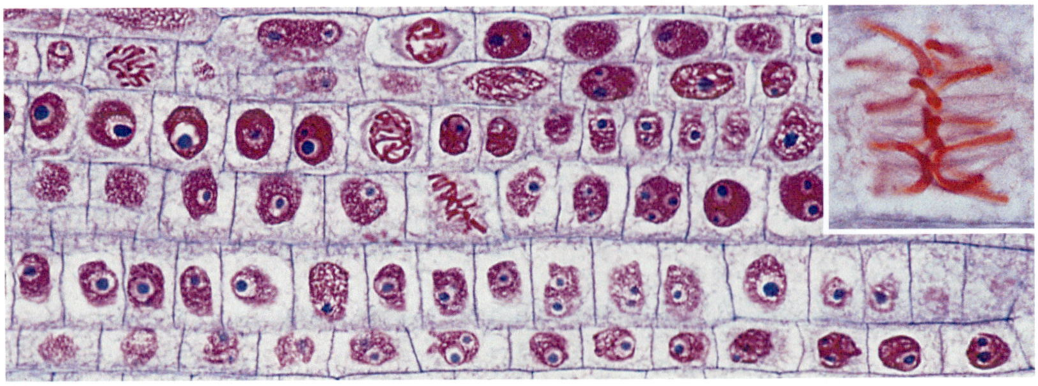

Figure 12 Stained cells in the root tip of an onion in different phases of mitosis. Inset: the separation of chromatids (magnification 1500x).

1 mature *voll ausgebildet, ausgereift*
2 adjacent *angrenzend*
3 pinch in *sich zusammenziehen*
4 flowering plant *Blütenpflanze*

or sugar solution. The cells on the outside are brick-shaped with thicker cell walls to protect the root against injuries.
The same processes of **differentiation** happen to young cells in humans and animals, but most of their tissues, in contrast to those of plants, are still able to divide by mitosis.

1 Describe and compare the processes that occur in plant and animal cells as they divide.
2 Point out the differences in the division of body cells of plants and animals.
3 Explain why small wounds in the skin and other organs heal relatively quickly in humans and animals.
4 Explain the processes which are responsible for the growth and development of the young bean plant. Another look at p. 176 might help you.

2.5 Cells work together to form tissues and organs

In humans, plants and animals, cells of the same kind are mostly arranged in larger groups in which all cells carry out the same function. These groups of cells are called **tissues**.

Key terms

Tissues, organs and organ systems
- organ *Organ*
- organ system *Organsystem*
- organism *Organismus, Lebewesen*
- tissue *Gewebe*

Humans and animals

In humans and vertebrate[1] animals the skeleton consists of **cartilage** and bone tissues, which are made up of cartilage cells and bone cells. Muscle cells form the muscle tissue, nerve cells the nerve tissue and glandular cells the **glandular tissues**, which are responsible for the **secretion** and **excretion** of fluids. The simplest type of tissue is the **epithelium**, which consists of square or cube-shaped cells that fit closely together like the bricks in a wall. These cells line[2] the spaces and tubes inside the body and are also found on the surface of the skin. The task of epithelia is to protect the tissues that lie

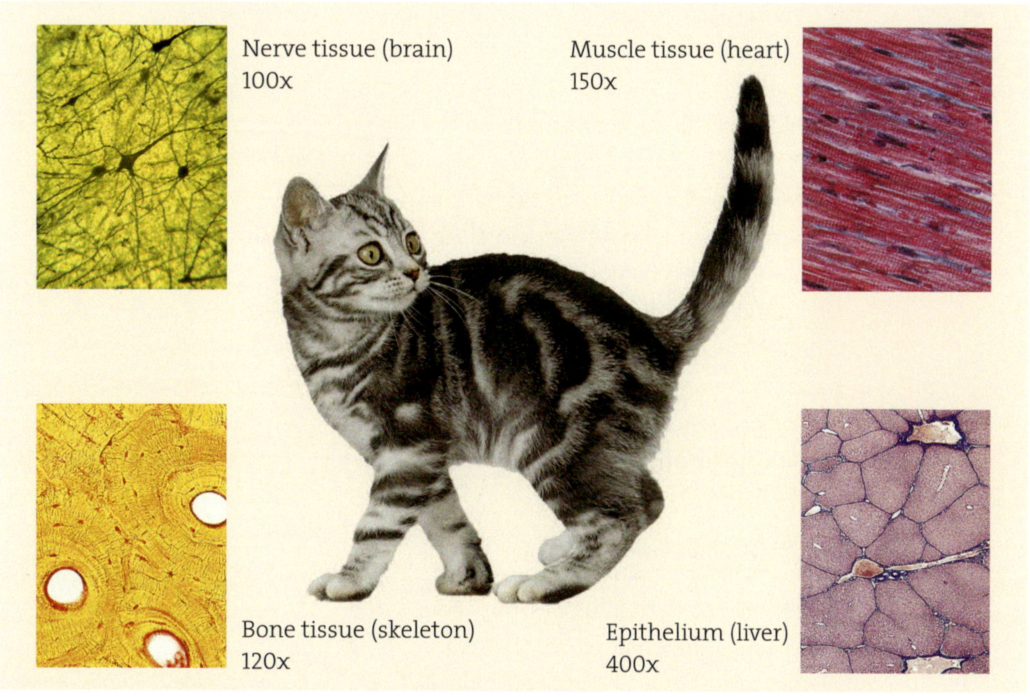

Figure 13 Four types of tissues in the body of a cat

1 vertebrate *mit einer Wirbelsäule*
2 line sth. *etwas auskleiden*

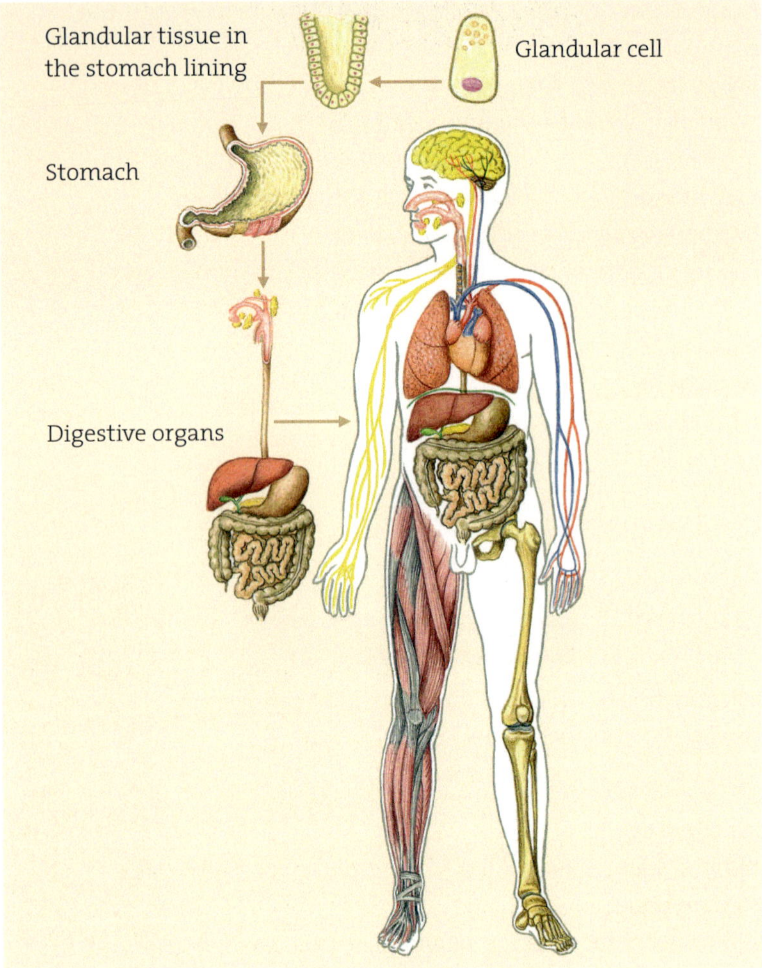

Figure 14 Organ systems in the human organism

underneath[1], and they all look different depending on which organ they cover.

An **organ** is a complex structure made up of many different tissues which work together to fulfil specific tasks. Your heart is made up of muscle and nerve tissue, covered by epithelia. The heart has only one task: to pump the blood through the body. By contrast, the **liver** and the **small intestine** both have a large number of vital functions to carry out and so consist of many different types of cells. The tissues of the small intestine secrete[2] **digestive juices** and transfer the digested food particles into the blood and the **lymph**. The liver, the biggest organ inside your body, is responsible for the metabolism of the food you eat, and for detoxifying any poisonous substances you consume, such as alcohol.

Certain tasks can only be fulfilled by the combined work of several organs. This is called an **organ system**. A good example is the **digestive system**. The process of digestion starts in the **mouth cavity** with the combined work of the **mucous glands**, the teeth and the tongue, and continues in the stomach with the work of the glandular tissue in the stomach lining. It comes to an end in the small intestine, when the **pancreatic juices**, as well as the juices of the liver and the intestine itself, split the food particles into molecules[3], which can be absorbed into the body. For more on the digestive system, see Unit 1, Chapter 5.

It is essential for the life of humans and animals that their organ systems work together in harmony[4]. The different organs with their specific tasks can only benefit[5] the organism when they are well coordinated. This coordination is mainly controlled by two systems working closely together: the **nervous system** and the **hormonal system**.

Plants

In plants, just as in animals, the different types of cells are arranged in tissues, but there is one important difference: In mature plants, only certain tissues have cells that are able to divide at any time. These **meristematic tissues** with

1 underneath *darunter (gelegen)*
2 secrete sth. *etwas absondern*
3 molecule *Molekül*
4 harmony *Harmonie*
5 benefit sb./sth. *jdm./etwas nützen, zugute kommen*

Key terms

Animal tissues
- cartilage *Knorpel*
- epithelium (pl. epithelia) *Deckgewebe, Epithel*
- excretion *Ausscheidung*
- glandular tissue *Drüsengewebe*
- secretion *Abgabe ins Körperinnere, Sekretion*

Animal organs
- digestive juice *Verdauungssaft*
- digestive system *Verdauungsapparat*
- hormonal system *Hormonsystem*
- liver *Leber*
- lymph *Lymphe*
- mouth cavity *Mundhöhle*
- mucous gland *Schleimdrüse*
- nervous system *Nervensystem*
- pancreas *Bauchspeicheldrüse*
- pancreatic juice *Saft der Bauchspeicheldrüse*
- small intestine *Dünndarm*

their small embryonic cells are found at the tips of shoots and roots (see Figures 11 and 15), in the **buds** of leaves and flowers, and in the **cambium** in woody plants (which enables the stems of trees to grow in diameter). Meristematic cells can differentiate into specialized cells, which mainly grow to form **permanent tissues**. In these tissues the cells lose their ability to divide when the tissues reach their final shape and size. In between the different plant tissues are bundles[1] of tough fibres. These **strengthening tissues** give stability to a plant and allow it to grow upright.

A good example of a plant organ is the leaf, which is responsible for photosynthesis and transpiration. Leaves contain various tissues whose combined activities enable these vital functions to take place. The cells with chloroplasts, where photosynthesis takes place, form the **photosynthetic tissue**. The tubes which run through the veins in the leaf, transporting water, minerals and the products of photosynthesis, are part of the **conducting tissue**, while the closely packed cells on the upper and lower surface of the leaf form the **epidermis**.

A flower is an organ system in which different organs work together. The **sepals** protect the growing flower and the **petals** attract insects. **Stamens** and **carpels** are the sexual organs of a flower, in which specific tissues produce the male and female sex cells.

As in humans and animals, the combined functions of cells, tissues, and organs make the life of a plant possible. The increase in the grade of specialization means an increase in efficiency. For more on plants, their structure and functions, see Unit 2, Chapter 1.

The principle of division of labour[2] plays an important role in **organisms** at every level. As one organ cannot fulfil[3] all the tasks necessary for an organism to function, many organs work together by forming organ systems. Each organ contains several tissues for the various processes that take place within the organ. A tissue can only function because it consists of many cells doing the same job. In unicellular organisms, the functions of the different organelles within the cell also reflect this principle (see 'Further Facts' p. 227).

Key terms

Plant tissues

- cambium *Kambium*
- conducting (vascular) tissue *Leitgewebe*
- epidermis *Hautgewebe*
- meristematic tissue *Bildungsgewebe*
- permanent tissue *Dauergewebe*
- photosynthetic tissue *Photosynthesegewebe*
- strengthening tissue *Stützgewebe*

Parts of a flower

- bud *Knospe*
- sepal *Kelchblatt*
- petal Kronblatt, *Blütenblatt*
- stamen *Staublatt*
- carpel *Fruchtblatt*

Parts of a vascular bundle

- phloem *Siebtiel*
- vascular bundle *Leitbündel*
- xylem *Gefäßteil*

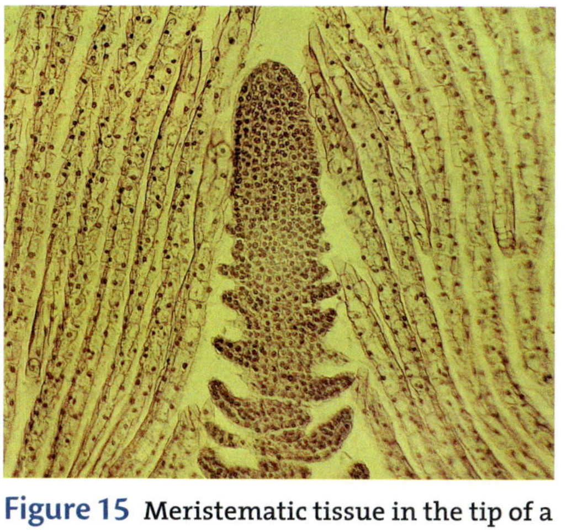

Figure 15 Meristematic tissue in the tip of a developing shoot (longitudinal section)

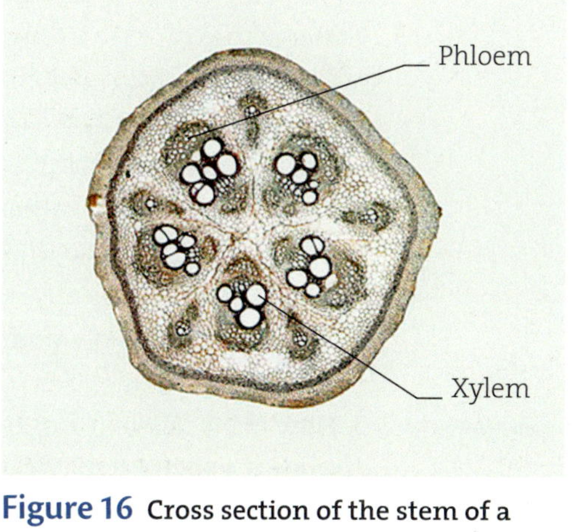

Figure 16 Cross section of the stem of a herbaceous plant with five vascular bundles

1 bundle *Bündel*
2 division of labour *Arbeitsteilung*
3 fulfil sth. *etwas erfüllen, ausführen*

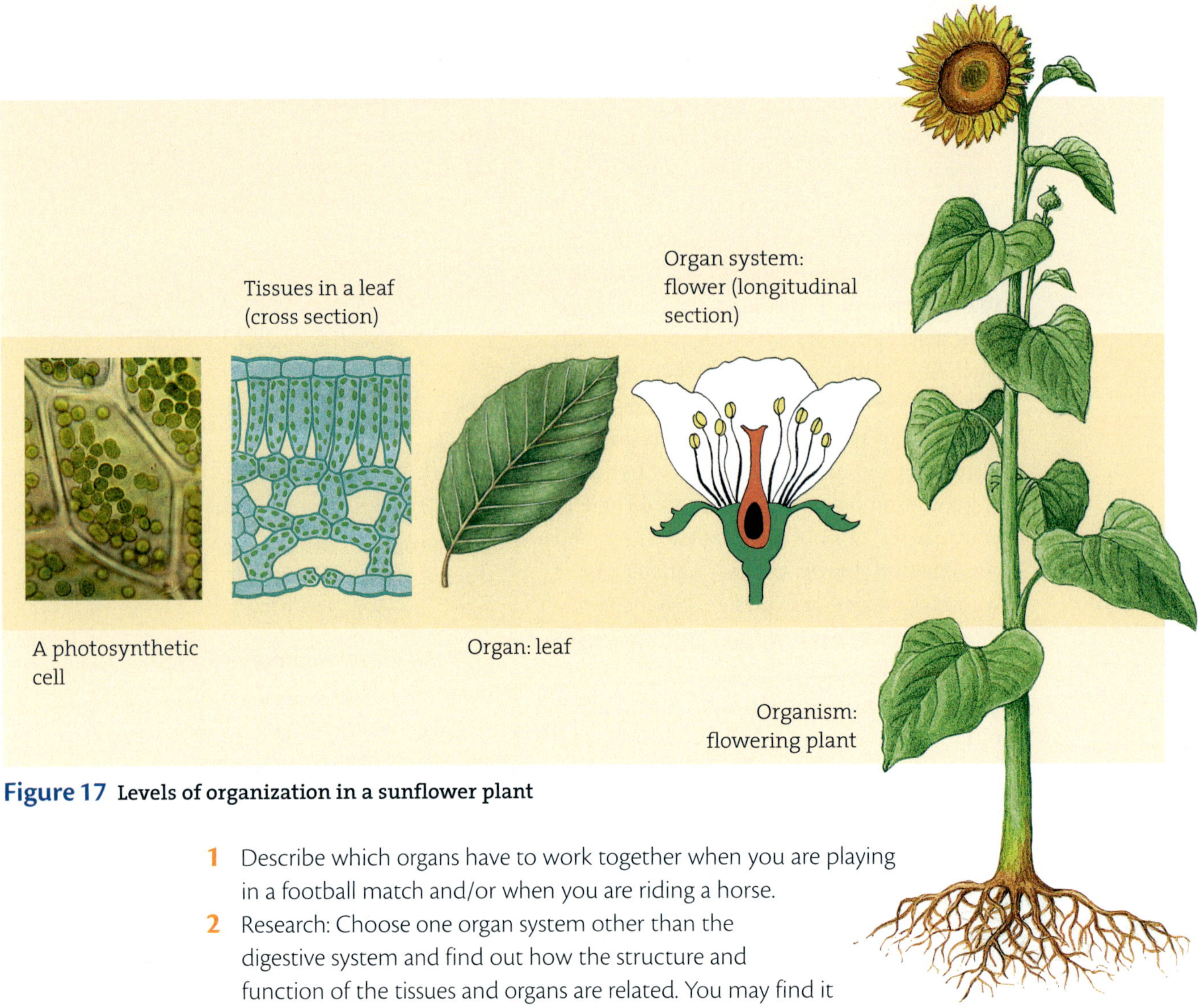

Figure 17 **Levels of organization in a sunflower plant**

1 Describe which organs have to work together when you are playing in a football match and/or when you are riding a horse.
2 Research: Choose one organ system other than the digestive system and find out how the structure and function of the tissues and organs are related. You may find it easiest to show the relationships between different tissues and organs using a flow chart or cause-and-effect diagram (see 'Working with flow charts', p. 84). Present your research to the class.
3 Using the sunflower plant as an example (see Figure 17), describe the different levels of organization from cells to tissues to organs. Make use of the information provided in this chapter.
4 Explain the main differences between plants and animals in the development and specialization of their tissues. You may find it helpful to work with a partner, with one of you describing the tissues of plants and the other describing the tissue of animals.
5 Observe the tissues in a longitudinal section of the root tip of an onion on prepared slides.
   a Use the low- and medium-power objectives to get an overview of the whole specimen.
   b Focus on the cells in the root tip using the medium- and high-power objectives and make a drawing of these cells.
   c Now look at tissues further away from the root tip. Draw some of the cells.
   d Describe and compare the cells near the root tip and further along the root.
6 Use prepared slides to observe cross sections of the stems of different flowering plants (e.g. a herbaceous plant and woody plants that were one and three years old when the slides were made).
   a How many different tissues do you recognize? How do they differ from one another?
   b Make a rough sketch of the whole cross section.
   c Now use the medium- or high-power objective and make labelled drawings of the tissues that you can identify.

Drawing a microscopic object

It is not easy to make an exact drawing of a microscopic object. The following points will help you to learn this skill:

1. Use **plain white paper**, size A4. Place the sheet of paper beside the microscope.
2. Always **draw in pencil**. With a pencil you can draw in more detail, erase things and shade in certain areas if necessary. Use different grades of pencil for very thin and bolder lines.
3. Make sure that your drawing is as close as possible to an **exact representation**[1] of what you see when you look through the eyepiece. If you see two fine lines, try to draw two fine lines, and if a line under the microscope is not straight, do not draw a straight line.
4. For each drawing you should **reserve half a page** to allow enough space for details.
5. **Label** all structures you can identify by drawing a horizontal line from the structure to the side of the drawing (the line or arrow must just touch the structure that is named). Write the correct label next to the end of the line (not on it), as shown in Figure 18.
6. Every good drawing has a **title** that tells you the name of the specimen, the total magnification, the name of the person who made the drawing and the date that it was made. This information is normally found in the upper right-hand corner of the page (see Figure 18).

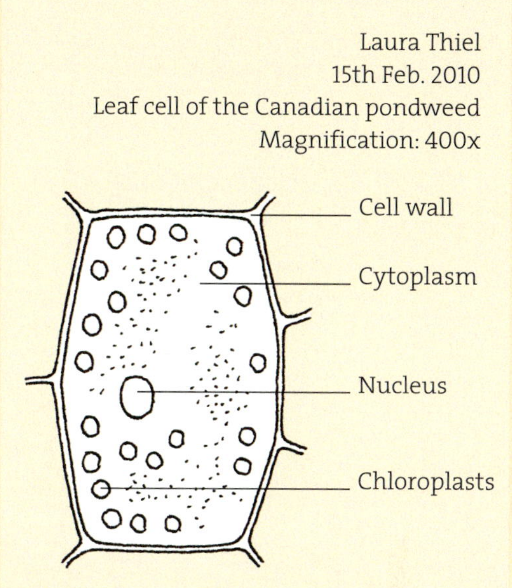

Figure 18 **A drawing of a single plant cell**

Learn to **keep both eyes open** while looking through the eyepiece. Any other technique will make you tired quickly. With practice you will be able to draw the object while you are looking at it through the microscope.
While you are still learning how to draw microscopic objects, it can be useful to draw the **field of sight** as seen through the eyepiece. The circle will help you to observe closely and keep everything in proportion.

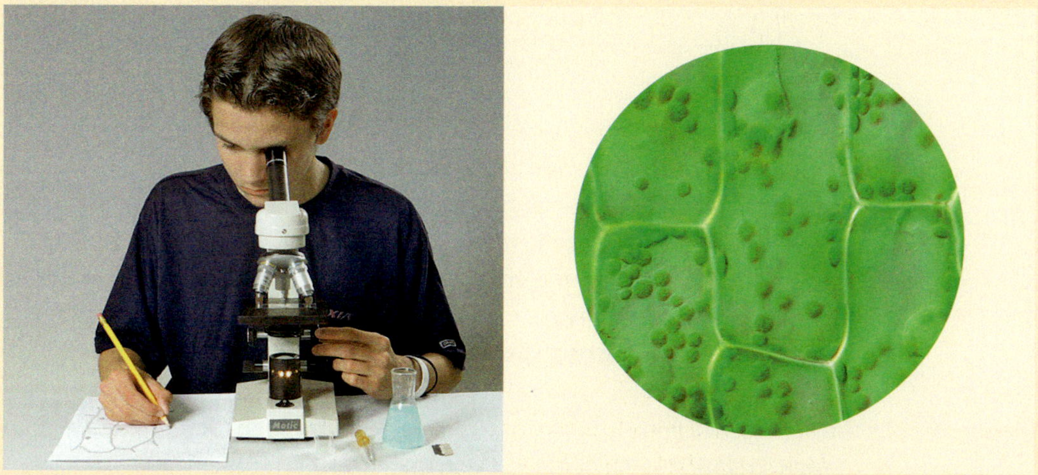

Figure 19 **The correct way to make a microscopic drawing (left); mesophyll[2] of the Canadian pondweed as seen through the light microscope (right)**

1 representation *Wiedergabe, Darstellungsweise*
2 mesophyll *Mesophyll, Grundgewebe im Inneren eines Blattes*

Activity 1 Observing living cells

A Observe the epidermis cells in the scale of an onion bulb

The scale of an onion bulb is a useful object for the beginner of microscopic observations.

Materials

- onion bulb
- scalpel, sharp knife or razor blade
- tweezers
- microscope slide
- cover slip
- pipette
- beaker (60 ml)
- strip of filter paper (5 cm x 1 cm) or a piece of paper towel
- water
- iodine solution (Lugol's solution)
- compound microscope

Figure 20 **Preparing a wet mount of onion cells**

Procedure

1. Slice the onion lengthwise (i.e. from top to bottom) into two pieces.
2. Take out one of the fleshy scales and make some parallel cuts lengthwise and at an angle of 45 degrees into the loose, almost transparent layer of cells on the inside of the scale.
3. Use the tweezers to grip one of the small squares of epidermis and place it into a drop of water on a clean slide.
4. Add a clean cover slip.
5. Observe the cells under low and then under medium and high power and try to locate the nucleus and other structures in each cell.
6. Stain the cells with iodine solution using the irrigation method.

For more information on the techniques needed for steps 1–6, see Chapter 1, p. 220.

1 Make a drawing of a section of onion tissue as observed through a microscope at medium power (see also 'Drawing a microscopic object', p. 235).

2 After step 5, draw a single onion cell in water as observed at high power and label the structures you can identify.

3 After adding iodine solution in step 6, make another drawing of a single cell. Ask your teacher what the tiny dots or eye-like structures are which you can see within the nucleus at high-power magnification.

4 Which structures separate the cell contents from each other?

5 Describe the positive and negative effects of iodine solution on your specimen. Which structures in particular absorb the most iodine solution?

B Observe the cells inside the leaf of a Canadian pondweed

The Canadian pondweed is a plant well known to people who keep fish in an aquarium or have a garden pond. It is a rather hardy[1] freshwater plant that comes from North America. The tender leaves are an easy object for microscopic observations of typical plant cells. In this activity you will study its cells under a compound microscope.

Materials

- branch of Canadian pondweed
- beaker (60 ml)
- microscope slide
- cover slip
- pipette
- tweezers
- strip of filter paper (5 cm x 1 cm)
- water
- compound microscope

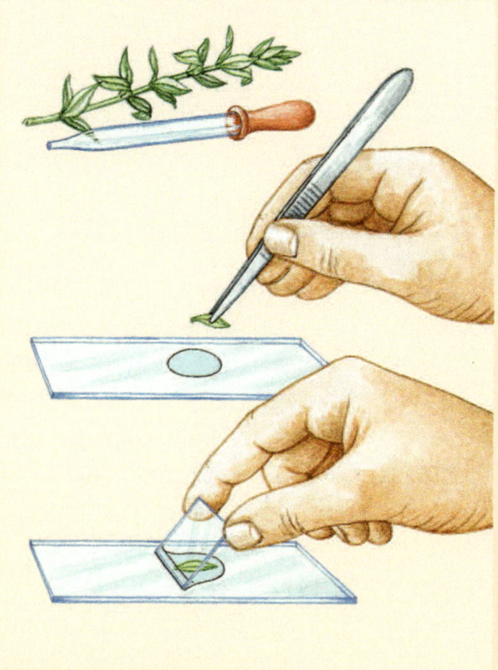

Figure 21 Preparing a wet mount of a leaf of the Canadian pondweed

Procedure

1. Break off a small leaf near the tip of a branch of pondweed. Here the leaves are thinner, allowing more light to pass through.
2. Prepare a wet mount of the leaf on a clean slide.
3. View the whole leaf under low power and locate the midrib[2] of the leaf.
4. Focus on some cells near the midrib under medium power and then under high power. You may be able to see the slow movement of the chloroplasts.
5. Remove the slide and bring it under a bright lamp or warm it between the palms of your hand, being very careful not to disturb the cover slip.
6. Then observe the cells near the midrib again.

1 hardy *widerstandsfähig*
2 midrib *Hauptader*

1 Make a labelled drawing of one or two cells under medium or high power.

2 Which organelles are visible in the Canadian pondweed cells that could not be found in the onion scale (see Activity 1a)? Explain this difference. Which organelles are difficult to see in the cells of the pondweed?

3 Describe the movement of the chloroplasts. For help see **Activate your English** below.

4 Point out and explain the effect that warming the slide had on the specimen.

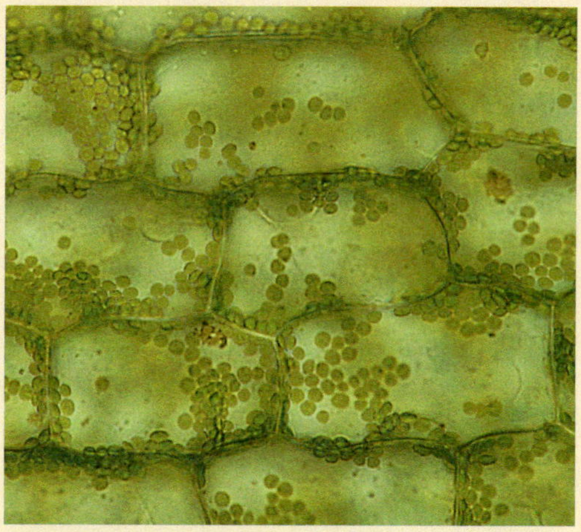

Figure 22 Cells in the leaf of a Canadian pondweed (magnification 100x)

Activate your English

Talking about chloroplasts
They (tend to) move/drift/float/...
in circles / in a circular motion
along the cell wall / around the edge
in a long row / one after the other
in opposite directions in adjoining cells

C Observe the cells of the human cheek lining

Human cheek cells are excellent subjects for observing basic structures in animal cells.

Materials
- microscope slide
- cover slip
- pipette
- beaker (60 ml)
- water
- 0.5 % methylene-blue solution
- compound microscope
- teaspoon or wooden spatula[1] (clean and sterile)
- 2 human hairs

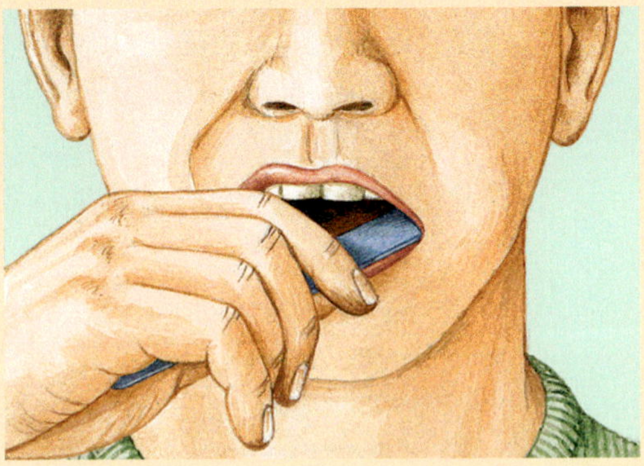

Figure 23 Taking a sample of cheek cells

Procedure
1. Using the edge of the teaspoon or spatula, gently[2] scrape the inside of your cheek to remove a small amount of surface material from this area (see Figure 23). Be very careful not to hurt yourself – you do not need to scrape very hard.
2. Spread the scrapings onto two clean slides and place a human hair into both scrapings.
3. Add a drop of water to the scraping on one slide and a drop of methylene-blue solution to the other.
4. Place cover slips on both specimens in the usual manner.

1 Compare the two wet mounts (one in water and the other in methylene-blue solution) under all magnifications, starting with the low-power objective lens.

2 Locate and examine cells that are separated from one another rather than those that are in clumps. Which cell structures are more clearly visible in the stained wet mount?

3 Draw two or three cheek cells as they appear under high magnification.

4 Label the nucleus, nucleolus (the dark spot in the nucleus), cell membrane, cytoplasm, and any other structures you recognize.

5 Describe the shape of cheek cells. Why are they different in shape from onion cells? For language help see **Activate your English** on the right.

6 What is the purpose[3] of adding a human hair to the cells on the slide?

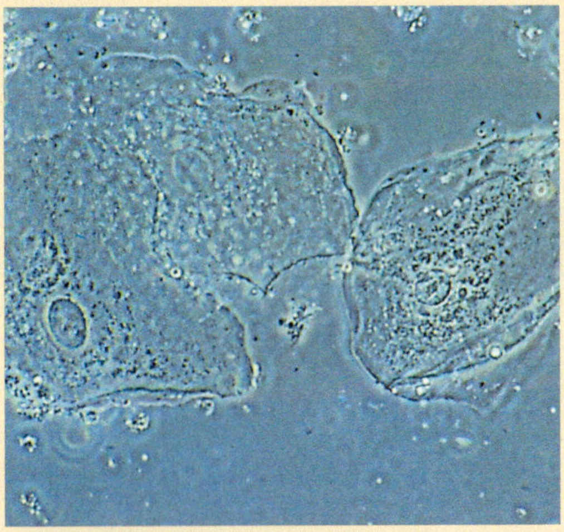

Figure 24 Cells of the cheek lining in methylene-blue solution (magnification: 400x)

Conclusion
Compare the cells of the Canadian pondweed with the cells of the cheek lining. Sum up the similarities and differences in a table. Then, together with a partner, take turns describing the key features of each kind of cell, and add any extra information to your table.

Activate your English

Talking about cell shapes

They / the cells have a ... shape / are shaped like ....

They are different in shape from onion cells because ...

The reason for their different shape is (probably) / may be that ...

1 spatula *Spatel*
2 gently *zart*
3 purpose *Zweck*

Activity 2 Build a three-dimensional model of a plant cell

Before you start, consider what materials and tools you might need to make a model of a plant cell that shows the structures and organelles you can see under the light microscope. Figure 25 presents two options[1] for such a model, but there are other possibilities you can discuss in class before you begin.

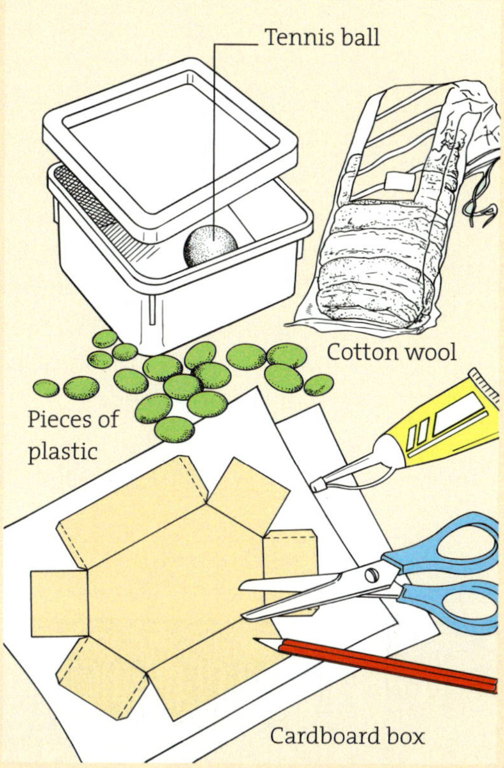

Figure 25 **Materials for making models of plant cells**

Option 1:

Materials

- transparent plastic box (so that you can see the structures inside without having to lift the lid)
- small ball representing the nucleus
- oval-shaped green bodies representing the chloroplasts (balls of modelling clay or small round green sweets work well)
- cotton wool representing cytoplasm

Procedure

Arrange the different parts in such a way that your model looks as realistic as possible.

Option 2:

Materials

- small cardboard box with a lid or a sheet of thin cardboard
- same materials for nucleus, cytoplasm and chloroplasts as in Option 1
- pair of scissors
- glue

Procedure

Figure 25 gives you an idea of how to complete the model. Although it requires more effort to model the cell out of a sheet of cardboard than to use a ready-made box, this technique will give you a model which resembles the cell you saw under the microscope much more closely than the ready-made box would.

Figure 26 **A model of a plant cell**

1 option *Alternative, Variante*

1 Discuss in class how you could build a model of an animal cell. Figure 26 may give you some ideas.

2 With a partner, discuss how your model has or has not helped you understand plant cells and their different structures and organelles.

2.6 A leaf consists of different tissues

A cross section through the leaf of a typical broad-leafed tree shows that this plant organ has a complicated inner structure.

Square-shaped and closely fitting cells form the upper and lower epidermis – the 'skin' of the leaves. The function of the epidermis is to protect the inner layers of cells against injuries, fungi, harmful germs and dust. The upper epidermis is often covered by a shiny layer of a waxy[1] material, the **cuticle**, which prevents the leaf from losing too much water in hot and dry weather. In most plants the lower epidermis has tiny holes, the **stomata**, which allow the exchange of gases and the release of water vapour[2]. The two layers of epidermis enclose the **mesophyll**, which consists of two types of tissue. Below the upper epidermis there is a layer of closely packed cells with no spaces between them. This layer looks like a palisade[3] (another word for a fence), and is therefore called the **palisade layer**. The cells in this layer contain lots of chloroplasts filled with the green pigment chlorophyll. The space between the palisade layer and the lower epidermis is loosely filled with irregularly shaped, thin-walled cells, which also contain chloroplasts. These cells create a large inner surface through which the exchange of oxygen, carbon dioxide and water vapour can easily take place. The system of air spaces between the spongy cells is connected to larger air chambers behind the stomata. This allows for the interchange of gases between the different areas of the leaf and the outside air. The veins of the leaf, with their xylem and phloem tubes, are embedded in this **spongy layer** of cells. The cells of the upper and lower epidermis do not contain chloroplasts, apart from the **guard cells** of the stomata, which are responsible for the opening and closing process.

Key terms

Leaf structure
- cuticle *Kutikula*
- guard cell *Schließzelle*
- mesophyll *Mesophyll*
- palisade layer *Palisadenschicht*
- spongy layer *Schwammschicht*
- stoma (pl. stomata) *Spaltöffnung*

1 waxy *wachsartig*
2 water vapour *Wasserdampf*
3 palisade *Lattenzaun*

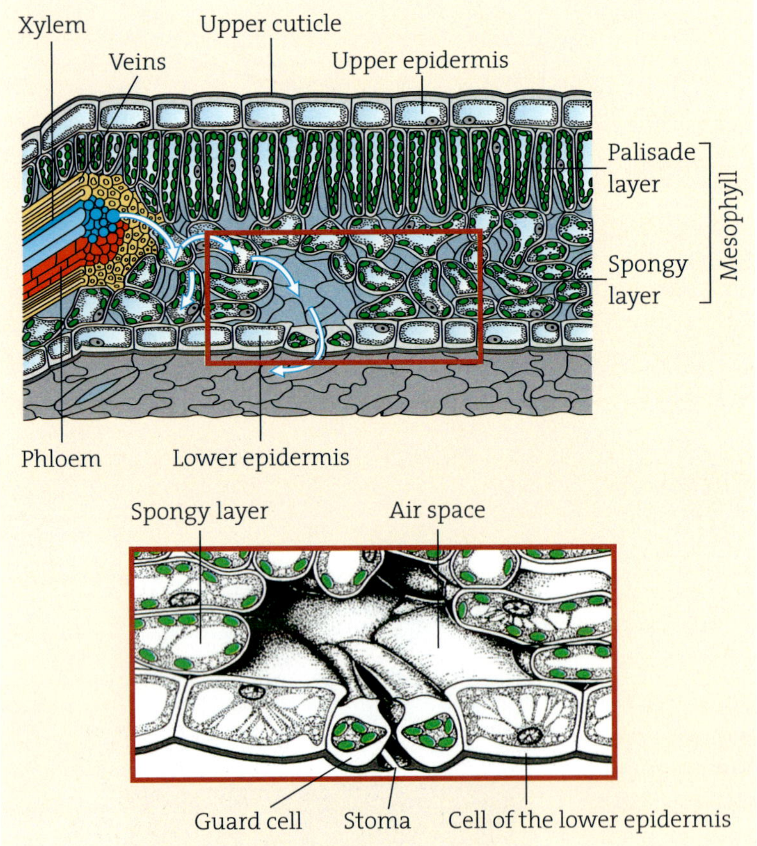

Figure 27 **Cross section of a leaf from a broad-leafed tree**

1 Using Figure 27 and the information in the text, describe the tissues in a leaf starting at the top and moving downwards. Point out the location and role of each type of tissue.
2 Explain why it is an advantage for the plant to have a shiny cuticle on the upper epidermis of its leaves and stomata in the lower epidermis.

Tip:
If you want to study the arrangement of cells and tissues in a leaf, you can either use prepared slides of the surface and the inside, or you can make your own wet mounts of the epidermis and of a cross section of a leaf.

Further activities

Activity 3

Preparing a piece of leaf epidermis for observation under the microscope

Materials

- microscope slides
- cover slips
- tweezers
- pipette
- small brush
- dissection needle or a pencil with a sharp tip
- single-edge razor blade or scalpel
- leaves of a tulip, beech or lime tree
- compound microscope

Procedure

1. Use the pipette to put a drop of water onto the centre of the slide.
2. Bend the upper side of the leaf round the pencil and make a slight diagonal cut into the surface of the leaf.
3. Grip the edge of the cutting with your tweezers and pull off a small piece of tissue from the underside of the leaf.
4. Put the piece of tissue into the drop of water.
5. Use the small brush to spread the epidermis out, trying not to let any part of it fold over.
6. Place the cover slip carefully on top of the object (starting on one side and lowering it slowly, supported by the dissection needle or the tip of your pencil).
7. Now put the slide onto the stage of the microscope.
8. If there are air bubbles under the cover slip which obscure your vision, use the tip of your pencil or dissection needle and tap carefully on the cover slip. This makes the air bubbles move slowly to the edge of the cover slip.

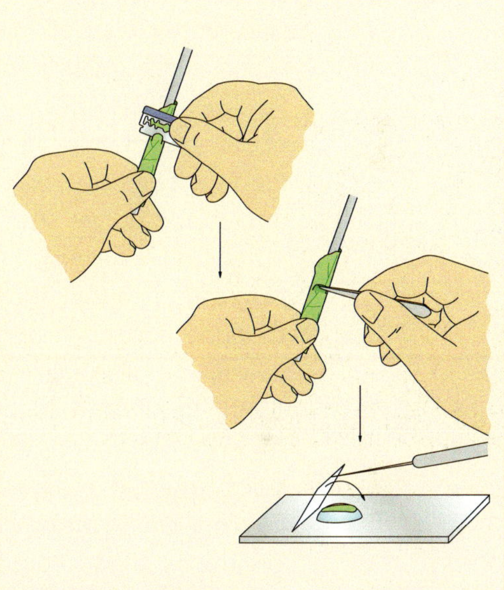

Figure 28 How to make a wet mount of a piece of leaf epidermis

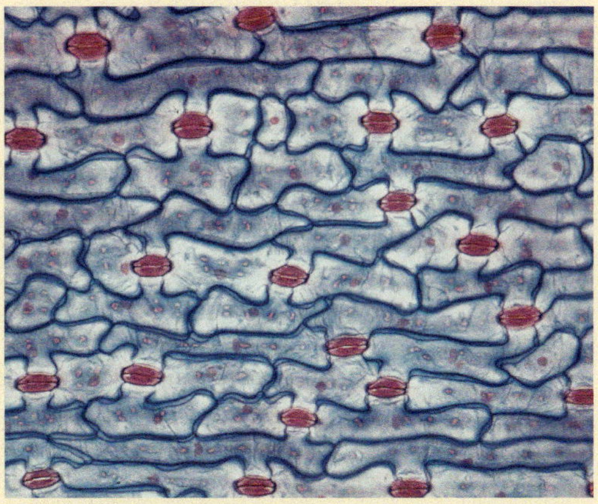

Figure 29 Microscopic picture of the lower epidermis of a tulip leaf (stomata stained red)

1 Observe the sample of epidermis tissue under the microscope. Figure 29 on this page and Figure 14 in Unit 2, Chapter 1, p. 166, show you what you can expect to see.

2 Make a labelled drawing of a section of the epidermis with stomata and another drawing without.

Activity 4 Preparing a cross section of a leaf

Materials
- microscope slides
- cover slips
- tweezers
- pipette or medicine dropper
- small brush
- dissection needle or a pencil with a sharp tip
- single-edge razor blade or scalpel
- piece of elder pith or a small, square block of polystyrene[1]
- leaves of a beech or lime tree, lilac or busy Lizzie
- compound microscope

Procedure
1. Use the pipette to put a drop of water onto the centre of the slide.
2. Cut the piece of elder pith in half lengthwise or cut the polystyrene block in half.
3. Sandwich the leaf between the two pieces of elder pith or polystyrene.
4. Press the two halves together and use the razor blade or the scalpel to produce a plane[2] surface for a cut.
5. Now start a horizontal cut 1–2 mm below the upper edge of the block of elder pith or polystyrene, keeping the blade tilted[3] slightly upwards while moving it through the material. This technique will produce a slice of the leaf which is increasingly thin from one end to the other.
6. Use the tweezers to put the cross section of the leaf into the drop of water.
7. Use the small brush to spread the piece of tissue out, trying not to let any part of it fold over.
8. Place the cover slip carefully on top of the object (starting on one side and slowly lowering it, supported by the dissection needle or the tip of your pencil).
9. Now place the prepared slide on the stage of the microscope.

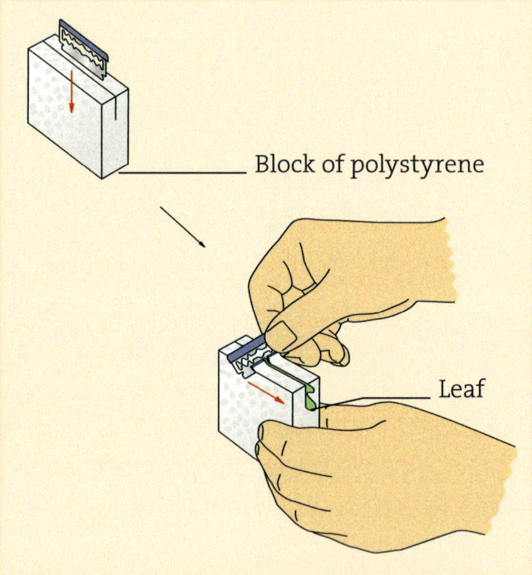

Figure 30 How to make a wet mount of a cross section of a leaf

1 polystyrene *Styropor*
2 plane *eben, flach*
3 tilted *geneigt*

1 Observe the cross section of the leaf under the microscope. Concentrate on the thin areas of the object.

2 First make a labelled drawing of the whole cross section at low or medium magnification. Then try to find an area similar to that shown in Figure 31 and make a labelled drawing of a few cells from each type of tissue. Then compare your findings with Figure 27, p. 240.

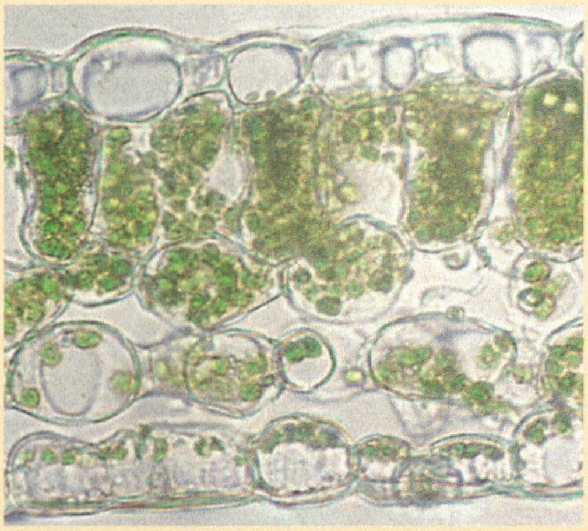

Figure 31 Cross section of a leaf under the light microscope

Summary

All cells of higher organisms (humans, animals and plants) have a nucleus and cytoplasm enclosed by a plasma membrane. The nucleus contains the genetic information and controls all activities of the cell. The cytoplasm is the substance in which metabolic processes take place.

Cells of animals and plants show characteristic differences. Animal cells are enclosed by a cell membrane and plant cells have an additional cell wall. The cytoplasm of plant cells contains chloroplasts for photosynthesis and large cell sap vacuoles. The mitochondria provide all cells with the necessary energy for life processes by means of cellular respiration.

All the cells of a multicellular organism originate from a fertilized egg cell that divides by mitosis. The result is a pair of identical daughter cells. These body cells can divide again and again or start to differentiate for specific functions. Mitosis is the basis for growth and repair.

Plants, multicellular animals and humans are made up of tissues consisting of large numbers of cells of the same kind. Different tissues work together in organs and all organs of the body work together in harmony in a healthy organism. Cells, tissues and organs function according to the principle of division of labour. The more specialized the different levels of organisation are, the greater is the efficiency of the whole organism.

Key terms

See the boxes:

A Special tasks

1. Draw a mind map based on the information in the summary above (for help see 'Improve your skills', p. 22).
2. Make a diagram of the different levels of organisation in the body of an animal of your choice, giving examples for each level: cells > tissues > organs > organ systems > organism. Present your diagram in class as a transparency or on a computer.

B True or false?

1. Plant cells have a cell wall, while animal cells have a cell membrane instead.
2. A typical animal cell differs from a plant cell in that it has large vacuoles.
3. Before mitosis starts, identical copies of the chromosomes are produced.
4. Tissues are made up of cells with different functions.
5. Nerve cells can change their shape when they carry out their functions.
6. Plant and animal cells have mitochondria, but only plant cells have chloroplasts.
7. Chloroplasts are the organelles of cellular respiration.
8. Microscopic drawings should always be done in ink.

C Critical thinking

1 The scales that make up the bulb of an onion are modified leaves. Explain why they do not contain chloroplasts, which are the typical organelles in leaves.

2 Draw conclusions from the comparison of the numbers of chromosomes in the body cells of humans and the different species of plants and animals in Table 1, p. 229.

3 **Say it in German:** Using a specific physical activity as an example, explain the advantage of a division of labour between the different tissues and organs of your body.

English-German vocabulary

A

absorb sth. [əb'sɔ:b ☆ -'zɔ:b] *etwas aufsaugen, etwas absorbieren, etwas abfangen*
absorbent [əb'sɔ:bənt ☆ -'zɔ:b-] *saugfähig*
absorption [əb'sɔ:pʃn ☆ -'zɔ:b-] *Aufnahme, Resorption, Stoffaufnahme*
abstinence ['æbstɪnəns] *Enthaltsamkeit*
accommodation [əˌkɒmə'deɪʃn ☆ əˌkɑ:m-] *Akkommodation*
account for sth. *etwas ausmachen; verantwortlich sein für etwas, etwas bedingen*
accumulated [ə'kju:mjəleɪtɪd] *angesammelt*
acidic [ə'sɪdɪk] *sauer, säurehaltig*
acidity [ə'sɪdəti] *Säure, Säuregehalt*
acorn *Eichel*
acoustic echo [ə'ku:stɪk 'ekəʊ ☆ 'ekoʊ] *hörbares Echo*
acquired immunodeficiency syndrome (AIDS) [ə'kwaɪəd ɪˌmju:nəʊdɪ'fɪʃnsi] *erworbenes Immunschwäche Syndrom*
active transport *aktiver (Energie verbrauchender) Transport*
acute [ə'kju:t] *scharf(sinnig), fein*
addiction [ə'dɪkʃn] *Sucht*
behavioural addiction [bɪ'heɪvjərəl ə'dɪkʃn] *stoffungebundene Sucht*
Adam's apple *Adamsapfel*
adaptation [ˌædæp'teɪʃn] *Adaptation, Anpassung*
adequate ['ædɪkwət] *geeignet, passend*
adequately *angemessen*
adjacent [ə'dʒeɪsnt] *angrenzend*
adjust sth. *etwas anpassen*
adjust to sth. *sich an etwas anpassen*
adjustment knob [nɒb ☆ nɑ:b] *Einstellknopf*
adolescence [ˌædə'lesns] *Jugendzeit*
advantage [əd'vɑ:ntɪdʒ] *Vorteil*
advice *Ratschlag*
afterbirth *Nachgeburt*
air pollutant [pə'lu:tənt] *Luftschadstoff*
airtight *luftdicht*
alcohol poisoning *Alkoholvergiftung*
alcoholism ['ælkəhɒlɪzm ☆ -hɔ:l-] *Alkoholabhängigkeit, Alkoholismus*
alder *Erle*
align with sth. [ə'laɪn] *sich ausrichten auf etwas*
alimentary canal [ˌælɪmentəri kə'næl] *Verdauungskanal*
alter sth. *etwas verändern*
alternately *abwechselnd*
alveolus (pl. alveoli) [ˌælvi'əʊləs] *Lungenbläschen*
amino acid [əˌmi:nəʊ 'æsɪd] *Aminosäure*
ammonia [ə'məniə] *Ammoniak*
amniotic fluid [ˌæmniɒtɪk 'flu:ɪd ☆ -ɑ:tɪk] *Fruchtwasser*
amoeba (pl. amoebae) [ə'mɪ:bə] *Amöbe, Urtierchen*
amount *Menge*
amplify sth. ['æmplɪfaɪ] *etwas verstärken*
ampulla [am'pʊlə] *Ampulle*
amputation [æmpju'teɪʃn] *Amputation*
amylase ['æmɪləɪz ☆ 'æməleɪs] *Amylase*
ancestor ['ænsestə] *Vorfahre*
anchor sth. *etwas verankern*
anemone [ə'neməni] *Anemone, Windröschen*
angle ['æŋgl] *Winkel*
annual [ˌænjuəl] *jährlich; einjährig*
anorexia nervosa [ˌænə'reksiə nɜ:vəʊsə] *Magersucht*
anorexic [ˌænə'reksɪk] *magersüchtig*
antagonist [æn'tægənɪst] *Gegenspieler*
anther *Staubbeutel*
antiallergic *nicht allergisch*
antibiotic [ˌæntibaɪ'ɒtɪk ☆ -ɑ:tɪk] *Antibiotikum*
antibody ['æntibɒdi ☆ -bɑ:di] *Antikörper*
anticlockwise *entgegen dem Uhrzeigersinn*
antidrug *Gegendroge*
antigen ['æntɪdʒən] *Antigen*
antiperspirant [ˌænti'pɜ:spərənt] *Antitranspirant, Deodorant*
antispasmodic *krampflösend*
anus ['eɪnəs] *After*
anvil ['ænvɪl] *Amboss*
anxiety [æŋ'zaɪəti] *Angst*
anxious [ˌæŋkʃəs] *ängstlich*
aorta [eɪ'ɔ:tə] *große Körperarterie, Schlagader*
apart *getrennt, auseinander*
appearance [ə'pɪərəns] *Aussehen*
appendix [ə'pendɪks] *Anhang, Blinddarmfortsatz, Wurmfortsatz*
apply sth. [ə'plaɪ] *etwas anlegen, etwas anwenden*
appreciation [əˌprɪ:ʃi'eɪʃn] *Wertschätzung*
aqueous humour ['eɪkwiəs 'hju:mə] *Kammerwasser*
arch [ɑ:tʃ] *Bogen, Wölbung*
armpit *Achselhöhle*
arrangement [ə'reɪndʒmənt] *Anordnung*
artery ['ɑ:təri] *Arterie*
artificial [ˌɑ:tɪ'fɪʃl] *künstlich*
artificial propagation [ˌprɑpə'geɪʃn] *künstliche Vermehrung*
artificial selection *künstliche Auslese*
asexual [ˌeɪ'sekʃuəl] *vegetativ, ungeschlechtlich*
ashore [ə'ʃɔ:] *ans Land, ans Ufer*
ask sb. out *jdn. einladen, mit jdm. auszugehen*
assess sth. *etwas bewerten, feststellen*
assign sb./sth. to sth. [ə'saɪn] *jdn./etwas etwas zuordnen*
astigmatism [ə'stɪgmətɪzəm] *Astigmatismus, Stabsichtigkeit*
athlete ['æθli:t] *Sportler, Leichtathlet*
atmosphere ['ætməsfɪə] *Atmosphäre*
atom ['ætəm] *Atom*
atrium (pl. atria) ['eɪtriəm] *Vorkammer, Vorhof*
attach sth. *etwas befestigen*
aubretia [ɔ:'brɪ:ʃə] *Blaukissen*
audible ['ɔ:dəbl] *hörbar*
auditory canal ['ɔ:dətri kə'næl ☆ -təri kə'næl] *Gehörgang*

auditory nerve ['ɔːdətri ☆ -tɔri] *Gehörnerv*
auricle ['ɔːrɪkl] *Ohrmuschel*
automatically *automatisch*
average ['ævərɪdʒ] *durchschnittlich*
on average ['ævərɪdʒ] *im Durchschnitt*
awareness [ə'weənəs] *(hier) Bewusstsein*
awn [ɒn ☆ ɑːn] *Granne*
axis (pl. axes) *Achse*

B

backbone *Wirbelsäule*
bacterial infection [bæk'tɪəriəl] *bakterielle Infektion*
bacterium (pl. bacteria) [bæk'tɪəriəm] *Bakterium, Bakterie*
baking powder ['beɪkɪŋ] *Backpulver*
baking tray *Backblech, Kuchenblech*
bale [beɪl] *Ballen*
ball and socket joint *Kugelgelenk*
bamboo *Bambus*
bandage ['bændɪdʒ] *Verband, Pflaster*
bank clerk [bæŋk klɑːk ☆ klɜːrk] *Bankangestellte/r*
bark [bɑːk] *Borke*
barley ['bɑːli] *Gerste*
basal metabolic rate ['beɪsl ˌmetə'bɒlɪk] *Grundumsatz*
basic *basisch, alkalisch; grundlegend*
basic skill *grundlegende Fähigkeit, Fertigkeit*
basilar membrane ['basɪlə 'membreɪn] *Grundmembran*
bat *Fledermaus*
addictive [ə'dɪktɪv] *abhängig machend, süchtig machend*
beak *Schnabel eines Vogels*
beaker *Becherglas, Becher*
beam of electrons [ɪ'lektrɒns] *Elektronenstrahl*
become loose [luːs] *sich lockern*
bedding *Lagerstreu*
beech *Buche*
beehive ['biːhaɪv] *Bienenstock*
bell jar *glockenförmiges Gefäß*
bend *beugen*
bend down *sich niederbeugen*
bend sth. *etwas beugen, etwas biegen*
benefit *Nutzen, Vorteil*
benefit from sth. *profitieren von etwas, Nutzen aus etwas ziehen*
beriberi *Beriberi*
beverage ['bevərɪdʒ] *Getränk*
beyond [bɪ'jɒnd] *jenseits*
biceps ['baɪseps] *Armbeuger*
bile [baɪl] *Gallenflüssigkeit*
binder ['baɪndə] *Ordner*
binge drinking [bɪndʒ] *Besäufnis, Komasaufen*
binge eating attack *Essattacke*
binocular microscope [bɪ'nɒkjələ 'maɪkrəskəʊp] *binokulares Mikroskop*
biofuel ['baɪəʊfjuːəl] *Biobrennstoff*
biogas *Biogas*
biological research [ˌbaɪə'lɒdʒɪkl ☆ -'lɑːdʒ-] *biologische Forschung*
biological system *biologisches System*
biomass *Biomasse*
birch [bɜːtʃ] *Birke*
bird of prey *Raubvogel*
birth-control pill *Antibabypille*
biscuit (BE) *süßer Keks*
blackberry *Brombeere*
blackout ['blækaʊt] *Ohnmachtsanfall*
bladder *(Harn)blase*
blade [bleɪd] *Blattspreite*
blank sample *Blindprobe*
bleeding *Blutung*
blender *Mixer*
blind spot [blaɪnd] *blinder Fleck*
blink *blinzeln, zwinkern*
blood alcohol concentration (BAC) *Blutalkoholkonzentration*
blood capillary [blʌd kə'pɪləri ☆ 'kæpəleri] *Blutkapillare*
blood clot ['blʌd klɒt ☆ klɑːt] *Blutgerinnsel*
blood group [blʌd] *Blutgruppe*
blood pressure ['blʌd preʃə] *Blutdruck*
blood transfusion [blʌd træns'fjuːʒn] *Bluttransfusion*
blood type [blʌd] *Blutgruppe*
blood vessel [blʌd] *Blutgefäß*
bloodstream ['blʌdstriːm] *Blutstrom, Blutbahn, Blutkreislauf*
blotting paper *Löschpapier*
bluebell *Glockenblume*
blueweed *Natterkopf*
blunt [blʌnt] *stumpf*
blurred [blɜːd] *unscharf, verschwommen*
blurred vision [ˌblɜːd 'vɪʒn] *verschwommene Sicht*
blurry ['blɜːri] *unklar, undeutlich, verschwommen*
boa ['bəʊə ☆ 'boʊə] *Boa*
body circulation ['bɒdi ˌsɜːkjə'leɪʃn ☆ 'bɑːdi] *Körperkreislauf*
body mass index (BMI) *Körpermassezahl*
body-specific *körpereigen*
bond *Bindung*
bone marrow ['mærəʊ] *Knochenmark*
bone membrane *Knochenhaut*
botanical guidebook *Pflanzenführer, Pflanzenbestimmungsbuch*
bottom *Hintern*
bounce off sth. [baʊns] *von/an etwas abprallen*
bowl [bəʊl] *Schüssel*
bra *BH*
bran *Kleie*
branch [brɑːntʃ] *Ast, Zweig*
branch off *abzweigen*
branched *verzweigt*
brant *äußere Schicht des Getreidekorn, Kleie*
break sth. down [breɪk] *etwas abbauen*
breast [brest] *Brust*
breastbone *Brustbein*
breath [breθ] *Atem, Atemzug*
breathing rate ['briːðɪŋ reɪt] *Zahl der Atemzüge pro Zeiteinheit*
breeding *Zucht*
breeding ground *Brutstätte*
brick *Ziegel*
brick-red *ziegelrot*
brim *Krempe*
bristle ['brɪsl] *Borste*
broad *breit*
broad bean [brɔːd] *Saubohne*

broad-leafed tree *Laubbaum*
broccoli *Brokkoli*
bronchi (sing. bronchus) ['brɒŋkaɪ ☆ 'brɑːŋ-] *Bronchien*
bronchiole ['brɒŋkiəʊl ☆ 'brɑːŋ-] *Bronchiole*
bronchitis [brɒŋ'kaɪtɪs ☆ brɑːŋ-] *Bronchitis*
broom *Besenginster*
brown radish *Rettich*
Brussels sprout ['brʌslz 'spraʊt] *Rosenkohl*
bryophyllum *Bryophyllum*
bud *Knospe*
building block ['bɪldiŋ blɒk] *Baustein*
bulb [bʌlb] *Zwiebel*
bulblike *knollenartig*
bulge [bʌldʒ] *ausbeulen, Schwellung*
bulimia [bu'lɪmiə] *Bulimie, Esssucht*
bump [bʌmp] *Erschütterung*
bumpy ['bʌmpi] *uneben*
bundle ['bʌndl] *Bündel*
Bunsen burner [ˌbʌnsn 'bɜːnə] *Bunsenbrenner*
burst – burst – burst [bɜːst] *platzen*
busy Lizzie [ˌbɪzi 'lɪzi] *Fleißiges Lieschen*
by-product [baɪ 'prɒdʌkt] *Nebenprodukt*

C

cabbage family [kæbɪdʒ] *Familie der Kreuzblütler*
cabbage variety [kæbɪdʒ və'raɪəti] *Kohlsorte*
cactus (pl. cacti) ['kæktəs / 'kæktai] *Kaktus*
caecum ['siːkəm] *Blinddarm*
Caesarean section [si'zeəriən 'sekʃn] *Kaiserschnitt*
calcareous [kæl'keəriəs] *kalkhaltig*
calcium ['kælsiəm] *Kalzium*
cambium ['kæmbiəm] *Kambium*
camomile ['kæməmaɪl] *Kamille*
can *Dose*
Canadian pondweed *Wasserpest*
cancer ['kænsə] *Krebs(erkrankung)*
candytuft *Schleifenblume*
canine *Eckzahn*
canned *aus der Dose*
capability [ˌkeɪpə'bɪləti] *Fähigkeit*
capillary [kə'pɪləri ☆ 'kæpəleri] *Kapillare*
capillary action *Kapillarwirkung*
capsule *Kapsel*
carbohydrate [ˌkɑːbəʊ'haɪdreɪt] *Kohlenhydrat*
carbon dioxide [daɪ'ɒksaɪd ☆ daɪ'ɑːksaɪd] *Kohlendioxid, Kohlenstoffdioxid*
carbon monoxide [mə'nɒksaɪd ☆ mə'nɑːksaɪd] *Kohlenstoffmonoxid*
cardboard *Pappe*
cardiac ['kɑːdiæk] *Herz-*
cardiac muscle *Herzmuskel*
caries *Zahnfäule, Karies*
carnivorous [kɑː'nɪvərəs] *fleischfressend*
carpal *Handwurzelknochen*
carpel *Fruchtblatt*
cartilage ['kɑːtəlɪdʒ] *Knorpel*
cartilage cell *Knorpelzelle*
case *Fall*
cast *Gipsverband*
caster sugar *Streuzucker*
cataract ['kætərækt] *grauer Star*
catkin *Blütenkätzchen*
cauliflower ['kɒliflaʊə] *Blumenkohl*
cavity *Höhle, Raum, Hohlraum*
cell *Zelle*
cell division *Zellteilung*
cell membrane *Zellmembran*
cell respiration *Zellatmung*
cell sap *Zellsaft*
cell sap vacuole [sel sæp 'vækjuəʊl] *Zellsaftraum*
cell theory *Zelltheorie*
cell wall *Zellwand*
cellular respiration ['seljʊlə] *Zellatmung*
cellulose *Zellulose*
cement *Zement; Zahnzement*
central nervous system *zentrales Nervensystem*
centromere ['sentrəʊmɪə] *Zentromer*
cereal ['sɪəriəl] *Getreideprodukt*
cereal plant *Getreidepflanze*
cerebral cortex ['serəbrəl 'kɔːteks ☆ sə'riːbrəl] *Großhirnrinde*
cerebrum [sə'riːbrəm, 'serəbrəm] *Großhirn*
cervix ['sɜːvɪks] *Muttermund*
chamber ['tʃeɪmbə] *Kammer*
charlock *Ackersenf*
charming *hübsch*
check-up *Kontrolluntersuchung*
chemical *chemisch*
chemical compound ['kemɪkl 'kɒmpaʊnd ☆ 'kɑːmpaʊnd] *chemische Verbindung*
chemical energy *gebundene Energie*
chemical messenger ['kemɪkl 'mesɪndʒə] *chemischer Botenstoff*
chemist's (BE) *Apotheke, Drogerie*
chest hair *Brusthaare*
chew [tʃuː] *kauen*
Chinese cabbage *Chinakohl*
chips (BE) *Pommes Frites*
chirp *zwitschern, zirpen*
chisel ['tʃɪzl] *Meißel*
chloride *Chlor*
chlorophyll ['klɒrəfɪl ☆ 'klɔːr-] *grüner Blattfarbstoff, Chlorophyll*
chloroplast *Blattgrünkörper, Chloroplast*
cholesterol [kə'lestərɒl ☆ -rɑːl] *Cholesterin*
chopped *gehackt*
choroid ['kɔːrɔɪd] *Gefäßhaut*
chromatid *Chromatid*
chromoplast *Chromoplast*
chromosome *Chromosom*
ciliary muscle ['sɪliəri mʌsl] *Ziliarmuskel*
cilium (pl. cilia) ['sɪliəm] *Wimperhärchen*
circular ['sɜːkjələ] *kreisförmig*
circumcise sb. ['sɜːkəmsaɪz] *jdn. beschneiden*
circumcision [ˌsɜːkəm'sɪʒn] *Beschneidung*
clamp *Klammer, Klemme*
clamp sth. *etwas abklemmen*
clamp sth. to sth. *etwas zusammenklammern*

cleanse sth. *etwas säubern*
clematis *Clematis, Waldrebe*
cling to sth. *an etwas haften, an etwas hängen*
clitoris *Klitoris, Kitzler*
clockwise *im Uhrzeigersinn*
close-up *Nahaufnahme*
clouded *trüb, unklar*
clove [kləʊv] *Gewürznelke*
clump [klʌmp] *Ballen, Klumpen*
clump together *verklumpen*
cluster *Büschel*
coarse *grob*
(maize) cob [kɒb ☆ kɑːb] *Maiskolben*
cochlea [kɒkliə ☆ 'kɑːk-] *Schnecke (im Innenohr)*
cochlear fluid [kɒkliə 'fluːɪd ☆ 'kɑːk-] *Flüssigkeit in den Schneckengängen*
coconut palm *Kokosnusspalme*
coil sth. up *etwas aufrollen*
coiled [kɔɪld] *aufgerollt*
coleoptile [kɒli'ɒptaɪl ☆ kɑːli'ɑːp-] *Keimscheide (beim Graskeimling)*
coleus plant *Buntnessel*
collarbone (clavicle) *Schlüsselbein*
collide with sb./sth. *mit jdm./etwas zusammenstoßen*
column ['kɒləm ☆ 'kɑːl-] *Spalte, Säule*
come to terms with sth. *mit etwas klarkommen*
comet *Komet*
commitment *bewusste Entscheidung, Zusage*
common *üblich, häufig*
common cold *Schnupfen, Erkältung*
compact bone tissue *hartes Knochengewebe*
compass needle *Kompassnadel*
compatible [kəm'pætəbl] *verträglich, passend*
competition *Wettkampf*
complete fracture *vollständiger Bruch*
complex *komplex, kompliziert*
compound eye *Komplexauge*
compound leaf *zusammengesetztes Blatt*
compound microscope *zusammengesetztes Mikroskop*
compulsive *zwanghaft*
concave lens [kɒn'keɪv ☆ kɑːn-] *Zerstreuungslinse*
concentric *konzentrisch*
condition *Befinden, Zustand; Leiden*
condom *Kondom*
conduct sth. *etwas durchführen*
conducting tissue *Leitgewebe*
cone *Zäpfchen (im Auge)*
coniferous tree [kə'nɪfərəs] *Nadelbaum*
conjunctiva [ˌkɒndʒʌŋ'tɪvə ☆ ˌkɑːn-] *Bindehaut*
connective tissue *Bindegewebe*
consciousness ['kɒnʃəsnəs ☆ 'kɔːn] *Bewusstsein*
constipation *Verstopfung*
consult sth. *etwas heranziehen, etwas zu Rate ziehen*
consumer *Verbraucher, Konsument*
consumption *Verzehr*
contain sth. *etwas enthalten*
continuous *andauernd, ununterbrochen*
contraceptive [ˌkɒntrə'səptiv] *Verhütungsmittel*
contract *(sich) anspannen*
contract [kən'trækt ☆ kɑːn-] *sich kontrahieren, sich zusammenziehen*
contractions (pl.) *Wehen*
contradict sb./sth. *jdm./etwas widersprechen*
contrast *Gegensatz*
control (experiment) *Kontrolle, Kontrolluntersuchung, Kontrollgruppe*
controversial [ˌkɒntrə'vɜːʃl ☆ ˌkɒːntrə'vɜːrʃl] *umstritten, kontrovers*
convert sth. *etwas umwandeln, umbauen*
convex lens ['kɒnveks ☆ 'kɑːn-] *Sammellinse*
cookie (AE) *süßer Keks*
coordination [kəʊˌɔːdɪ'neɪʃn ☆ koʊˌɔːrd-] *Abstimmung, Koordination*
cope with sth. *etwas bewältigen*
copper spiral ['spaɪrəl] *Kupferspirale*
core *Kerngehäuse*
cork *Kork*
corm *Sprossknolle*
cornea ['kɔːniə] *Hornhaut*
corolla *Blütenkrone*
coronary artery ['kɒrənri 'ɑːtəri ☆ 'kɔːrəneri] *Herzkranzgefäß*
correlate ['kɒrəleɪt ☆ 'kɔːr-] *zueinander in Beziehung setzen*
correspond to sth. *übereinstimmen mit etwas*
cortex *Wurzelrinde*
cotton *Baumwolle*
cotton stick *Wattestäbchen*
cotton wool *Watte*
cotyledon [kɒtɪ'liːdn ☆ kɑːt-] *Keimblatt*
cough [kɒf ☆ kɔːf] *Husten*
counselling centre *Beratungszentrum*
cover slip *Deckglas*
cover sth. *etwas behandeln, sich mit etwas befassen*
crack *Riss*
cramp *Krampf*
crave sth. *sich nach etwas sehnen*
cress *Kresse*
crop *Ernte, Ernteertrag*
cross section *Querschnitt*
cross-pollination *Fremdbestäubung*
crown *Krone; Zahnkrone*
cruciate ligament ['kruːʃiəɪt 'lɪgəmənt] *Kreuzband*
crumbly *krümelig, brüchig*
cube *Würfel*
cubic millimetre (mm^3) ['kjuːbɪk 'mɪlimiːtə] *Kubikmillimeter*
cuckoo flower *Wiesen-Schaumkraut*
cucumber *Gurke*
cultivate sth. *etwas anbauen, kultivieren*
cultivation *Anbau, Bebauung*
cultural background *kulturelle Herkunft*
cup-shaped *kelchförmig*
cupula ['kjuːpjuːlə] *Cupula*
curl [kɜːl ☆ kɜːrl] *rollen*
curvature ['kɜːvətʃə] *Wölbung*
curve *sich biegen*

curved *gewölbt*
curvy *kurvenreich*
cut *Einschnitt*
cuticle **['kju:tɪkl]** *Kutikula*
cutting *Steckling*
cylindrical **[sə'lɪndrɪkl]** *zylinderförmig*
cytology *Zelllehre, Zytologie*
cytoplasm *Zellplasma*

D

daffodil **['dæfədɪl]** *Osterglocke, Narzisse*
dahlia **['deɪliə]** *Dahlie*
dairy product **['deəri]** *Milchprodukt, Molkereiprodukt*
daisy *Gänseblümchen*
damage sth. *etwas beschädigen*
damaged **['dæmɪdʒd]** *beschädigt*
dandelion **['dændɪlaiən]** *Löwenzahn*
dazzle sb. *jdn. blenden*
debt **[det]** *Schuld*
decay *faulen, zerfallen*
decibel **['desɪbel]** *Dezibel*
decisive *entscheidend*
decline *Rückgang*
decrease *abnehmen, sinken*
deficiency **['dɪfɪʃnsi]** *Mangel*
deficiency disease *Mangelkrankheit*
delicate *empfindlich*
deliver a baby *ein Kind zur Welt bringen, ein Kind entbinden*
dendrite *Dendrit*
dense *dicht*
dental floss *Zahnseide*
dentine **['denti:n]** *Zahnbein*
deoxygenated blood **[ˌdi:'ɒksɪdʒəneɪtɪd blʌd]** *sauerstoffarmes Blut*
deposit **[dɪ'pɒzɪt ☆ -'pɑ:z-]** *Ablagerung*
dermis **['dɜ:mɪs]** *Haut*
descendant **[dɪ'sendənt]** *Nachkomme*
desirable *wünschenswert*
destarch sth. **[dɪ'sta:rtʃ]** *die Stärke aus etwas entfernen*
destroy sth. *etwas zerstören*
detect sth. *etwas aufdecken, aufspüren*
deteriorate **[dɪ'tɪəriəreɪt]** *verschlechtern*
determine sth. **[dɪ'tɜ:mɪn]** *etwas feststellen, bestimmen*
detoxification *Entgiftung*
devastating *verheerend, vernichtend*
diabetes **[ˌdaɪə'bi:ti:z]** *Zuckerkrankheit*
dial **['daɪəl]** *(am Telefon eine Nummer) wählen*
diameter *Durchmesser*
diaphragm **['daɪəfræm]** *Diaphragma, Blende, Zwerchfell*
diarrhoea **[ˌdaɪə'rɪə]** *Durchfall*
dicot **['dɪkɒt ☆ 'daɪkɑ:t]** *zweikeimblättrige Pflanze*
diet **['daɪət]** *Ernährung*
balanced diet *ausgewogene Ernährung*
dietary fibre **['daɪətri ☆ -teri]** *Ballaststoff, Faserstoff*
differentiate *differenzieren*
differentiation *Differenzierung*
digest sth. **[daɪ'dʒest]** *etwas verdauen*
digested **[daɪ'dʒestɪd]** *verdaut*
digestion **[daɪ'dʒestʃən]** *die Verdauung*
digestive **[daɪ'dʒestɪv]** *die Verdauung betreffend*
digestive juice **[daɪ'dʒestɪv]** *Verdauungssaft*
digestive system **[daɪ'dʒestɪv]** *Verdauungsapparat*
dilate **[daɪ'leɪt]** *dehnen, erweitern*
dilute hydrochloric acid **[daɪ'lu:t ˌhaɪdrəklɒrik 'æsɪd]** *verdünnte Salzsäure*
dim *dämmerig*
disaccharide *Zweifachzucker*
disadvantage *Nachteil*
disc *Scheibe*
discard sth. *etwas abwerfen*
disc-shaped *diskusförmig*
dish *Gericht*
dislocation *Verrenkung*
disorder *Funktionsstörung*
display *Ausstellung*
dispersal *Verbreitung*
disperse *sich verteilen*
disperse sth. *etwas verbreiten, etwas verstreuen*
display sth. *etwas zeigen, zur Schau stellen*
dissatisfying *unbefriedigend*
dissect sth. **[dɪ'sekt]** *etwas zerlegen*
dissection needle **[dɪ'sekʃn]** *Präpariernadel*
dissolve **[dɪ'zɒlv ☆ -'zɑ:lv]** *sich auflösen*
dissolve sth. **[dɪ'zɒlv ☆ -'zɑ:lv]** *etwas auflösen*
distort sth. *etwas verzerren*
distorted *verzerrt*
distribute sth. *etwas verteilen*
divide into sth. **[dɪ'vaɪd]** *sich in etwas aufteilen*
divide sth. by sth. *etwas teilen durch etwas*
division of labour *Arbeitsteilung*
division of the nucleus *Kernteilung*
dizziness *Schwindel*
dome-shaped *gewölbt, kuppelartig*
donor **['dəʊnə]** *Spender*
dormancy **['dɔ:mənsi]** *Samenruhe, Keimruhe*
dormant **['dɔ:mənt]** *ruhend*
dormouse **['dɔ:maʊs]** *Haselmaus*
dose *Dosis*
double sugar *Doppelzucker*
dough **[doʊ]** *Teig*
downward (AE) *abwärts*
downwards (BE) *abwärts*
downy *flaumig*
draw air *Luft ziehen*
drill *Bohrer*
droplet *Tröpfchen*
drought **[draʊt]** *Dürre*
drug *Droge; Arzneimittel, Medikament*
over-the-counter drug nicht verschreibungspflichtiges Medikament
drugstore (AE) *Apotheke, Drogerie*
due date *Geburtstermin*
duodenum **[ˌdju:ə'di:nəm]** *Zwölffingerdarm*
dusty *staubig*

dye [daɪ] *Färbemittel, Farbstoff*
dyer's woad ['daɪəʒ wəʊd ☆ woʊd] *Färberwaid*

E

eagle *Adler*
ear *Ähre, Kornähre*
ear canal [kə'næl] *Gehörgang*
ear ossicles ['ɒsɪkls ☆ 'ɑːs-] *Hörknöchelchen*
ear plug *Ohrstöpsel*
eardrum *Trommelfell*
earmuff *Ohrenschützer*
earwax *Ohrenschmalz*
eating disorder *Essstörung*
echolocate [ˌekəʊləʊ'keɪt ☆ ˌekoʊ'loʊkeɪt] *sich durch Echoortung orientieren*
economically [ˌiːkə'nɒmɪkli] *wirtschaftlich*
ecosystem ['iːkəʊsɪstəm ☆ 'iːkoʊ-] *Ökosystem*
ecstasy *Ecstasy*
edible *essbar, verzehrbar*
egg cell *Eizelle*
ejaculation [iˌdʒækju'leɪʃn] *Samenerguss*
electromagnet *Elektromagnet*
electron microscope *Elektronenmikroskop*
elodea [e'ləʊdiə, ˌeləʊ'diːə] *Wasserpest*
elongated ['ilɒŋgeɪtɪd] *länglich, lang gestreckt*
embed sth. in sth. [ɪm'bed] *etwas in etwas einbetten*
embedded [ɪm'bedɪd] *eingebettet*
embryo *Embryo*
embryonic cell *embryonale Zelle*
emergency [i'mɜːdʒənsi] *Notfall*
emphysema [ˌemfɪ'siːmə] *Emphysem*
enable sb. to do sth. *jdm. ermöglichen, etwas zu tun*
enamel [ɪ'næml] *Zahnschmelz*
enclose sth. *etwas umschließen*
endolymph ['ɛndəʊɪmf] *Endolymphe*
endosperm ['endəʊspɜːm] *Nährgewebe, Endosperm*
endure sth. *etwas ertragen, etwas aushalten*
energy ['enədʒi] *Energie*
energy requirement ['enədʒi] *Energiebedarf*
energy-rich ['enədʒi] *energiereich*
engrave sth. *etwas einprägen*
environment [ɪn'vaɪrənmənt] *Umgebung*
enzyme ['enzaɪm] *Enzym, Ferment*
eosin *Eosin, ungiftiger roter Farbstoff*
epidermis [ˌepɪ'dɜːmɪs] *Epidermis, Oberhaut, Hautgewebe*
epididymis (pl. epididymides) [ˌepɪ'dɪdəmɪs] *Nebenhoden*
epiglottis [ˌepɪ'glɒtɪs ☆ -'glɑːtɪs] *Kehldeckel*
epithelium (pl. epithelia) [epɪ'thiːliʌm] *Deckgewebe, Epithel*
equalize sth. ['iːkwəlaɪz] *etwas ausgleichen*
equation *Gleichung*
erect *aufgerichtet, aufrecht*
erectile tissue [ɪ'rektaɪl 'tiʃu] *Schwellkörper*
Erlenmeyer flask *Erlenmeierkolben*
erythrocyte [ɪ'rɪθrəsaɪt] *rotes Blutkörperchen*
essential [ɪ'senʃl] *unerlässlich*
essential fatty acid [ɪ'senʃl] *essentielle Fettsäure*
essential oil [ɪ'senʃl] *ätherisches Öl*
estimate sth. *etwas schätzen, etwas abschätzen*
eucalyptus tree [ˌjuːkə'lɪptəs] *Eukalyptusbaum*
euphoric [juː'fɒrɪk] *euphorisch*
Eustachian tube [juː'steɪʃn tjuːb ☆ tuːb] *Ohrtrompete*
evaluation *Auswertung*
evaporate *verdampfen*
evaporation *Verdunstung (über eine freie Oberfläche)*
event *Ereignis*
evident *deutlich, offensichtlich*
examination [ɪgˌzæmɪ'neɪʃn] *Untersuchung*
examine sb./sth. [ɪgˌzæmɪn] *jdn./etwas untersuchen*
excrete sth. *etwas ausscheiden*
excretion *Ausscheidung*
exercise *Bewegung*
exhale [eks'heɪl] *ausatmen*
expand *sich ausdehnen*
expend sth. *etwas ausgeben*
expose sb./sth. to sth. *jdn./etwas einer Sache aussetzen*
exposure [ɪk'spəʊʒə] *Belastung, Einwirkung*
extensor *Streckmuskel*
extinguish sth. [ɪk'stɪŋgwɪʃ] *etwas auslöschen*
extract sth. *etwas extrahieren, etwas gewinnen*
extraordinary [ɪk'strɔːdnri ☆ ɪk'strɔːrdəneri] *außergewöhnlich*
eye socket *Augenhöhle*
eye strain *Augenbelastung*
eye wash *Augendusche*
eyeball *Augapfel*
eyebrow ['aɪbraʊ] *Augenbraue*
eyelash *Augenwimper*
eyelid *Augenlid*

F

fabric *Stoff*
facial hair ['feɪʃl] *Gesichtsbehaarung*
faeces [ˌfiːsiːz] *Kot*
faint [feɪnt] *schwach*
fallen arches ['fɔːlən ɑːtʃəʒ] *Senkfuß*
Fallopian tube [fəˌləʊpiən 'tjuːb] *Eileiter*
far-sightedness [fɑː 'saɪtɪdnəs] *Weitsichtigkeit*
fat *Fett*
fatal ['feɪtl] *tödlich*
fatigue *Schwäche, Müdigkeit*
fatty acid *Fettsäure*
femur ['fiːmə] *Oberschenkelknochen*
ferment *fermentieren*
fermenter *Fermentierer, Gärbehälter*
fern *Farn*
fertile *fruchtbar*
fertilization [ˌfɜːtəlaɪ'zeɪʃn] *Befruchtung*

fertilize **['fɜːtəlaɪz]** *befruchten*
fibre **['faɪbə]** *Faser*
fibrin **['faɪbrɪn]** *Fibrin*
fibrinogen **['faɪbrɪnədʒən]** *Fibrinogen*
fibrous root **['faɪbrəs]** *Faserwurzel*
fibula **['fɪbjələ]** *Wadenbein*
field of view *Gesichtsfeld*
field of vision **['vɪʒn]** *Gesichtsfeld*
field penny-cress *Acker-Hellerkraut*
filament **['fɪləmənt]** *Staubfaden*
finger bones *Fingerknochen*
fir **[fɜː]** *Tanne*
flank sb./sth. *jdn./etwas flankieren*
flap Klappe
flask **[flɑːsk ☆ flæsk]** *Glaskolben*
flat feet *Plattfüße*
flatbread *Fladenbrot*
fleshy *fleischig*
flexor *Beugemuskel*
fling sth. out *etwas hinauswerfen, etwas herausschleudern*
flip book *Daumenkino*
flooded *überflütet, überschwemmt*
flour **['flaʊə]** *Mehl*
flow chart *Fließdiagramm*
flower bud *Blütenknospe flower diagram Blütendiagramm*
flower stalk **[stɔːk]** *Blütenstiel*
flowering plant *Blütenpflanze*
flu *Grippe*
fluorescent screen **[ˌflɔː'resnt]** *Leuchtschirm*
fluoride **['flɔːraɪd]** *fluorhaltig*
fluorine **['flɔːriːn]** *Fluor*
flush sth. *etwas spülen*
foam **[fəʊm]** *Schaum(stoff)*
focus (on) sth. *etwas anvisieren, etwas scharf sehen, etwas scharf einstellen*
be in/out of focus *scharf/nicht scharf eingestellt sein*
focusing distance *Bezugssehweite*
fodder *Viehfutter*
foetus **['fiːtəs]** *Fötus*
folded *gefaltet*
foliage *Blätterdach; Blattwerk, Laub*
follow-up experiment *Anschlussexperiment*
food chain *Nahrungskette*
food remains (pl.) *Nahrungsreste*
forage **['fɒrɪdʒ ☆ 'fɔː-]** *Viehfutter*
forage crop **['fɒrɪdʒ ☆ 'fɔː-]** *Futterpflanze*
forceps (pl.) *Pinzette*
forefinger *Zeigefinger*
forehead *Stirn*
foreskin *Vorhaut*
forest *Wald*
forewing **[fɔːwɪŋ]** *Vorderflügel*
format *Format, Größe, Form*
fortified *angereichert*
fossil fuel **['fɒsl ☆ 'fɑːsl 'fjuːəl]** *fossiler Brennstoff*
foxglove *Fingerhut*
fractions of a second *Bruchteile einer Sekunde*
fracture **['fræktʃə]** *Knochenbruch*
open/closed fracture **['fræktʃə]** *offener/geschlossener Bruch*
fragrant **['freɪgrənt]** *duftend*
fraternal twins *zweieiige Zwillinge*
French fries (AE) **[fraɪʒ]** *Pommes Frites*
from underneath *von unten her*
fruit *Frucht*
fruit bat *Fruchtfledermaus*
functional defect *Funktionsstörung*
fungus (pl. fungi) **['fʌŋgəs]** *Pilz*
funnel *Trichter*
fur **[fɜː]** *Pelz*
furrow **['fʌrəʊ]** *Furche*
fuse **[fjuːz]** *miteinander verschmelzen*
fuzzy **['fʌzi]** *unklar, undeutlich, verschwommen*

G

gall bladder **[gɔːl]** *Gallenblase*
gambling *Glücksspiel*
gamete **['gæmiːt]** *Keimzelle*
garden bean *Gemüsebohne, Gartenbohne*
garden silverleaf *Garten-Silberblatt*
garlic mustard *Knoblauchhederich, Knoblauchsrauke*
gastric juice **['gæstrɪk]** *Magensaft*
gauze **[gɔːz]** *Gaze*
gender *Geschlecht*
generate sth. *etwas erzeugen*
genetic information *Erbinformation*
genital area *Genitalbereich, Intimbereich, Schambereich*
gently *zart*
geranium **[dʒə'reɪniəm]** *Geranie*
germ *Keim, Krankheitserreger*
germinate *keimen*
get rid of sth. *etwas loswerden*
get used to sth. *sich an etwas gewöhnen*
giant redwood *Mammutbaum*
give in (to sth.) *(einer Sache) nachgeben*
gland *Drüse*
glandular tissue **['glændjʊlə 'tiʃu]** *Drüsengewebe*
glans penis *Eichel*
glass tube **[glɑːs 'tjuːb ☆ glæs 'tuːb]** *Glasrohr*
glaucoma **[glɔː'kəʊmə ☆ glaʊ'koʊmə]** *grüner Star*
globe *Erdkugel, Globus*
glow *Leuchten*
glowing *glühend*
glucose **['gluːkəʊs ☆ -koʊz]** *Traubenzucker, Glukose*
glume *Hüllspelze, Hochblatt*
gluten *Kleber*
glycerol **['glɪsərɒl]** *Glyzerin*
glycogen **['glaɪkədʒin]** *Glycogen, 'tierische Stärke'*
go into labour *Wehen bekommen*
goggles *Schutzbrille*
gonorrhoea **[ˌgɒnə'rɪə ☆ ˌgɑːnə'riːə]** *Gonorrhö*
gooseberry **['gʊzbəri ☆ 'guːsberi]** *Stachelbeere*
goosegrass **['guːsˌgrɑːs]** *Klebkraut*
gorse **[gɔːs ☆ gɔːrs]** *Stechginster*
gospel **['gɒspl ☆ 'gɑːspl]** *Evangelium*
grade *Schulnote*
gradual *allmählich*
graft, grafting *Pfropfung*
graft sth. *etwas pfropfen*
grass *Gras*
grasshopper *Heuschrecke, Heupferd*
gravity **['grævəti]** *Schwerkraft*
graze *grasen*

grease sth. *etwas einfetten*
greenish *grünlich*
grind sth. (up) *etwas (zer)mahlen*
growth centre [grəʊθ] *Wachstumszentrum*
growth ring [grəʊθ] *Jahresring*
growth spurt [grəʊθ spɜːt] *Wachstumsschub*
guard cell [gɒːd] *Schließzelle, Spaltöffnungszelle*
guinea pig ['gɪni] *Meerschweinchen*
gullet ['gʌlɪt] *Speiseröhre*
gum disease [gʌm dɪ'ziːz] *Zahnfleischerkrankung*
gum(s) *Zahnfleisch*
gummy bear *Gummibärchen*
gut *Darm*
gynaecologist [ˌgaɪnə'kɒldzɪst] *Gynäkologe, Frauenarzt*
gypsum ['dʒɪpsəm] *Gips*

H

habit *Angewohnheit, Sucht, Gewohnheit*
haemoglobin [ˌhiːmə'gləʊbɪn ☆ -'gloʊ-] *Hämoglobin, roter Blutfarbstoff*
hand lens *Lupe*
handkerchief ['hæŋkətʃɪf, -tʃiːf] *Taschentuch*
hand-microtome [hænd 'maɪkrəˌtəʊm] *Handmikrotom*
hangover *Kater*
hard palate ['pælət] *harter Gaumen*
harden *härten*
hardy *widerstandsfähig*
harmful *schädlich*
harmony *Harmonie*
hawthorn ['hɔːθɔːn] *Weißdorn*
hay *Heu*
hazel *Hasel*
head (of a bone) *Gelenkkopf*
head (of a flowering plant) *Kopf, Korb*
headache ['hedeɪk] *Kopfschmerz*
heal *heilen*
health-centre staff [helθ 'sentə stɑːf ☆ stæf] *Angestellte/r einer Gesundheitsberatung*
healthy *gesund*
heart attack *Herzanfall*
heart disease [dɪ'ziːz] *Herzerkrankung*
heart rate *Herzfrequenz*
heartbeat *Herzschlag*
heat *Hitze*
heat-sensitive ['sensətɪv] *hitzeempfindlich*
hemp plant *Hanfpflanze*
herbaceous ['hɜː'beɪʃəs ☆ ɜːr'b-] *krautig*
herbarium *Herbar, Sammlung getrockneter Pflanzen*
herbivorous ['hɜː'bɪvərəs ☆ ɜːr'b-] *pflanzenfressend*
high-cholesterol [kə'lestərɒl ☆ -rɑːl] *mit viel Cholesterin*
high-pitched *hoch*
hinder sb./sth. from doing sth. *jdn./etwas davon abhalten, etwas zu tun*
hinge joint *Scharniergelenk*
HIV (human immunodeficiency virus) ['hjumən ɪˌmjuːnəʊdɪ'fɪʃnsi 'vaɪrəs] *HIV (humanes Immunschwäche Virus)*
hollow ['hɒləʊ ☆ 'hɑːloʊ] *hohl; Einbuchtung, Vertiefung, Mulde*
hollow back ['hɒləʊ ☆ 'hɑːloʊ] *Hohlrücken*
hooked *abhängig*
hormonal system ['hɔːməʊnl] *Hormonsystem*
hormone ['- -] *Hormon, Botenstoff*
horse chestnut *Rosskastanie*
horseradish ['hɔːsrædɪʃ] *Meerrettich*
hotplate *Heizplatte*
hover *schweben*
human consumption [kən'sʌmpʃn] *menschlicher Verzehr*
human papillomavirus [ˌpæpɪ'ləʊmə'vaɪrəs] *Human-Papilloma-Virus*
humerus ['hjuːmərəs] *Oberarmknochen*
hummingbird *Kolibri*
husk *Spelze*
hyacinth ['haɪəsɪnθ] *Hyazinthe*
hydrochloric acid [ˌhaɪdrəˌklorɪk 'æsɪd] *Salzsäure*
hymen ['haɪmən] *Jungfernhäutchen*
hypothesis (pl. hypotheses) [haɪ'poθəsɪs] *Hypothese*

I

identical twins *eineiige Zwillinge*
illegal *illegal*
image ['ɪmɪdʒ] *Bild, Abbild*
image distance *Bildweite*
immense [ɪ'mens] *unermesslich, ungeheuer*
impact *Aufprall*
impaired *beeinträchtigt*
impermeable [ɪm'pɜːmiəbl] *undurchlässig*
implant *sich einnisten*
implantation *Einnistung*
impression *Eindruck*
incisor [ɪn'saɪzə] *Schneidezahn*
incomplete fracture *unvollständiger Bruch*
inconspicuous *unauffällig, unscheinbar*
increase *zunehmen, steigern*
incredible *unglaublich*
indefinite [ɪn'defɪnət] *unbegrenzt, unbestimmt*
in-depth [ˌɪn'depθ] *detailliert*
index finger *Zeigefinger*
indicate sth. *etwas nachweisen*
indicator *Indikator*
industrial plant [ɪn'dvstriəl] *Industriebetrieb*
infatuation [ɪnˌfætʃu'eɪʃn] *Verliebtheit*
infect sb./sth *jdn./etwas infizieren, jdn./etwas anstecken*
infected *infiziert*
infection *Ansteckung, Infektion, Infekt, ansteckende Krankheit*
inflamed *entzündet*
inflammation *Entzündung*
inflorescence [ˌɪnflə'resəns] *Blütenstand, Blütezeit*
influenza [ˌɪnflu'enzə] *Grippe*
ingredient [ɪn'griːdiənt] *Zutat*
inhalation ['ɪndʒəri] *Inhalieren*

inhale (sth.) *(etwas) einatmen*
inherit sth. **[ɪn'herɪt]** *etwas erben*
inhibit sth. *etwas behindern, hemmen*
injured **['ɪndʒəd]** *verletzt*
injury *Verletzung*
inkjet-printer *Tintenstrahldrucker*
inner ear *Innenohr*
inner lining *Schleimhaut*
inner lining of the uterus **['juːterəs]** *Gebärmutterschleimhaut*
inner scale *Vorspelze*
insert sth. *etwas einführen, einlegen*
instruction *Anweisung*
insulation *Isolierung*
intensity *Stärke*
interact with sb./sth. *aufeinander einwirken, mit jdm./etwas interagieren*
intercostal muscles **[ˌɪntə'kɒstl ☆ -'kɑːstl 'mʌsls]** *Zwischenrippenmuskeln*
internode *Stängelabschnitt, Stängelglied*
interval *Zeitabstand*
intervertebral disc *Bandscheibe*
intestinal juice *Dünndarmsaft*
intimacy *Intimität*
intoxicating *berauschend*
intoxication *Rausch, Berauschtsein*
intra-ocular pressure (IOP) **[ˌɪntrə'ɒkjələ 'preʃə ☆ -'ɑːk-]** *Augeninnendruck*
intrauterine device (IUD) **[ˌɪntrə'jutraɪn dɪ'vaɪs]** *intrauterines Pessar, 'Spirale'*
invade sth. *in etwas eindringen*
investigate sth. *etwas untersuchen*
investigation *Untersuchung*
involuntary muscle *unwillkürlicher Muskel*
iodine *Jod*
iodine solution *Jodlösung*
iris **['aɪrɪs]** *Regenbogenhaut, Iris; Schwertlilie*
iron **['aɪən]** *Eisen*
iron oxide **['aɪən 'ɒksaɪd ☆ 'ɑːk-]** *Eisenoxid*
irregular **[ɪ'regjələ]** *unregelmäßig*
irrigation method *Durchzugstechnik (in der Mikroskopie)*
issue **['ɪʃuː]** *(Streit-)Frage; Thema; Ausgabe,*
itch *jucken*
item **['aɪtəm]** *Gegenstand, Objekt*

J

jackhammer *Presslufthammer*
jaw **[dʒɔː]** *Kiefer*
jawbone **['dʒɔːbəʊn]** *Kieferknochen*
jay *Eichelhäher*
jelly bear **[beə]** *Gummibärchen*
jellylike *geleeartig*
joint *Gelenk*
joint capsule **['kæpsjuːl]** *Gelenkkapsel*
judge sth. **[dʒʌdʒ]** *etwas beurteilen*
juice *Saft*
juicy *saftig*

K

kale *Grünkohl*
keep sth. afloat *etwas schwimmfähig halten*
keep sth. in store *etwas bevorraten*
Kentucky bluegrass *bläuliches Wiesenrispengras*
key term *Schlüsselwort*
kick (the ball) *(den Ball) schießen*
kidney *Niere*
kohlrabi **[ˌkəʊl'rɑːbi]** *Kohlrabi*

L

lab *Labor*
label sth. *etwas beschriften*
labia majora **['leɪbiə]** *große Schamlippen*
labia minora **['leɪbiə]** *kleine Schamlippen*
lacteal **['laːktiːl]** *Lymphkapillare*
lancet *Lanzettmesser, Lanzettnadel*
large intestine **[ɪn'testɪn]** *Dickdarm*
larva (pl. larvae) **['lɑːvə]** *Larve*
larynx **['lærɪŋks]** *Kehlkopf*
lateral root *Seitenwurzel*
latex rubber **['leɪteks]** *Latexgummi*
laurel **['lɒrəl]** *Lorbeer*
layer *Schicht*
layer of cartilage **['kɑːtɪlɪdʒ]** *Knorpelschicht*
leaf base *Blattgrund*
leaf margin **['mɑːdʒɪn]** *Blattrand*
leaf stalk **[stɔːk]** *Blattstiel*
left and right lungs *Lungenflügel*
legume **['legjuːm]** *Hülsenfrucht*
length **[leŋθ]** *Dauer, Länge*
lens *Linse*
lethal **['liːθl]** *tödlich*
lettuce **['letɪs]** *Salat*
leucocyte **['luːkəsaɪt]** *weißes Blutkörperchen*
leucoplast **['luːkoplæst]** *Leukoplast*
level out *sich ausgleichen*
lever *Hebel*
life-threatening *lebensbedrohlich*
ligament **['lɪgəmənt]** *Gelenkband*
light energy *Lichtenergie*
light microscope *Lichtmikroskop*
light-sensitive **['sensətɪv]** *lichtempfindlich*
ligule **['lɪgjuːl]** *Blatthäutchen*
lily of the valley *Maiglöckchen*
limb **[lɪm]** *Gliedmaße*
lime tree *Linde*
lime water *Kalkmilch, Kalkwasser*
line sth. *etwas auskleiden*
line graph *Kurvendiagramm*
linen **['lɪnɪn]** *Leinen, Leinwand*
lipase **['laɪpeɪz]** *Lipase*
liquid organ **['lɪkwɪd 'ɔːgən]** *flüssiges Organ*
liver *Leber*
liver cirrhosis **[sə'rəʊsɪs]** *Leberzirrhose*
livestock *Vieh, Viehbestand*
locust **['ləʊkəst ☆ 'loʊ-]** *Heuschrecke*
log sheet *Laufzettel*
lollipop **['lɒlipɒp ☆ 'lɑːlipɑːp]** *Lutscher*
long bone *Röhrenknochen*
longitudinal muscle **[ˌlɒŋgɪ'tjuːdɪnl ☆ ˌlɑːndʒətuːdənl]** *Längsmuskel*
longitudinal section **[ˌlɒŋgɪ'tjuːdɪnl ☆ ˌlɑːndʒətuːdənl]** *Längsschnitt*
long-sighted **['saɪtɪd]** *weitsichtig*

long-sightedness ['saɪtɪdnəs] *Weitsichtigkeit*
loosen ['lu:sn] *sich lockern, sich lösen*
loudness *Lautstärke*
love [lʌv] *Liebe*
in love *verliebt*
lower jaw *Unterkiefer*
lubricate sth. *etwas schmieren, etwas einfetten, etwas geschmeidig machen*
lung artery *Lungenarterie*
lung cancer ['lʌŋ kænsə] *Lungenkrebs*
lung circulation [ˌsɜ:kjə'leɪʃn] *Lungenkreislauf*
lung vein [veɪn] *Lungenvene*
lupin ['lu:pɪn] *Lupine*
lymph [lɪmf] *Lymphe, Gewebsflüssigkeit*
lymph capillary [lɪmf kə'pɪləri ☆ 'kæpəleri] *Lymphkapillare*
lymph gland [lɪmf] *Lymphdrüse*
lymph node [lɪmf] *Lymphknoten*
lymphatic system [lɪm'fætɪk] *Lymphsystem*
lymphocyte ['lɪmfəsaɪt] *Lymphozyt, Lymphzelle*

M

magnesium [mæg'ni:ziəm] *Magnesium*
magnetite ['magnɪtaɪt] *Magneteisenstein*
magnification [ˌmægnɪfɪ'keɪʃn] *Vergrößerung*
magnify sth. ['mægnɪfaɪ] *etwas vergrößern*
magnifying power ['mægnɪfaɪŋ] *Vergrößerungskraft*
main root *Hauptwurzel*
main vein [veɪn] *Hauptvene*
maintain sth. *etwas beibehalten, etwas aufrecht erhalten*
maize [meɪz] *Mais*
malt whisky *Malzwhisky*
maltose ['mɔ:ltəʊz] *Maltose*
marijuana [ˌmærə'wɑ:nə] *Marihuana*
marrow *Mark*
marrow cavity *Markhöhle*
marsh marigold [mɑ:ʃ 'mærigəʊld ☆ -goʊld] *Sumpfdotterblume*
mashed *zerdrückt, zerstampft*
masturbation [ˌmæstə'beɪʃn] *Masturbation, Selbstbefriedigung*
match *Streichholz*
mature [mə'tʃʊə] *ausgereift, voll entwickelt*
mature [mə'tʃʊə] *heranreifen*
mauve [məʊv] *hellviolett*
meadow ['medəʊ ☆ 'medoʊ] *Wiese*
meadow clary ['medəʊ 'kleɪri] *Wiesen-Salbei*
meadow grass *Wiesenrispengras*
measure sth. ['meʒə] *etwas messen*
measuring pipette ['meʒərɪŋ pɪ'pet] *Messpipette*
mechanical [mə'kænɪkl] *mechanisch*
mechanical injury [mə'kænɪkl 'ɪndʒəri] *mechanische Verletzung*
mechanism ['mekənɪzəm] *Mechanismus*
median canal ['mi:diən kə'næl] *Schneckengang*
medical care *medizinische Hilfe*
medical treatment *medizinische Behandlung*
medication *Medikamente*
medicine *Medikament*
medium *Milieu, Nährboden*
meniscus (pl. menisci) [mə'nɪskəs] *Meniskus*
menopause ['menəpɔ:z] *Wechseljahre*
menstrual cycle ['menstruəl] *Menstruationszyklus*
menstrual flow ['menstruəl] *Menstruationsblut*
menstruation [ˌmenstru'eɪʃn] *Menstruation, monatliche Blutung, Regelblutung*
mental retardation *geistige Behinderung*
meristematic tissue [ˌmerə'stəmætik tɪʃu] *Bildungsgewebe*
mesophyll ['mezəfɪl] *Mesophyll, Grundgewebe im Inneren des Blattes*
metabolic process [ˌmətə'bɒlɪk 'prəʊses ☆ -'bɑ:- 'prɑ:-] *Stoffwechselprozess*
metabolism [mə'tæbəlɪzəm] *Stoffwechsel*
metacarpal [ˌmetə'kɑ:pl] *Mittelhandknochen*
methylated spirit [ˌmeθəleɪtɪd 'spɪrɪt] *Brennspiritus*
methylene-blue solution ['meθəli:n] *Methylenblau-Lösung*
micrograph *mikroskopisches Foto, mikroskopisches Bild*
microscopic slide *Objektträger*
microscopic structure *Feinstruktur*
middle ear *Mittelohr*
migratory bird ['maɪgrətri, maɪ'greɪtəri ☆ 'maɪgrətɔ:ri] *Zugvogel*
milk teeth *Milchzähne*
mill *mahlen*
millet *Rispenhirse*
mind-altering *bewusstseinsverändernd*
mineral *Mineral*
mitochondrion (pl. mitochondria) [ˌmaɪtəʊ'kɒndriən ☆ -'kɑ:n] *Mitochondrium*
mitosis [maɪ'təʊsɪs] *Mitose*
mixture ['mɪkstʃə] *Mischung*
modelling clay *Knetmasse*
moderate *mäßig*
moist [mɔɪst] *feucht*
moisten sth. ['mɔɪsn] *etwas anfeuchten*
molar *hinterer Backenzahn*
molecule ['mɒlɪkju:l ☆ 'mɑ:l-] *Molekül*
monocot ['mɒnəkɒt ☆ 'mɑ:nokɑ:t] *einkeimblättrige Pflanze*
monoculture ['mɒnəkʌltʃə ☆ 'mɑ:n-] *Monokultur, Anbau nur einer Art*
monosaccharide *Einfachzucker*
mosaic [məʊ'zeɪɪk ☆ moʊ-] *Mosaik*
moss *Moos*

moth [mɒθ ☆ mɑːθ] *Nachtschmetterling*
motion sickness *Reisekrankheit*
motor neuron [ˌməʊtə'njʊərɒn ☆ ˌmoʊtə'nʊrɑːn] *motorische Nervenzelle*
mould [məʊld] *Schimmel*
mouldy [məʊldi] *schimmelig*
mount a specimen ['spesɪmən] *ein Präparat auf den Objektträger bringen*
mount sth. *etwas befestigen*
mountain ash *Eberesche, Vogelbeere*
mouth cavity *Mundhöhle*
mucous gland ['mjuːkəs] *Schleimdrüse*
mucous membrane ['mjuːkəs 'membreɪn] *Schleimhaut*
mucus ['mjuːkəs] *Schleim*
muffin tin *Muffinform*
multiples of (a number) *ein Vielfaches von (einer Zahl)*
muscle *Muskel*
muscle fibre ['mʌsl 'faɪbə] *Muskelfaser*
muscle strain ['mʌsl] *Muskelzerrung*
mustard *Senf, Ackersenf*

N

narration *Erzählung*
nasal cavity ['neɪzl 'kævəti] *Nasenhöhle*
nasal passage ['neɪzl 'pæsɪdʒ] *Nasengänge*
nasturtium [nə'stɜːʃəm] *Kapuzinerkresse*
nausea ['nɔːziə, 'nɔːsiə] *Übelkeit*
navel *Bauchnabel*
navigation [ˌnævɪ'geɪʃn] *Orientierung, Navigation*
near-sightedness ['saɪtɪdnəs] *Kurzsichtigkeit*
nectar *Nektar*
nectary *Nektarium*
neglect sth. *etwas vernachlässigen; etwas unerledigt lassen*
nerve *Nerv*
nerve cell *Nervenzelle*
nerve impulse [ˌnɜːv 'ɪmpʌls] *Nervenimpuls*
nervous system *Nervensystem*
network *Netzwerk*
neuron ['njʊərɒn ☆ 'nʊrɑːn] *Nervenzelle*
neutral ['njuːtrəl ☆ 'nuː-] *neutral*
neutral red solution ['njuːtrəl ☆ 'nuː-] *Neutralrot-Lösung*
nicotine ['nɪkətiːn] *Nikotin*
night blindness *Nachtblindheit*
nitrogen ['naɪtrədʒən] *Stickstoff*
node *Knoten*
noise level *Geräuschpegel*
nosebleed *Nasenbluten*
in note form *in Stichworten*
nucleus ['njuːkliəs ☆ 'nuːk-] *Zellkern*
numb *taub*
number sth. *etwas nummerieren*
numbness *Gefühllosigkeit*
nutmeg *Muskatnuss*
nutrient ['njuːtriənt ☆ 'nuː-] *Nährstoff*
nutritional value [nju'trɪʃənl ☆ nu-] *Nährwert*
nutritious [nju'trɪʃəs ☆ nu-] *nahrhaft*
nylon stocking *Nylonstrumpf*

O

oak *Eiche*
oats *Hafer*
obese [əʊ'biːs] *fettleibig*
obesity [əʊ'biːsəti] *Fettleibigkeit*
obtain sth. *etwas erhalten; etwas gewinnen*
occur *vorkommen*
odour ['əʊdə] *Geruch*
offspring *Nachwuchs, Nachkommenschaft*
omnivorous [ɒm'nɪvərəs ☆ ɑːm-] *allesfressend*
onset *Anfang, Beginn*
optic nerve *Sehnerv*
optical illusion *optische Täuschung*
orchid ['ɔːkɪd] *Orchidee*
organ *Organ*
organ system *Organsystem*
organelle *Zellorganell*
organism *Organismus*
orgasm *Orgasmus*
ornamental plant *Zierpflanze*
ostrich ['ɒstrɪtʃ ☆ 'ɑːs-] *Strauß*
outcome *Ergebnis*
outer ear *äußeres Ohr*
outer scale *Deckspelze*
oval window *ovales Fenster*
ovary (pl. ovaries) ['əʊvəri] *Eierstock, Fruchtknoten*
oven *Backofen*
overcome sth *etwas überwinden*
overlapping turns *Überkreuzungen*
overstretch *überdehnen*
overwhelming *überwältigend*
ovulation [ˌɒvju'leɪʃn ☆ ˌɑːv-] *Eisprung*
ovule ['ɒvjuːl ☆ 'oʊ-] *Samenanlage*
ox bile [ɒks ☆ ɑːks] *Ochsengalle*
oxygen ['ɒksɪdʒən ☆ 'ɑːks-] *Sauerstoff*
oxygenated ['ɒksɪdʒəneɪtɪd ☆ 'ɑːks-] *mit Sauerstoff angereichert*
oxygenated blood ['ɒksɪdʒəneɪtɪd blʌd ☆ 'ɑːks-] *sauerstoffreiches Blut*

P

pace *Schritt, Geschwindigkeit, Tempo*
pacemaker *Schrittmacher*
package *Verpackung*
pain receptor [rɪ'septə] *Schmerzrezeptor*
painkiller *Schmerzmittel*
palisade layer [ˌpælɪ'seɪd] *Palisadenschicht*
palm [pɑːm] *Handfläche; Palme*
pancreas ['pæŋkriəs] *Bauchspeicheldrüse*
pancreatic juice [ˌpæŋkri'ætɪk] *Bauchspeicheldrüsensaft*
panel discussion *Podiumsdiskussion*
panicle *Rispe*
panties (pl.) *Schlüpfer*
paper fastener *Rundkopfklammer*
parachute ['pærəʃuːt] *Fallschirm*
parallel *parallel*
paralysis [pə'ræləsɪs] *Lähmung*

paramecium (pl. paramecia) *Pantoffeltierchen, Paramaecium*
particle *Teilchen*
partition *Trennwand*
pass out *ohnmächtig werden*
passage ['pæsɪdʒ] *Gang, Trakt*
passive smoking ['pæsɪv 'sməukɪŋ] *Passivrauchen*
pasta ['pæstə, 'pɑːstə] *Nudeln, Teigwaren*
patch *Pflaster*
pathway *Übertragungsweg*
pave the way for sb./sth. *jdm./etwas den Weg ebnen*
pea [piː] *Erbse*
peel sth. *etwas schälen*
peer (usu. pl.) *Gleichaltrige/r*
peer group *Gruppe von Gleichaltrigen*
peer pressure ['preʃə] *Gruppenzwang*
pelvis *Becken*
penetrate sth. *etwas durchdringen, in etwas eindringen*
penis *Penis*
pepsin *Pepsin*
peptidase *peptidspaltendes Enzym*
perceive sth. [pə'siːv] *etwas wahrnehmen, etwas erkennen*
perception [pə'sepʃn] *Wahrnehmung*
perforate sth. ['pɜːfəreɪt] *etwas durchlöchern*
performance [pə'fɔːməns] *Leistungsfähigkeit, Leistungsvermögen*
period *Zeitraum; Periode, Menstruation*
peristalsis [ˌperɪ'stælsɪs] *peristaltische Bewegung*
permanent mount *Dauerpräparat*
permanent teeth *bleibende Zähne*
permanent tissue ['tɪʃuː, 'tɪsjuː] *Dauergewebe*
permission to do sth. *Erlaubnis, etwas zu machen*
petal ['petl] *Kronblatt, Blütenblatt*
petiole *Blattstiel*
Petri dish ['petri dɪʃ] *Petrischale*
petting *Petting, Streicheln der Sexualorgane*
phagocyte ['fægəsaɪt] *Fresszelle, Phagozyt*
pharynx ['færɪŋks] *Rachen*
phase *Abschnitt, Phase*
phenomenon [fə'nɒmɪnən ☆ -nɑːm-] *Erscheinung, Phänomen*
phloem ['fləuəm] *Siebteil*
phosphorus ['fɒsfərəs ☆ 'fɑːs-] *Phosphor*
photoreceptor cell ['fəutəurɪseptə ☆ 'foutou-] *Sehzelle, Lichtsinneszelle*
photosynthesis [ˌfəutəu'sɪnθəsɪs] *Photosynthese, Fotosynthese*
photosynthetic tissue [ˌfəutəu'sɪnθətɪk 'tɪʃuː, 'tɪsjuː] *Photosynthesegewebe*
phrase *Gruppe von Wörtern*
physical exercise ['fɪzɪkl 'eksəsaɪz] *körperliche Aktivität(en)*
physical ['fɪzɪkl] *körperlich*
phytoplankton *pflanzliches Plankton, Phytoplankton*
pickled cabbage ['kæbɪdʒ] *Sauerkraut*
pierce sth. ['piːəs] *etwas durchdringen*
pigeon ['pɪdʒɪn] *Taube*
pigment ['pɪgmənt] *Farbstoff*
pinch *Prise*
pinch sth. together *etwas abschnüren*
pine *Kiefer*
pinna ['pɪnə] *Ohrmuschel*
pipette [pɪ'pet] *Pipette*
pistil ['pɪstɪl] *Stempel*
pit *Grube*
pith *Mark*
pit organ *Grubenorgan*
pit viper *Grubenotter*
pitch *Tonhöhe, Stimmlage*
pitch *(bei einem Schiff) sich auf und ab bewegen,*
pith *Mark*
placenta [plə'sentə] *Plazenta, Mutterkuchen*
plant press *Pflanzenpresse*
plant succession [sək'seʃn] *Sukzession; Abfolge der Pflanzen, das nacheinander Auftreten von Pflanzenarten (in einem bestimmten Gebiet)*
plant-based *pflanzenbasiert*
plaque [plæk] *Zahnbelag*
plasma ['plæzmə] *Plasma*
plaster *Pflaster, Gipsverband*
plastid *Plastid*
platelet ['pleɪtlət] *Blutplättchen*
pleural membrane ['pluərəl 'membreɪn] *Lungen- und Rippenfell*
pliers (pl.) ['plaɪəz] *Zange*
plot sth. *etwas eintragen*
plumule *Sprossknospe*
pneumonia [njuː'məuniə ☆ nuː'mou-] *Lungenentzündung*
pointed scissors (pl.) ['sɪzəz] *Spitzenschere*
poisonous ['pɔɪzənəs] *giftig*
polenta [pə'lentə] *Polenta, Maisgericht*
polish sth. *etwas polieren, abschleifen*
pollen basket *Pollenkörbchen*
pollen sac *Pollensack*
pollen tube [tjuːb ☆ tuːb] *Pollenschlauch*
polypeptide [ˌpɒli'peptaɪd ☆ ˌpɑːl-] *Polypeptid*
polysaccharide *Vielfachzucker*
poplar ['pɒplə ☆ 'pɑːp-] *Pappel*
poppy ['pɒpi ☆ 'pɑːpi] *Mohn*
porridge ['pɒrɪdʒ ☆ 'pɔːr-] *Haferbrei*
postural defect ['pɒstʃərəl ☆ 'pɑːs-] *Haltungsschaden*
posture ['pɒstʃə ☆ 'pɑːs-] *Körperhaltung*
potassium [pə'tæsiəm] *Kalium*
potting soil *Topferde*
poultry ['pəultri] *Geflügel*
precipitation [prɪˌsɪpɪ'teɪʃn] *Niederschlag*
predator ['predətə] *Räuber*
prediction *Voraussage*
pregnancy ['pregnənsi] *Schwangerschaft*
premolar [priː'məulə] *vorderer Backenzahn*
prescribe sth. [prɪ'skraɪb] *etwas (ein Medikament) verschreiben*

prescription drug *verschreibungspflichtiges Medikament*
preserve sth. *etwas aufbewahren, konservieren*
pressure ['preʃə] *Druck*
pressure bandage ['preʃə 'bændɪdʒ] *Druckverband*
pressure receptor ['preʃə rɪ'septə] *Druckrezeptor*
prevalence ['prevələns] *Verbreitung*
prevent sth. *etwas verhindern, einer Sache vorbeugen*
previous ['pri:viəs] *vorhergehend*
prey *Beute*
primary teeth *Milchzähne*
procedure *Vorgehensweise, Ablauf*
process sth. ['prəʊses ☆ 'prɑ:ses] *etwas verarbeiten*
processing *Verarbeitung*
producer [prə'dju:sə] *Produzent*
product ['- -] *Produkt; Reaktionsprodukt*
project *hervorstehen*
project sth. onto sth. [-'-] *etwas auf etwas projizieren*
projection *Ausstülpung, Vorwölbung*
propel sth. [-'-] *etwas vorwärts treiben*
the pros and cons *das Pro und Kontra*
prostate gland *Prostata*
protease *eiweißspaltendes Enzym*
protein *Protein*
proverb *Sprichwort*
pruning shears (pl.) [ʃɪəz] *Gartenschere, Baumschere*
psychological [ˌsaɪkə'lɒdʒɪkl ☆ -'lɑ:dʒ-] *psychisch*
psychosis [saɪ'kəʊsɪs] *Psychose*
puberty ['pju:bəti] *Pubertät*
pubic hair *Schambehaarung*
puff of wind *Windstoß*
pulmonary artery ['pʌlmənəri ☆ -neri] *Lungenarterie*
pulmonary vein ['pʌlmənəri ☆ -neri] *Lungenvene*
pulp *Zahnmark*
pulp cavity *Zahnhöhle*
pulse [pʌls] *Puls(schlag)*
pulse rate at rest [pʌls] *Ruhepuls*
pulse-wave [pʌls] *Pulswelle*
pump [pʌls] *Pumpe*
pumpkin pie *Kürbiskuchen*
pupil ['pju:pl] *Pupille*
purpose ['pɜ:pəs] *Zweck*
put a strain on sth. *etwas belasten*
python ['paɪθən ☆ -θɑ:n] *Python*

Q

query ['kwɪəri ☆ 'kwɪri] *Suchanfrage*

R

radially ['reɪdiəli] *strahlenförmig*
radicle *Keimwurzel*
radio wave *Funkwelle*
radioactivity *Radioaktivität*
radish ['rædɪʃ] *Radieschen*
radius ['reɪdiəs] *Speiche*
range *Spannbreite; Angebot, Auswahl*
rape *Raps; Vergewaltigung*
rapeseed oil *Rapsöl*
rapeseed plant *Raps*
rash *Ausschlag*
raspberry ['rɑ:zbəri ☆ 'ræzberi] *Himbeere*
rate of consumption *Geschwindigkeit, mit der etwas konsumiert wird*
ratio ['reɪʃiəʊ] *Verhältnis*
ray *Strahl*
razor blade *Rasierklinge*
react *reagieren*
reactant [ri'æktənt] *Reaktionspartner*
receive sth. [rɪ'si:v] *etwas erhalten*
receptacle [rɪ'septəkl] *Blütenboden*
reception [rɪ'sepʃn] *Aufnahme*
receptor (cell) [rɪ'septə] *Rezeptor(zelle); Sinneszelle, die einen Reiz aufnimmt*
recipe ['resəpi] *Rezept*
recipient [rɪ'sɪpiənt] *Empfänger*
reckless *leichtsinnig*
recognize sth. *etwas erkennen*
recommendation *Empfehlung*
record ['rekɔ:d ☆'rekərd] *Nachweis, Beleg*
record sth. [rɪ'kɔ:d] *etwas aufschreiben, notieren*
rectum ['rektəm] *Mastdarm*
red blood cell [blʌd] *rotes Blutkörperchen*
red cabbage ['kæbɪdʒ] *Rotkohl*
red-green colour blindness *Rotgrünblindheit*
refine sth. *etwas veredeln, aufbereiten, raffinieren*
reflect on sth. *über etwas nachdenken*
regenerate [rɪ'dʒenəreɪt] *sich regenerieren*
regeneration [rɪˌdʒenə'reɪʃn] *Erneuerung, Regeneration*
regulate sth. *etwas regulieren*
relapse *Rückfall*
in relation to sth. *im Verhältnis zu etwas*
relax [rɪ'læks] *sich entspannen*
release sth. *etwas freisetzen*
reliability *Zuverlässigkeit*
relief [rɪ'li:f] *Erleichterung*
relieve [rɪ'li:v] *erleichtern*
rely on sb./sth. [rɪ'laɪ] *sich auf jdn./etwas verlassen, auf jdn./etwas angewiesen sein*
remain *übrig bleiben*
remedy *Heilmittel, Arzneimittel*
remove sth. [-'-] *etwas entfernen*
renewable energy [rɪ'nju:əbl 'enədʒi] *erneuerbare/nachwachsende Energie*
replace sth. by sth. *etwas mit etwas ersetzen*
reproduce [ˌri:prə'dju:s ☆ -'du:s] *sich fortpflanzen*
resin *Harz*
resist sth. *etwas widerstehen*
resistance *Widerstand*
resolving power *Auflösungsvermögen*
resource [rɪ'sɔ:s ☆ 'ri:sɔ:rs] *Rohstoffquelle, Bodenschatz*
respiration [ˌrespə'reɪʃn] *Atmung*
external respiration [ɪk'stɜ:nl ˌrespə'reɪʃn] *äußere Atmung*
internal respiration *innere Atmung*

internal respiration *innere Atmung*
respiratory system [re'spɪrətri ☆ 'respərətɔ:ri] *Atmungssystem*
respire [rɪ'spaɪə] *atmen*
respond to sb./sth. *auf jdn./etwas ansprechen*
retain sth. *etwas beibehalten*
retina ['retɪnə ☆ 'retənə] *Netzhaut*
review sth. [rɪ'vju:] *etwas überprüfen, nochmals durchsehen*
revision *Überprüfung*
rhesus factor ['ri:səs] *Rhesusfaktor*
rhizome ['raɪzəʊm] *Rhizom*
rhubarb ['ru:bɑ:b] *Rhabarber*
ribcage *Brustkorb*
ribs *Rippen*
rice *Reis*
right away *sofort*
rigid ['rɪdʒɪd] *rigide, unflexibel*
ring muscle *Ringmuskel*
rinse sth. [rɪns] *etwas spülen, etwas abspülen, etwas ausspülen*
rockery *Steingarten*
rod *Stäbchen*
rod-shaped *stabförmig*
root *Wurzel; Zahnwurzel; sich bewurzeln, wurzeln*
root cap *Wurzelhaube*
root hair *Wurzelhaar*
root tip *Wurzelspitze*
rooting powder *Mittel zur Förderung der Wurzelbildung*
rootstock *Unterlage*
rose family *rosenartige Gewächse*
rosette *Rosette*
rotate *kreisen, rotieren*
roughage ['rʌfɪdʒ] *Ballaststoffe*
round back *Rundrücken*
round window *rundes Fenster*
row [rəʊ ☆ roʊ] *Zeile, Reihe*
rub *reiben*
rubber bung *Gummistopfen*
rubber sheet *Gummihaut*
rubber tubing ['tju:b- ☆ 'tu:b-] *Gummischlauch*
runner *Ausläufer*
runner bean *Stangenbohne*
rustle *rascheln*
rye [raɪ] *Roggen*

S

saccule ['sakju:l] *Vorhofsäckchen*
saclike *sackartig*
saddle joint *Sattelgelenk*
safety goggles ['seɪfti ˌɡɒglz ☆ ˌɡɑ:glz] *Schutzbrille, Sicherheitsbrille*
saliva [sə'laɪvə] *Speichel*
salivary gland [sə'laɪvəri] *Speicheldrüse*
sample *Probe*
sanitary pad *Monatsbinde*
sanitary towel ['taʊəl] *Monatsbinde*
satisfaction *Befriedigung*
saturated fatty acid ['sætʃəreɪtɪd] *gesättigte Fettsäure*
savoy cabbage [sə'vɔɪ 'kæbɪdʒ] *Wirsing*
saying *Sprichwort*
scab *Kruste*
scalpel ['skælpəl] *Skalpell*
scanning electron microscope (SEM) *Raster-Elektronenmikroskop*
scar *Narbe*
scar tissue ['tɪʃu:, 'tɪsju:] *vernarbtes Gewebe*
scarlet runner *rotblühende Stangenbohne, Feuerbohne*
scatter sth. *etwas verstreuen*
scent [sent] *Duft, Duftstoff*
scented ['sentɪd] *duftend*
schematic drawing [ski:'mætɪk] *schematische Zeichnung*
scion ['saɪən] *Edelreis*
sclera ['sklɪərə] *weiße Augenhaut*
scoliosis [ˌskəʊli'əʊsɪs] *seitliche Verkrümmung der Wirbelsäule, Skoliose*
score a goal *ein Tor schießen*
scrape sth. off *etwas abkratzen*
screen *Schirm*
scrotum ['skrəʊtəm] *Hodensack*
seasickness *Seekrankheit*
secrete sth. [sɪ'kri:t] *etwas absondern, ausscheiden*
secretion [sɪ'kri:ʃn] *Sekretion, Abgabe ins Körperinnere*
sectioning razor *Rasiermesser für mikroskopische Schnitte*
seed *Samen*
seed fibre ['faɪbə] *Samenfaser*
seed head *Ähre, Fruchtstand*
seedling *Keimling*
seizure ['si:ʒə] *Anfall*
selective dye [daɪ] *selektive Färbung*
selective stain *selektive Färbung*
self-deception *Selbsttäuschung*
self-pollination *Selbstbestäubung*
semen ['si:men] *Sperma*
semicircular [ˌsemi'sɜ:kjələ] *halbkreisförmig*
semicircular canals [ˌsemi'sɜ:kjələ kə'næls] *Bogengänge*
semi-fluid ['semi 'flu:ɪd] *halbflüssig*
seminal vesicles ['semɪnl 'vesɪklz] *Bläschendrüsen*
sense organ *Sinnesorgan*
sensitive ['sensətɪv] *empfindlich*
sensory cell ['sensəri] *Sinneszelle*
sensory neuron ['sensəri 'njʊərɒn ☆ 'nʊrɑ:n] *Sinnesnervenzelle*
sepal ['sepl] *Kelchblatt*
sequence ['si:kwəns] *Filmausschnitt*
serving *Portion*
set of teeth *Gebiss*
setting *Szenario, Situation*
settle *sich niederlassen*
severe [sɪ'vɪə] *schwer, heftig, schwerwiegend*
sex *Geschlecht*
sexual abuse ['sekʃuəl ☆ ə'bju:s] *sexueller Missbrauch*
sexual arousal ['sekʃuəl] *sexuelle Erregung*
sexual harassment ['sekʃuəl] *sexuelle Nötigung*
sexual intercourse ['sekʃuəl] *Geschlechtsverkehr*
sexually transmitted disease (STD) ['sekʃuəli] *Geschlechtskrankheit*
shaft (of a bone) *Knochenschaft*
shape *Form*
sheath [ʃi:θ] *Blattscheide, Hülle, Scheide*
shed sth. *etwas ausscheiden, etwas abwerfen*
shepherd's purse ['ʃepədz pɜ:s] *Hirtentäschelkraut*
shin bone *Schienbeinknochen*
shock absorber *Stoßdämpfer*

shoe insert *Schuheinlage*
shoot *Spross*
shoot tip *Sprossspitze*
short cut *Abkürzung*
short-sighted **['saɪtɪd]** *kurzsichtig*
short-sightedness **['saɪtɪdnəs]** *Kurzsichtigkeit*
short-term *kurzfristig*
shoulder blade *Schulterblatt*
shred sth. *etwas hechseln, zerkleinern*
shrink *schrumpfen*
side root *Seitenwurzel*
sieve **[sɪv]** *Sieb*
sieve tubes **[sɪv tju:bz ☆ tu:bz]** *Siebröhren*
silage **['saɪlɪdʒ]** *Silage, fermentiertes Tierfutter*
silo *Silo, Getreidespeicher*
simulate sth. **['sɪmjuleɪt]** *etwas simulieren*
single sugar *Einfachzucker*
single-edge razor blade *Rasierklinge mit einer scharfen Seite*
skeletal muscle *Skelettmuskel*
ski goggles **['ski: ˌgɒglz ☆ ˌgɑ:glz]** *Skibrille*
skill *Fertigkeit, Fähigkeit*
skull *Schädel*
slender *schlank, schmal*
slice sth. *etwas in Scheiben schneiden*
slippery *gleitfähig, schlüpfrig*
small intestine **[ɪn'testɪn]** *Dünndarm*
small red radish *Radieschen*
smooth meadow grass *bläuliches Wiesenrispengras*
smoothie *cremiges Getränk mit frischen Früchten*
snail *Schnecke*
snapshot *Momentaufnahme*
snort sth. *etwas (eine Droge wie Kokain) schnupfen*
snowdrop *Schneeglöckchen*
sober *nüchtern*
socket *Gelenkpfanne; Hohlraum für den Zahn im Kiefer*
sodium **['səʊdiəm]** *Natrium*
sodium hydroxide solution **['səʊdiəm haɪ'drɒksaɪd ☆ -'drɑ:k-]** *Natronlauge*
soft drink *Limonade*
soft palate **['pælət]** *weicher Gaumen, Gaumensegel*
solid **['sɒlɪd ☆ 'sɑ:l-]** *fest*
soluble **['sɒljəbl ☆ 'sɑ:l-]** *löslich*
solution *Lösung*
solvent *Lösungsmittel*
soothe sth. **[su:ð]** *etwas beruhigen, etwas lindern*
sore throat **[sɔ: θrəʊt ☆ θroʊt]** *Halsschmerzen, Halsentzündung*
sorghum **['sɔ:gəm]** *Mohrenhirse*
sound energy *Schallenergie*
sound intensity *Lautstärke*
sound level *Geräuschpegel*
source *Quelle*
soya bean (BE) soy bean (AE) *Sojabohne*
spatula **['spætʃələ]** *Spatel*
specimen **['spesɪmən]** *Untersuchungsobjekt, Einzelprobe, Exemplar*
spectacles (pl.) *Brille*
sperm *Samen, Spermie, Spermium; Samenflüssigkeit mit Spermien*
sperm duct *Spermienleiter*
spice *Gewürz*
spikelet *Ährchen*
spinal cord *Rückenmark*
spindle-shaped *spindelförmig*
spine *Stachel*
spine of a book *Buchrücken*
spleen *Milz*
splint *Schiene*
split sth. *etwas teilen, etwas aufspalten*
spoke *Speiche*
spongy **['spʌndʒi]** *schwammig, weich*
spongy bone tissue **['spʌndʒi 'tɪʃu:]** *schwammiges Knochengewebe*
spongy layer **['spʌndʒi]** *Schwammschicht*
spontaneous **[spɒn'teɪniəs ☆ spɑ:n]** *spontan*
sprain *verstauchen*
sprain *Verstauchung*
spruce *Fichte*
squeeze sth. *(sich oder etwas) durchzwängen*
squirrel **['skwɪrəl ☆ 'skwɜ:rəl]** *Eichhörnchen*
stability *Festigkeit*
stage *Stadium*
stage of dormancy *Ruhestadium*
stalk *Halm, Stiel, Stängel*
stamen **['steɪmən]** *Staubblatt*
stand out *hervortreten*
staple food *Grundnahrungsmittel*
starch **[stɑ:tʃ]** *Stärke*
starch grain **[stɑ:tʃ]** *Stärkekorn*
state *Zustand*
stationary *stationär, unbeweglich*
stem *Stamm*
stereoscopic vision **[ˌsteriə'skɒpɪk ☆ -'skɑ:pɪk]** *räumliches Sehen*
stereotype *Klischee, Stereotyp*
sterile **['steraɪl]** *keimfrei, steril*
stick to sth. *an etwas festhalten*
stigma **['stɪgmə]** *Stigma*
stimulate sth. **['stɪmjuleɪt]** *etwas reizen*
stimulus (pl. stimuli) **['stɪmjələs]** *Reiz*
stir sth. *etwas rühren*
stirrup **['stɪrəp]** *Steigbügel*
stock *Levkoje*
stoma (pl. stomata) *Spaltöffnung*
stomach **['stʌmək]** *Magen*
upset stomach *verdorbener Magen*
stomach acid *Magensäure*
stomach lining *Magenschleimhaut*
stopper *Stöpsel, Stopfen*
store *Vorrat*
store sth. *etwas speichern, etwas lagern*
straighten **['streɪtn]** *(sich) aufrichten; strecken*
strand *Strang, Faden*
straw *Stroh*
strengthening tissue *Stützgewebe*
stroke *Gehirnschlag, Schlaganfall*
style *Griffel*
substance addiction *stoffgebundene Sucht*
subtropical *subtropisch*
succession *Abfolge*
suck sth. *etwas saugen*
suction bulb **['sʌkʃn bʌlb]** *Saugball*

suction pump ['sʌkʃn pʌmp] *Saugpumpe*
sufficient [sə'fɪʃnt] *ausreichend*
sugar solution *Zuckerlösung*
suggestion [sə'dʒestʃən ☆ səg'dʒ-] *Vorschlag*
supply sth. [sə'plaɪ] *etwas liefern, versorgen*
surface area ['sɜ:fɪs] *Oberfläche, Grundfläche*
surgical ['sɜ:dʒɪkl] *chirurgisch*
surplus ['sɜ:pləs] *überschüssig*
surround sth. *etwas umgeben*
survey *Umfrage*
survive *überleben*
suspensory ligament [sə'spensəri 'lɪgəmənt] *Linsenband*
swallow ['swɒləʊ ☆ 'swɑ:loʊ] *schlucken*
swap ideas *Ideen austauschen*
swap roles [swɒp ☆ swɑ:p] *die Rollen tauschen*
sweat [swet] *Schweiß*
sweat glands [swet] *Schweißdrüsen*
sweet pea [pi:] *Wicke*
swell *anschwellen*
sycamore ['sɪkəmɔ:] *Ahorn*
symmetry ['sɪmətri] *Symmetrie*
synovial fluid [saɪ'nəʊviəl 'flu:ɪd] *Gelenkschmiere*
syphilis ['sɪfɪlɪs] *Syphilis*

T

table *Tabelle*
tablespoon *Esslöffel*
tag *Anhänger*
take in sth. *etwas aufnehmen, etwas in sich aufnehmen*
talk sth. over with sb. *etwas mit jdm. besprechen*
tampon ['tæmpɒn ☆ -pɑ:n] *Tampon*
tap *Absperrhahn*
tap sth. *etwas klopfen*
taproot *Pfahlwurzel*
tar *Teer*
tarsals ['tɑ:sls] *Fußwurzelknochen*
tear [teə] *reißen*
tear duct [tɪə ☆ tɪr] *Tränennasengang*
tear fluid [tɪə ☆ tɪr] *Tränenflüssigkeit*
tear gland [tɪə ☆ tɪr] *Tränendrüse*
technique [tek'ni:k] *Methode, Verfahren*
tectorial membrane *Deckmembran*
temperate *gemäßigt*
temperature receptor ['temprətʃə rɪ'septə] *Temperaturrezeptor*
temporary mount *Frischpräparat, zeitlich begrenzt haltbares Präparat*
tendon *Sehne*
tendril *Blattranke*
tension ['tenʃn] *Spannung*
test tube *Reagenzglas*
test-tube holder *Reagenzglashalter*
test-tube rack *Reagenzglasständer*
testa *Samenschale*
testis (pl. testes) ['testɪs] *Hoden*
testosterone [te'stɒstərəʊn ☆ te'stɑ:stəroʊn] *Testosteron*
thatched roof *reedgedecktes Dach*
'the pill' *Antibabypille*
thigh bone [θaɪ] *Oberschenkelknochen*
thistle ['θɪsl] *Distel*
thoroughly ['θʌrəli ☆ 'θɜ:r-] *gründlich*
thresh sth. *etwas dreschen*
threshold of hearing ['θreʃhəʊld ☆ -hoʊld] *Hörgrenze*
threshold of noise pain ['θreʃhəʊld ☆ -hoʊld] *Schmerzgrenze*
thrombocyte ['θrɒmbəʊsʌɪt ☆ 'θrɑ:mboʊ-] *Blutplättchen*
thrombosis [θrɒm'bəʊsɪs ☆ θrɑ:mboʊ-] *Thrombose*
thump [θʌmp] *pochen*
tibia ['tɪbiə] *Schienbeinknochen*
tighten ['taɪtn] *sich straffen*
tilted *geneigt*
in time *rechtzeitig*
tiny *winzig*
on tiptoes *auf Zehenspitzen*
tissue ['tɪʃu:, 'tɪsju:] *Gewebe*
toe bones *Zehenknochen*
tofu *Tofu*
tolerance *Toleranz*
tong *Zange*
tonsils ['tɒnsls ☆ 'tɑ:nsls] *Mandeln*
tool *Werkzeug, Gerät*
tooth decay *Zahnfäule, Karies*
topple sth. over *etwas umkippen*
touch receptor [rɪ'septə] *Tastrezeptor*
touch-me-not *Springkraut*
tough [tʌf] *zäh*
toxic shock syndrome (TSS) ['sɪndrəʊm ☆ -droʊm] *Toxic-Shock-Syndrom*
toxin ['tɒksɪn ☆ 'tɑ:ksən] *Giftstoff*
trachea [trə'ki:ə ☆ 'treɪkiə] *Luftröhre*
transmission *Übertragung*
transmission electron microscope (TEM) *Transmissions-Elektronenmikroskop*
transparent [træns'pærent] *durchsichtig*
transpiration *Verdunstung (über die Spaltöffnungen)*
trap sb./sth. *jdn./etwas fangen*
treat *Leckerbissen*
treat sb./sth. *jdn./etwas behandeln*
treatment *Behandlung*
tremendous *enorm*
triceps ['traɪseps] *Armstrecker*
tripod ['traɪpɒd ☆ -pɑ:d] *Dreifuß*
trunk *Baumstamm*
trustworthy ['trʌstwɜ:ði] *vertrauenswürdig*
tube-like *röhrenförmig*
tuber *Knolle*
tuberculosis [tju:ˌbɜ:kju'ləʊsɪs ☆ tu:ˌbɜ:rkjə'loʊsɪs] *Tuberkulose*
tulip ['tju:lɪp ☆ 'tu:-] *Tulpe*
tumour ['tju:mə ☆ 'tu:mə] *Geschwulst, Tumor*
tuning fork ['tju:nɪŋ fɔ:k ☆ 'tu:nɪŋ] *Stimmgabel*
turnip ['tɜ:nɪp] *weiße Rübe*
tweezers (pl.) *Pinzette*
twig [twɪg] *dünner Zweig*
twisted *gebogen, gedreht*
two-dimensional *flächig, zweidimensional*

U

ulna ['ʌlnə] *Elle*
ultrasonic [ˌʌltrə'sɒnɪk ☆ -'sɑːn-] *mit Ultraschallqualität, Ultraschall-*
ultrasound scan *Ultraschalluntersuchung*
ultraviolet [ˌʌltrə'vaɪələt] *ultraviolett*
ultraviolet radiation [ˌʌltrə'vaɪələt ˌreɪdi'eɪʃn] *ultraviolette Strahlung*
umbilical cord [ʌmˌbɪlɪkl] *Nabelschnur*
unbearable [ʌn'beərəbl] *unerträglich*
uncoiled *abgerollt*
undergo sth. *etwas durchmachen, etwas erleben*
underneath *darunter (gelegen)*
underweight [ˌʌndə'weɪt] *Untergewicht*
unfavourable [ʌn'feɪvərəbl] *ungünstig, negativ*
unicellular organism [ˌjuːnɪ'seljələ] *Einzeller*
unimaginable [ˌʌnɪ'mædʒɪnəbl] *unvorstellbar*
unique [ju'niːk] *einzigartig*
unknown *unbekannt*
unpredictable *unberechenbar*
unsaturated fatty acid [ˌʌn'sætʃəreɪtɪd] *ungesättigte Fettsäure*
unscented [ˌʌn'sentɪd] *ohne Duft*
upper jaw [dʒɔː] *Oberkiefer*
uproot sth. *etwas entwurzeln*
upside-down *umgedreht*
upwards ['ʌpwədz] *aufwärts*
urea [jʊ'riːə] *Harnstoff*
urethra [jʊ'riːθrə] *Harnröhre, Harnspermienröhre*
urge [ɜːdʒ] *(starker) Drang*
urinary bladder ['jʊərɪnəri] *Harnblase*
uterus ['juːtərəs] *Gebärmutter*
utricle ['juːtrɪkl] *Utriculus*
utter sth. *etwas ausstoßen*

V

vacuole ['vækjuəʊl] *Zellsaftraum, Vakuole*
vagina [və'dʒaɪnə] *Vagina, Scheide*
value ['væljuː] *Wert*
values (pl.) *Werte, Überzeugungen*
valve [vælv] *Ventil*
variegated ['veəriəgeɪtɪd] *panaschiert*
variety *Vielfalt*
vary *variieren*
vascular bundle ['væskjələ] *Gefäßbündel*
vegetative ['vedʒɪtətɪv] *vegetativ*
vein [veɪn] *Vene*
vein of a leaf [veɪn] *Blattader*
Velcro™ fastener *Klettverschluss*
vena cava [ˌviːnə 'keɪvə] *Hauptvene*
ventricle ['ventrɪkl] *Herzkammer*
versatile ['vɜːsətaɪl ☆ -tl] *vielseitig*
vertebra (pl. vertebrae) ['vɜːtɪbrə] *Wirbelkörper*
vertebral column ['vɜːtɪbrəl 'kɒləm ☆ 'kɑːl-] *Wirbelsäule*
vertebrate ['vɜːtɪbrət] *Wirbeltier*
vertical ['vɜːtɪkl] *senkrecht, vertikal*
vessel *Gefäß, Ader*
vestibular apparatus [vɛ'stɪbjʊlə ˌæpə'reɪtəs ☆ -'rætəs] *Vestibularapparat*
vestibular nerve [vɛ'stɪbjʊlə nɜːv] *Vestibularnerv*
vibrate [vaɪ'breɪt ☆ 'vaɪbreɪt] *vibrieren*
victim *Opfer*
villus (pl. villi) ['vɪləs] *Darmzotte*
virus ['vaɪrəs] *Virus*
visible *sichtbar*
visor ['vaɪzə] *Schutzschirm, Blendschutz*
visual resolution ['vɪʒuəl] *Auflösungsvermögen, Sehschärfe*
visualize sth. ['vɪʒuəlaɪz] *etwas veranschaulichen, bildlich darstellen*
vital ['vaɪtl] *lebenswichtig, vital*
vital capacity ['vaɪtl kə'pæsəti] *maximaler Lungeninhalt*
vital stain ['vaɪtl steɪn] *Vitalfärbung, Färbung eines lebenden Objekts*
vitamin ['vɪtəmɪn ☆ 'vaɪt-] *Vitamin*
vitreous humour ['vɪtrɪəs 'hjuːmə] *Glaskörper*
vocal cord/s ['vəʊkəl] *Stimmband/Stimmbänder*
voice box *Kehlkopf*
volume ['vɒljuːm ☆ 'vɑːljəm] *Lautstärke; Volumen*
voluntary muscle *willkürlicher Muskel*
vomit ['vɒmɪt ☆ 'vɑːm-] *sich erbrechen*
vulva ['vʌlvə] *weibliche Scham*

W

wallflower *Goldlack*
walnut *Walnuss*
waste [weɪst] *Abfall(stoff)*
waste ground *Brachland, Ödland*
waste material *Abfallstoff, Reststoff*
waste product *Abfallprodukt*
waste substance ['sʌbstəns] *Abfallstoff*
water *Wasser*
water lily *Wasserlilie*
water resistant *wasserfest, wasserundurchlässig*
water vapour ['veɪpə] *Wasserdampf*
wavelength ['weɪvleŋθ] *Wellenlänge*
wax *Wachs*
waxy *wachsartig*
wear sth. down [weə] *etwas abnutzen*
wet mount *Feuchtpräparat*
wet pack *feuchter Wickel*
wheat [wiːt] *Weizen*
white blood cell [blʌd] *weißes Blutkörperchen*
white cabbage ['kæbɪdʒ] *Weisskohl*
white radish ['rædɪʃ] *Rettich*
wholegrain product *Vollkornprodukt*
wholemeal *Vollkorn*
whorl [wɜːl] *Blattquirl*
wild cabbage ['kæbɪdʒ] *Wildkohl*
wild carrot *wilde Möhren*
wild radish ['rædɪʃ] *Hederich*
willow *Weide*
wilt *welken*
windpipe *Luftröhre*
wire ['waɪə] *Metalldraht*

with the naked eye *mit bloßen Auge*
withdrawal symptoms [wɪð'drɔːəl] *Entzugserscheinungen*
wither ['wɪðə] *welken, verwelken*
withstand sth. *etwas aushalten, widerstehen*
womb [wuːm] *Gebärmutter*
wood *Wald*
wooden splint *Holzspan*
woody plant *Holzpflanze*
woolly burdock ['wʊli 'bɜːdɒk] *Filz-Klette*
wound [wuːnd] *Wunde*

X

x-axis (pl. axes) *x-Achse*
x-ray *Röntgenstrahl*
xylem ['zaɪləm] *Holzteil, Gefäßteil*

Y

y-axis (pl. axes) *y-Achse*
yeast [jiːst] *Hefe*
yellow spot *gelber Fleck*

German-English vocabulary

A

etwas abbauen break sth. down
Abfallprodukt waste product
Abfallstoff waste, waste substance, waste material
etwas abfangen absorb sth.
abhängig addicted, hooked
abhängig machen be addictive
etwas abklemmen clamp sth.
abnehmen decrease
etwas abnutzen wear sth. down
Abschnitt phase
etwas abschnüren pinch sth. together
etwas absondern secrete sth.
Absonderung secretion
etwas abspülen rinse sth.
abwechselnd alternately
etwas abwerfen discard sth., shed sth.
Achse axis (pl. axes)
Achselhöhle armpit
Adaptation adaptation
alkalisch basic, alkaline
Alkohol am Steuer driving under the influence of alcohol (DUI)
Alkoholismus alcoholism
Alkoholvergiftung alcohol poisoning
allesfressend omnivorous
allmählich gradual
Anbau cultivation
etwas anbauen cultivate sth.
andauernd continuous
Anfall seizure
etwas anfeuchten moisten sth.
angemessen adequate
auf jdn./etwas angewiesen sein rely on sb./sth.
angrenzend adjacent
Angst anxiety
ängstlich anxious
etwas anlegen apply sth.
Anordnung arrangement
(sich an) etwas anpassen adjust (to) sth.
Anpassung adaptation
etwas ansammeln accumulate sth.
auf jdn./etwas ansprechen respond to sb./sth.
jdn./etwas anstecken infect sb./sth
ansteckend infectious
Ansteckung infection
Antibabypille birth-control pill, 'the pill'
Antibiotikum antibiotic
etwas anvisieren focus (on) sth.
Anweisung instruction
anwenden apply
Apotheke chemist's (BE); drugstore (AE)
Arzneimittel drug
Atem breath
Atemzug breath
atmen respire, breathe
Atmosphäre atmosphere
Atmung respiration, breathing
etwas aufbewahren preserve sth.
etwas aufdecken detect sth.
aufgerichtet erect
aufgerollt coiled
etwas auflösen dissolve sth.
sich auflösen dissolve
Aufnahme absorption
etwas in sich aufnehmen take sth. in
Aufprall impact
aufrecht erect
etwas aufrecht erhalten maintain sth.
etwas aufrollen coil sth. up
etwas aufsaugen absorb sth.
etwas aufschreiben record sth.
etwas aufspalten split sth.
etwas aufspüren detect sth.
aufwärts upwards
mit bloßem Auge with the naked eye
Augendusche eye wash
ausatmen exhale
sich ausdehnen expand
Ausgabe issue
mit jdm. ausgehen ask sb. out
ausgereift mature
etwas ausgleichen equalize sth.
etwas aushalten endure sth., withstand sth.

etwas auslöschen extinguish sth.
ausreichend sufficient
etwas ausscheiden excrete sth.
Ausscheidung excretion
Ausschlag rash
Aussehen appearance
unerlässlich essential
jdn./etwas einer Sache aussetzen expose sb./sth. to sth.
etwas ausspülen rinse sth.
Ausstellung display
Auswahl range
Auswertung evaluation
automatisch automatic(ally)

B

Backofen oven
Backpulver baking powder
Bakterium bacterium (pl. bacteria)
Ballaststoffe dietary fibre, roughage
basisch basic
Baumschere pruning shears (pl.)
Bebauung cultivation
Becher beaker
sich mit etwas befassen deal with sth.
etwas befestigen attach sth., mount sth.
Befinden condition
sich befinden be located
Befriedigung satisfaction
etwas befruchten fertilize sth.
Befruchtung fertilization
jdn./etwas behandeln treat sb./sth.
Behandlung treatment
etwas behindern inhibit sth.
etwas beibehalten retain sth.
etwas belasten put a strain on sth.
Belastung exposure
berauschend intoxicating
etwas beruhigen soothe sth.
Besäufnis binge drinking
etwas beschädigen damage sth.
beschädigt damaged
etwas beschriften label sth.
etwas mit jdm. besprechen talk sth. over with sb.
etwas bestimmen determine sth.
Bestimmungsbuch guidebook
etwas beugen bend sth.
etwas beurteilen judge sth.
Beute prey
Bewegung exercise
bewusste Entscheidung commitment
Bewusstsein consciousness; awareness
etwas biegen bend sth.
sich biegen curve
Bild image
Biobrennstoff biofuel
Biogas biogas
biologische Forschung biological research
Blatt leaf
Blattader vein (of a leaf)
Blattwerk foliage
Blendschutz visor
Blindprobe blank sample
blinzeln blink
Blutalkoholkonzentration blood alcohol concentration (BAC)
Blutbahn bloodstream
Blutdruck blood pressure
Blutgefäß blood vessel
Blutgerinnsel blood clot
Blutgruppe blood group/type
Blutkreislauf blood circulation
Blutstrom bloodstream
Blutung bleeding
Bodenschatz resource
Botenstoff hormone
breit broad
Brennstoff fuel
Brille spectacles (pl.)
Bronchitis bronchitis
Bruch fracture
brüchig crumbly
Bruchteil fraction
Bulimie bulimia
Bunsenbrenner Bunsen burner

C

chemisch chemical
chemische Verbindung chemical compound
chirurgisch surgically
Chlorophyll chlorophyll
Cholesterin cholesterol
Chromosom chromosome

D

darunter (gelegen) underneath
Dauer length
Dehnen (of the eye) dilate
den Ball schießen kick the ball
Deodorant antiperspirant
detailliert in-depth
deutlich evident
Dezibel decibel
dicht dense
die Rollen tauschen swap roles
differenzieren differentiate
Differenzierung differentiation
diskusförmig disc-shaped
Dosis dose
etwas dreschen thresh sth.
Droge drug
Drogerie chemist's (BE); drugstore (AE)
Druck pressure
Drüse gland
duftend fragrant, scented
dünn thin, slender
etwas durchdringen penetrate sth., pierce sth.
Durchfall diarrhoea
etwas durchführen conduct sth.
etwas durchlöchern perforate sth.
etwas durchmachen undergo sth.
Durchmesser diameter
Durchschnitt average
im Durchschnitt on average
durchschnittlich average
durchsichtig transparent
Dürre drought

E

Ecstasy ecstasy
etwas einatmen inhale sth.
etwas in etwas einbetten embed sth. in sth.
in etwas eindringen invade sth., penetrate sth.
Eindruck impression
eineiige Zwillinge identical twins
etwas einführen insert sth.
einjährig annual
jdn. einladen ask sb. out
etwas einlegen insert sth.
etwas (scharf) einstellen focus sth.

Einwirkung exposure
Einzelprobe specimen
einzigartig unique
Eisprung ovulation
Empfänger recipient
empfindlich delicate, sensitive
Energie energy
Energiebedarf energy requirement
energiereich energy-rich
(ein Kind) entbinden deliver (a baby)
etwas entfernen remove sth.
Entgiftung detoxification
etwas enthalten contain sth.
Enthaltsamkeit abstinence
etwas (von etwas) entnehmen extract sth. (from sth.)
entscheidend decisive
sich entspannen relax
etwas entwurzeln uproot sth.
Entzugserscheinungen withdrawal symptoms
entzündet inflamed
Entzündung inflammation
etwas erben inherit sth.
Erbinformation genetic information
sich erbrechen vomit
Ergebnis outcome
etwas erhalten receive sth.
Erkältung cold
etwas erkennen detect sth., perceive sth., recognize sth.
Erlaubnis permission
erleichtern relieve
Erleichterung relief
jdm. ermöglichen, etwas zu tun enable sb. to do sth.
Ernährung diet
ausgewogene Ernährung balanced diet
erneuerbare Energie renewable energy
Erneuerung regeneration
Ernte crop, harvest
etwas (mit etwas) ersetzen replace sth. (by sth.)
etwas ertragen endure sth.
erweitern (of the eye) dilate
etwas erzeugen generate sth.
Essattacke binge-eating attack
essbar edible
Esslöffel tablespoon
Essstörung eating disorder
Esssucht bulimia
euphorisch euphoric
Exemplar specimen
etwas (von etwas) extrahieren extract sth. (from sth.)

F

Fähigkeit capability, skill
jdn./etwas fangen trap sb./sth.
Farbstoff dye
Faser fibre
faulen decay
Ferment enzyme
fermentieren ferment
Fertigkeit skill
fest solid
an etwas festhalten stick to sth.
Festigkeit stability
etwas feststellen determine sth.
Fett fat
fettleibig obese
Fettleibigkeit obesity
feucht moist
fleischfressend carnivorous
Fließdiagramm flow chart
Form shape, form
Fortpflanzung reproduction
sich fortpflanzen reproduce
fossiler Brennstoff fossil fuel
Fotosynthese photosynthesis
Fötus foetus
Frauenarzt gynaecologist
etwas freisetzen release sth.
Frucht fruit
fruchtbar fertile
Funktionsstörung disorder, functional defect
Futterpflanze forage crop

G

Gaze gauze
Gebärmutter uterus, womb
gebogen twisted
Geburtstermin due date
gedreht twisted
gefaltet folded
Gefäß vessel
Gefühllosigkeit numbness
Gegendroge antidrug
Gegensatz contrast
im Gegensatz zu etwas in contrast to sth.
Gegenstand object, item
Gehirnschlag stroke
geistige Behinderung mental retardation
Gelenk joint
Genitalbereich genital area
Geräuschpegel noise level, sound level
Gericht dish
Geruch odour
gesättigt saturated
Geschlecht gender
Geschlechtskrankheit sexually transmitted disease (STD)
Geschlechtsverkehr sexual intercourse
Gesichtsfeld field of view, field of vision
gesund healthy
Getränk beverage, drink
Getreideprodukt cereal
getrennt apart
Gewebe tissue
Gewohnheit habit
gewölbt dome-shaped, curved
giftig poisonous
Giftstoff toxin
Gipsverband cast
Gleichung equation
gleitfähig slippery
Glücksspiel gambling
glühend glowing
Gras grass
grasen graze
Grippe influenza, flu
grob coarse
grundlegend basic
gründlich thorough(ly)
Grundnahrungsmittel staple food
grüner Blattfarbstoff chlorophyll
Gruppenzwang peer pressure
Gummischlauch rubber tubing
Gummistopfen rubber bung
Gynäkologe gynaecologist

H

an etwas haften cling to sth.
halbflüssig semi-fluid
Halsschmerzen sore throat
Haut dermis, skin
heilen heal
Heizplatte hotplate
etwas hemmen inhibit sth.
heranreifen mature
hervorstehen project
hervortreten stand out
Herzanfall heart attack
Herzschlag heartbeat
etwas hinauswerfen fling sth. out
hitzeempfindlich heat-sensitive
hohl hollow
Höhle cavity
hörbar audible
Hörgrenze threshold of hearing
Hormon hormone
Hülle sheath
Husten cough
Hypothese hypothesis (pl. hypotheses)

I

illegal illegal
Infektion infection
jdn./etwas infizieren infect sb./sth
infiziert infected
inhalieren inhale
mit jdm./etwas interagieren interact with sb./sth.
Intimbereich genital area

J

jährlich annual
jenseits beyond
jucken itch
Jugendzeit adolescence

K

Kammer chamber
Kapsel capsule
Kater hangover
kauen chew
Keim germ
keimen germinate
keimfrei sterile
Keimling seedling
mit etwas klarkommen come to terms with sth.
Klettverschluss Velcro™ fastener
Klumpen clump, clot
Knetmasse modelling clay
Knochenbruch fracture
Komasaufen binge drinking
kompliziert complex
Kondom condom
etwas konservieren preserve sth.
Konsument consumer
kontrahieren contract
Kontrolle control
Kontrolluntersuchung check-up
kontrovers controversial
konzentrisch concentric
Koordination coordination
Kopfschmerz headache
körpereigen body-specific
Körperhaltung posture
Körperkreislauf body circulation
körperlich physical
körperliche Aktivität(en) physical exercise, physical activity
körperliche Intimität physical intimacy
Krankheitserreger germ
Krebs(erkrankung) cancer
kreisen rotate
kreisförmig circular
Kuchenblech baking tray
etwas kultivieren cultivate sth.
Kultur culture
kulturelle Herkunft cultural background
künstlich artificial
kuppelartig dome-shaped
Kurvendiagramm line graph
kurzfristig short-term
kurzsichtig short-sighted, nearsighted
Kurzsichtigkeit short-sightedness, near-sightedness

L

Labor laboratory, lab
etwas lagern store sth.
länglich elongated
Laub foliage
Laubbaum broad-leafed tree
Laufzettel log sheet
Lautstärke loudness, sound intensity, volume
lebensbedrohlich life-threatening
lebenswichtig vital
leichtsinnig reckless
Leiden condition
Leistungsfähigkeit performance
Leuchten glow
lichtempfindlich light-sensitive
etwas liefern supply sth.
etwas lindern soothe sth.
sich lockern become loose, loosen
sich lösen loosen
löslich soluble
Lösung solution
etwas loswerden get rid of sth.
luftdicht airtight
Luftschadstoff air pollutant
Lungenkrebs lung cancer
Lupe hand lens

M

verdorbener Magen upset stomach
Magensäure stomach acid
Magersucht anorexia nervosa
magersüchtig anorexic
etwas mahlen grind sth.
Mangel deficiency
Mangelkrankheit deficiency disease
mäßig moderate
Masturbation masturbation
mechanisch mechanical
Mechanismus mechanism
Medikament drug, medicine, medication
verschreibungspflichtiges Medikament prescription drug
nicht verschreibungspflichtiges Medikament over-the-counter drug
medizinisch medical
medizinische Behandlung medical treatment
medizinische Hilfe medical care
Mehl flour
Menge amount

Menstruation menstruation, period
Menstruationszyklus menstrual cycle
etwas messen measure sth.
Methode technique
mikroskopisches Foto micrograph
Milchprodukt dairy product
Mineral mineral
Mischung mixture
Monatsbinde sanitary towel/pad
Müdigkeit fatigue
Muskel muscle
Muskelfaser muscle fibre
Muskelzerrung muscle strain

N

über etwas nachdenken reflect on sth.
(jdm./etwas) nachgeben give in (to sth.)
Nachkommenschaft offspring
Nachteil disadvantage
nachwachsende Energie renewable energy
etwas nachweisen indicate sth.
Nachwuchs offspring
Nadelbaum coniferous tree
Nahaufnahme close-up
nahrhaft nutritious
Nährstoff nutrient
Nebenprodukt by-product
Nervensystem nervous system
Netzwerk network
neutral neutral
nicht allergisch antiallergic
sich niederbeugen bend down
Niederschlag precipitation
Notfall emergency
etwas notieren record sth.
nüchtern sober
etwas nummerieren number sth.
Nutzen benefit
Nutzen aus etwas ziehen benefit from sth.

O

Objektträger microscope slide
offensichtlich evident
ohne Duft unscented
ohnmächtig werden pass out
Ohnmachtsanfall blackout
Ökosystem ecosystem
Opfer victim
optische Täuschung optical illusion
Organ organ
Organismus organism
Organsystem organ system
Orientierung navigation

P

parallel parallel
passend compatible
Periode period
Petrischale Petri dish
pflanzenfressend herbivorous
Pflaster plaster, patch, bandage
Phänomen phenomenon
Phase phase
Pinzette tweezers (pl.), forceps (pl.)
Pipette pipette
platzen burst
etwas polieren polish sth.
Portion serving
Präpariernadel dissection needle
Probe sample
Produkt product
profitieren von etwas benefit from sth.
etwas auf etwas projizieren project sth. onto sth.
psychisch psychological
pumpen pump

Q

Quelle source
Querschnitt cross section

R

Radioaktivität radioactivity
etwas raffinieren refine sth.
Rasierklinge razor blade
Rasiermesser für mikroskopische Schnitte sectioning razor
jdn./etwas zu Rate ziehen consult sb./sth.
Ratschlag advice
Räuber predator
Raubvogel bird of prey
Rausch intoxication
Reagenzglas test tube
Reagenzglashalter test-tube holder
Reagenzglasständer test-tube rack
reagieren react
sich regenerieren to regenerate
etwas regulieren regulate sth.
reiben rub
reif mature
etwas reißen tear sth.
Reiz stimulus (pl. stimuli)
etwas reizen stimulate sth.
Reststoff waste material
Rezept recipe
rigide rigid
röhrenförmig tube-like
Röntgenstrahl x-ray
Rohstoffquelle resource
rotieren rotate
Rückfall relapse
Rückgang decline
ruhend dormant
Ruhepuls pulse rate at rest
etwas rühren stir sth.
rund round

S

sackartig saclike
Saft juice
saftig juicy
etwas säubern cleanse sth.
sauer acidic
Sauerstoff oxygen
saugen suck
saugfähig absorbent
Saugpumpe suction pump
Säuregehalt acidity
säurehaltig acidic
schädlich harmful
etwas schälen peel sth.
scharf eingestellt sein be in of focus
nicht scharf eingestellt sein be out of focus
etwas scharf stellen focus sth.
etwas schätzen estimate sth.
etwas in Scheiben schneiden slice sth.
Schicht layer

Schirm screen
Schlaganfall stroke
schlucken swallow
schlüpfrig slippery
Schlüsselwort key term
Schmerzgrenze threshold of pain/ tolerance
Schmerzmittel painkiller
etwas schmieren lubricate sth.
Schnupfen (common) cold
etwas schnupfen snort sth.
schrumpfen shrink
Schutzbrille goggles, safety goggles
Schutzschirm visor
schwach faint
schwammig spongy
Schwangerschaft pregnancy
schweben hover
schwerwiegend severe
Schwindel dizziness
sich nach etwas sehnen crave sth.
Selbstbefriedigung masturbation
senkrecht vertical
sexuelle Erregung sexual arousal
sichtbar visible
etwas simulieren simulate sth.
sinken decrease
Skalpell scalpel
Spalte column
Spannbreite range
Spannung tension
etwas speichern store sth.
Spirale intrauterine device (IUD)
spontan spontaneous
Sportler athlete
spülen flush
Stadium stage
Stärke intensity; starch
stationär stationary
staubig dusty
steigen increase
Stoffaufnahme absorption
Stoffwechsel metabolism
Stöpsel stopper
etwas straffen tighten sth.
etwas strecken straighten sth.
Streichholz match
Streitpunkt issue
stumpf blunt
subtropisch subtropical
Sucht addiction
süchtig machen be addictive

T

Tabelle table
Tampon tampon
taub numb
teilen divide
etwas teilen durch etwas divide sth. by sth.
Tempo pace
Thema issue
tödlich fatal, lethal
Toleranz tolerance
Traubenzucker glucose
trüb clouded

U

etwas überdehnen overstretch sth.
übereinstimmen mit etwas correspond to sth.
überflütet flooded
überleben survive
etwas überprüfen review sth.
überschüssig surplus
überschwemmt flooded
Übertragung transmission
überwältigend overwhelming
etwas überwinden overcome sth.
üblich common
übrig bleiben remain
im Uhrzeigersinn clockwise
entgegen dem Uhrzeigersinn anti-clockwise
etwas umbauen convert sth.
Umfrage survey
umgeben surround
Umgebung environment
umgedreht upside-down
etwas umschließen enclose sth.
umstritten controversial
etwas umwandeln convert sth.
unauffällig inconspicuous
unbefriedigend dissatisfying
unbegrenzt indefinite
unbekannt unknown
unberechenbar unpredictable
unbestimmt indefinite
unbeweglich stationary
undurchlässig impermeable
uneben bumpy
unerlässlich essential
unermesslich immense
unflexibel rigid
ungeschlechtlich asexual
unglaublich incredible
ungünstig unfavourable
unklar blurry, clouded, fuzzy
unregelmäßig irregular
unscharf blurred
unscheinbar inconspicuous
Untergewicht underweight
jdn./etwas untersuchen examine sb./sth.
Untersuchung examination, investigation
Untersuchungsobjekt specimen
ununterbrochen continuous
unvorstellbar unimaginable

V

variieren vary
etwas verändern alter sth.
etwas verankern anchor sth.
etwas veranschaulichen visualize sth.
etwas verarbeiten process sth.
etwas verbinden fuse sth.
Verbraucher consumer
etwas verbreiten disperse sth.
verdampfen evaporate
etwas verdauen digest sth.
Verdauung digestion
verdünnen dilute
Verdunstung (über eine freie Oberfläche) evaporation
etwas veredeln refine sth.
Verfahren technique
Vergewaltigung rape
etwas vergrößern magnify sth.
Vergrößerung magnification
Verhalten behaviour
im Verhältnis zu etwas in relation to sth.
etwas verhindern prevent sth.
Verhütungsmittel contraceptive
etwas verklumpen clump sth. together, clot sth.

sich auf jdn./etwas verlassen rely on sb./sth.
verletzt injured
Verletzung injury
jdn./etwas vernachlässigen neglect sb./sth.
verschlechtern deteriorate
verschmelzen fuse
jdm. etwas verschreiben prescribe sb./sth.
verschwommen fuzzy, blurred, blurry
verschwommene Sicht blurred vision
etwas verstärken amplify sth.
etwas verstauchen sprain sth.
Verstopfung constipation
etwas verstreuen disperse sth., scatter sth.
etwas verteilen distribute sth.
vertikal vertical
Verzehr consumption
verzehrbar edible
etwas verzerren distort sth.
verzerrt distorted
verzweigt branched
vibrieren vibrate
Viehfutter fodder, forage
Vielfalt variety
Voraussage prediction
einer Sache vorbeugen prevent sth.
vorderer Backenzahn premolar
Vorfahre ancestor
vorhergehend previous
vorkommen occur
Vorrat store
Vorschlag suggestion
Vorteil advantage, benefit
etwas vorwärts treiben propel sth.

W

wachsartig waxy
etwas wahrnehmen perceive sth.
Wahrnehmung perception
Wald forest, wood
wasserfest water resistant
wasserundurchlässig water resistant
Watte cotton wool
Wattestäbchen cotton bud
jdm./etwas den Weg ebnen pave the way for sb./sth.
Wehen (pl.) contractions
Wehen bekommen go into labour
weitsichtig long-sighted, far-sighted
Weitsichtigkeit long-sightedness, far-sightedness
welken wilt
Werkzeug tool
Wert value
jdm./etwas widersprechen contradict sb./sth.
Widerstand resistance
widerstandsfähig hardy; resistant
etwas widerstehen resist sth., withstand sth.
sich wiederherstellen regenerate
Winkel angle
winzig tiny
Wirbeltier vertebrate
wirtschaftlich economical
Wölbung arch, curvature
Wunde wound
wünschenswert desirable

Z

zäh tough
Zahnseide dental floss
Zange pliers (pl.), tong
zart gentle
Zeile row
Zeitabstand interval
Zeitraum period
etwas zerdrücken mash sth.
etwas zerlegen dissect sth.
etwas zermahlen grind sth. up
etwas zerren tear sth.
etwas zerstampfen mash sth.
etwas zerstören destroy sth.
Zierpflanze ornamental plant
Zucht breeding
etwas züchten breed sth.
zunehmen increase
jdn./etwas etwas zuordnen assign sb./sth. to sth.
Zusage commitment
mit jdm./etwas zusammenstoßen collide with sb./sth.
sich zusammenziehen contract
Zustand condition, state
Zutaten ingredients
Zuverlässigkeit reliability
Zweck purpose
zweidimensional two-dimensional
zweieiige Zwillinge fraternal twins
zylinderförmig cylindrical

Abbreviations and symbols

AE	American English
BE	British English
°C	degree(s) Celsius, centigrade
cm	centimetre(s)
e.g.	(Latin) *exempli gratia* = for example
g	gram(s)
jdm.	*jemandem*
jdn.	*jemanden*
kg	kilogram(s)
kJ	kilojoule(s)
km	kilometre(s)
ml	milliliter(s)
mm	millimeter(s)
mm^2	square millimetre(s)
p., pp.	page, pages
pl.	plural
sb.	somebody
sing.	singular
sth.	something
x	times
☆	American pronunciation follows

Regular differences in American pronunciation – as for instance the [r] after vowels – have not been given.

Acknowledgements

Bildquellenverzeichnis

Fotos: adpic Bildagentur, Bonn: S. 120 u. (B. Leitner); S. 122 Fig. 16 (M. Baumann) | Agentur Focus, Hamburg: S. 122 Fig. 17 (Tipsimages); S. 129 Fig. 33 (SPL/Gschmeissner); S. 139 Fig. 52 (Scharf); S. 149 Fig. 4 a (Siu / Science Photo Library), b (CNRI / Science Photo Library); S. 153 Fig. 11 (Curtis / SPL); S. 174 Fig. 34 (Claude Nuridsany & Marie Perennou / Science Photo Library); S. 214 u. re. (SPL); S. 226 Fig. 5 a (SPL), Fig. 5 b (J. Berger/MPI/SPL) | Alamy, Abington: S. 10, 5. Bild (Jupiterimages/Goodshoot), S. 14 (Jim Lane); S. 15 (Mark Andersen / Rubberball); S. 33 re. o. (Paul Rapson), re. u. (Ashley Cooper); S. 34 Fig. 6 (Medical-on-Line); S. 66 Fig. 15 (Dinodia Images); S. 149 u. (Ace Stock Limited); S. 152 (topdog images); S. 175 u. (Wildlife GmbH) | Archiv Hoyer: S. 170 Fig. 22; S. 171 Fig. 25 (Leo, F.); S. 178 Fig. 43 | arifoto.de: S. 156 Fig. 14 (Michael Reichel) | H. Bellmann: S. 170 Fig. 21 | Bevilacqua: S. 103 Fig. 16 a (Cedri) | Blickwinkel: S. 159 o. (Layer); S. 194 Fig. 10 (J. Flohe) | Buff, W., Biberbach: S. 179 Fig. 47; S. 214 o., u. li.; S. 235 Fig. 19 re.; S. 237 Fig. 22 | Buhtz, A.: S. 162 Fig. 5 (BMU) | Bundesanstalt für Arbeitsschutz und Arbeitsmedizin (BAuA), Dortmund: S. 131 Fig. 38 a-b | Bundeszentrale für gesundheitliche Aufklärung (BzgA), Köln: S. 107 Fig. 22 | Corbis, Düsseldorf: Titel links (© Charles Krebs); Titel rechts (© Bill Varie); S. 9 o. (Renee Lynn); S. 65 Fig. 14-2 (David Stoecklein); S. 154 Fig. 13 li. (Lueger) | Cornelsen Archiv, Berlin: S. 9 Mi li. (Corel), Mi. re. (Corel); S. 138 Fig. 50; S. 141 Fig. 58-4; S. 211 o. re.; S. 218 Fig. 14; S. 215 Fig. 9; S. 216 Fig. 10 li., Fig. 11 | Crown copyright material is reproduced with the permission of the Controller of HMSO and Queen's Printer for Scotland: S. 62 Fig. 12 | V. Döring, Hohen Neuendorf: S. 23 Fig. 14; S. 25 Fig. 18; S. 28 Fig. 24; S. 33 Fig. 5; S. 37 Fig. 10; S. 52 Fig. 13; S. 54 Fig. 15; S. 71-72 (alle); S. 77 Fig. 5-6; S. 99 Fig. 11; S. 105 Fig. 20 r.; S. 124 Fig. 23, 24; S. 131 Fig. 37; S. 135 Fig. 44-45; S. 136 Fig. 47; S. 137 Fig. 49; S. 151 Fig. 6; S. 163 Fig. 8; S. 184 Fig. 55; S. 203 Fig. 32; S. 206 Fig. 40; S. 209 Fig. 43; S. 211 Mi.; S. 219 Fig. 15-17; S. 221 Fig. 20-22; S. 235 Fig. 19 li. | T. Eisner, Ithaca, NY (Cornell University): S. 139 Fig. 54 a-b | eye of science: S. 67 Fig. 16 (NIBSC/SPL) | Getty Images, München: S. 59 Fig. 4 (Tetra Images); S. 60 Fig. 6 (Teubner); S. 150 Fig. 5 (SSPL via Getty Images) | G. v. Hagens, Institut für Plastination, Heidelberg: S. 34 Fig. 7 (beide) | K. Heinzel: S. 186 Fig. 60 | J. Hollatz, Heidelberg: S. 120 Fig. 11 a-b; S. 136 Fig. 46; S. 185 Fig. 55 | iStockphoto, Calgary: S. 65 Fig. 14-4 (Dmitriy Shironosov); S. 153 Fig. 9 (robotboot); S. 159 Mi. (lubilub); S. 165 u. re. (hazel proudlove); S. 169 Fig. 19 (Roger Whiteway); S. 172 Fig. 27 re. (Georg Hanf); S. 178 Fig. 44 (Wojtek Kryczka); S. 187 Fig. 62 (TT); S. 191 Fig. 2 (Denis Jr. Tangney); S. 197 Fig. 20 brussels (David T. Gomez); S. 198 Fig. 21 (BasieB); S. 202 Fig. 29 (Bernhard Weber), Fig. 30 (Antema Photography) | IFA-Bilderteam: Titel Mitte (© Beauty Archive) I IWF Wissen und Medien gGmbH, Göttingen: S. 214 o. re. | IWF Institut für wissenschaftliche Fotografie Kage, Lauterstein: S. 95 Fig. 7; S. 214; S. 226 Fig. 5 d | Ishihara's Test for Colour-Blindness: S. 126 Fig. 29 | J. Jaenicke: S. 171 Fig. 26 | Joker, Bonn: S. 154 Fig. 13 re. (Gudrun Petersen) | Keystone, Hamburg: S. 144 li; S. 151 Fig. 7 (Volkmar Schulz) | David Kjaer (rspb-images.com): S. 175 Fig. 37 | W. Kleesattel, Schwäbisch Gmünd: S. 140 Fig. 57; S. 141 Fig. 58-6 | R. Kleinert, Hainburg: S. 12 | G. Klepel: S. 239 Fig. 26 | K. Kühntopp, Essen: S. 141 Fig. 58-3 | LEO Elektronenmikroskopische GmbH, Oberkochen: S. 217 Fig. 12 | J. Lieder: S. 11 o. re., o. Mi., o. li; S. 161 Fig. 3; S. 211 u.; S. 212 Fig. 1; S. 230 Fig. 12 (beide); S. 231 Fig. 13 u.li, u.re; S. 241 Fig. 29 | LifeART: S. 128 Fig. 31-32 (Lippincott Williams) | medicalpicture, Köln: S. 39 Fig. 13 | Nature & Science, Vaduz: S. 189 Fig. 63, S. 205 Fig. 37 | Naturfotografie Hecker: S. 226 Fig. 5 c (Sauer) | naturganznah.de: S. 195 Fig. 14; S. 196 Fig. 18; S. 198 Fig. 22 | Okapia, Frankfurt a.M.: S. 11 o. li (NAS/Biophoto); S. 45 Fig. 3; S. 102 Fig. 15; S. 103 Fig. 16 d (OSF/Bromhall); S. 118 Fig. 8; S. 144 re. (Weinhäupl); S. 176 Fig. 39 / S. 206 Fig. 41 (Naturbild/Geduldig); S. 178 Fig. 45; S. 217 Fig. 13 (Kage), S. 211 o. li./S. 228 Fig. 6 (NAS/Porter), S. 228 Fig. 7 (Biophoto Ass./Science Sov.); S. 231 Fig. 13 o. re. (NAS/Biophoto) | Oxford University Press: S. 86 Fig. 16; S. 101 Fig. 14 Ron Pickering, Complete Biology, OUP, 2000. Reprinted by permission of OUP | Pfletschinger/Angermayer: S. 171 Fig. 24 | PhotoAlto: S. 144 Mi. (Katarina Sundelin) | Photolibrary, London: S. 40 Fig. 14; S. 58 Fig. 1 (Corbis); S. 65 Fig. 14-1 (Hervé de Gueltzl / Photononstop), 14-3 (David Madison); S. 174 Fig. 32 (Satoshi Kuribayashi); S. 178 Fig. 46 (Philippe Bonduel); S. 195 Fig. 17 (De Agostini Editore) | picture-alliance, Frankfurt a.M.: S. 10, 4. Bild (Herbert Schwind / Okapia KG); 7. Bild (Aribert Jung/Klett); S. 59 Fig. 2, S. 104 Fig. 19 (dpa); S. 127 Fig. 30 (Medicalpicture) | Photoshot, London. S. 64 Fig. 13a | Pixelio.de: S. 200 Fig. 25 | Prentice Hall Health: S. 148 Fig. 3 a-b (Susan Tapert) | Prisma: S. 119 Fig. 10 a-b (Schröder) | Project Photos, Augsburg: S. 21 Fig. 12 (Reinhard Eisele); S. 64 Fig. 13b; S. 124 Fig. 22, S. 169 Fig. 18 (2 Fotos von Reinhard Eisele) | Reinhard-Tierfoto, Heiligkreuzsteinach: S. 141 Fig. 58-5; S. 172 Fig. 27 li.; S. 176 Fig. 40 o. re. / S. 181 Fig. 51; S. 193 Fig. 4; S. 195 Fig. 16 | Sargent-Welch: S. 220 Fig. 19 | Schering AG Berlin: S. 195 Fig. 15 | R. Schuster, Greifswald: S. 194 Fig. 9, Fig. 11; S. 195 Fig. 12-13 | Shutterstock images, N.Y.: S. 9, 4. Bild (Liga Lauzuma), 5. Bild (Kristian Sekulic), 6. Bild (Dragan Trifunovic); S. 10, 1. Bild (Monkey Business Images), 2. Bild (Albo), 3. Bild (Andrey Stratilatov), 6. Bild (Kiyoshi Takahase Segu / iDesign); S. 11, 5. Bild (James Steidl); S. 13 o. li, (Jamey Ekins), o. re. (Jaimie Duplass), Mi. (Mandy Godbehear), u. li. (Anita Patterson Peppers), u. re. (Losevsky Pavel); S. 60 Fig. 9 (Viktor1); S. 61 Fig. 10 (Christopher Edwin Nuzzaco); S. 68 o. (Liv Friis-Larsen), u. (Jeanne Hatch); S. 90 Fig. 1 (Tracy Whiteside); S. 92 Fig. 4 (Olena Mykhaylova); S. 93 Fig. 5 (Stephen Coburn); S. 94 Fig. 6, 1. Bild (Jason Stitt), 2. Bild (Galina Barskaya), 3. Bild (Rick Becker-Leckrone), 4. Bild (Phil Date); S. 100 Fig. 12 (Galina Barskaya); S. 103 Fig. 16b (Leptospira), Fig. 16c (Sebastian Kaulitzki); S. 105 Fig. 20 l (Karin Lau, Roman Sigaev); S. 108 Fig. 23 (Yakov Stavchansky); S. 109 Fig. 24 (Miodrag Gajic); S. 114 Fig. 1 (Juriah Mosin); S. 147 Fig 2 li. (Margit), Mi. (Anton Prado Photo), re. (Sergei Didyk); S. 151 Fig. 8 (Simone von den Berg); S. 154 Fig. 12 (Radu Razvan); S. 154 Fig. 13 Mi. (Galina Barskaya); S. 159 u. (Tusia); S. 165 u. li. (Vitor Costa); S. 173 Fig. 29 re. (Madlen); Fig. 30 (iofoto); S. 174 Fig. 33 (Georgy Markov), Fig. 35 (Brian A Jackson); S. 175 Fig. 36 (Sergey Chushkin), Fig. 38 (Maksym Gorpenyuk); S. 176 Fig. 40 o. li. (LazarevaEl); S. 180 Fig. 49 (Yurchyks); S. 186 Fig. 56 (Igor Dutina), Fig. 57 (Liv Friis-Larsen); S. 187 Fig. 60 o. (Elena Schweitzer), u. (Elena Elisseeva), Fig. 61 (Filip Fuxa); S. 193 Fig. 5 (Mircea Bezergheanu), Fig. 6 (Viktor1); S. 196 Fig. 19 (LianeM); S. 197 Fig. 20 broccoli (Stefanie Mohr Photography), kohlrabi (Rob Werfel Photography), cauliflower (George Green), cabbage (Jarrod Erbe), kale (CLM), savoy cabbage (Santje); S. 199 Fig. 23 (Jason Joelson); S. 200 Fig. 26 (szefei); S. 201 Fig. 27 (Tusia), Fig. 28 (humbak); S. 202 Fig. 31 (Sofia); S. 213 Fig. 6 (AVAVA); S. 214 Fig. 8 o. re. (francesco de marco); S. 225 re. (Hannamariah); S. 231 Fig. 13 Mi (Eric Isselée) | Silvestris Fotoservice, Kastl: S. 11 Mi re. (Robba); S. 104 Fig. 19 r (Kerscher); S. 141 Fig. 58-2; S. 231 Fig. 13 o. li. | Still Pictures, Berlin: S. 186 Fig. 58 (Martin Bond) | Stockfood: S. 60 Fig. 8 (Michael Brauner) | J. Stockhaus: S. 27 Fig. 23 | H. Streble: S. 44 Fig. 1; S. 213 Fig. 5; S. 238 Fig. 24 | Superbild, Berlin: S. 61 Fig. 11 (JAL – Jacques_Alexandre) | H. Theuerkauf, Gotha: S. 18 Fig. 7 li.; S. 50 Fig. 10; S. 160 Fig. 1; S. 162 Fig. 6; S. 163 Fig. 7; S. 166 Fig. 14; S. 179 Fig. 48; S. 182 Fig. 52; S. 194 Fig. 8; S. 197 Fig. 20 wild cabbage; S. 205 Fig. 38; S. 212 Fig. 2; S. 2/3 Fig. 4; S. 216 Fig. 10 re.; S. 224 Fig. 1-2; S. 230 Fig. 9-11; S. 233 Fig. 15-16; S. 234 Fig. 17 li.; | Ufop Bonn: S. 199 Fig. 24 | ullstein bild, Berlin: S. 153 Fig. 10 (Westphal) | vision photos, Berlin: S. 110 Fig. 25 | Visum, Hamburg: S. 191 Fig. 3 (Silke Reents) | Wechsler: S. 141 Fig. 58-1 | Weiß, Neckarsteinach: S. 125 Fig. 25 | K.H. Willer: S. 242 Fig. 31

Grafiker/Illustrationen: G. Biste, Schwäbisch Gmünd | H. Blümel, Mücka | Ulrike Braun, Berlin | Camilla Catania, Berlin | H. Haydin, Bensheim | Gabriele Heinisch, Berlin | P. Hesse, Berlin | K. Krischke, Marbach | J. Mair, München | K. Mall, Berlin | L.-E. Müller, Leipzig | L. H. Schnellbächer, Schlüchtern | Christine Wächter

Es war leider nicht möglich, zu allen Abbildungen die Inhaber der Rechte zu ermitteln. Berechtigte Ansprüche werden entsprechend den üblichen Vereinbarungen abgegolten.